SECOND-LEVEL
BASIC ELECTRONICS

SECOND-LEVEL BASIC ELECTRONICS

Prepared by the U. S. Navy

(Bureau of Naval Personnel)

Dover Publications, Inc., New York

Published in Canada by General Publishing Company, Ltd., 30 Lesmill Road, Don Mills, Toronto, Ontario.
Published in the United Kingdom by Constable and Company, Ltd., 10 Orange Street, London WC 2.

This Dover edition, first published in 1972, is an unabridged and unaltered republication of *Basic Electronics Volume 2*, originally published by the United States Government Printing Office in 1971 as Rate Training Manual NAVPERS 10087-C.

International Standard Book Number: 0-486-22841-X
Library of Congress Catalog Card Number: 79-189774

Manufactured in the United States of America
Dover Publications, Inc.
180 Varick Street
New York, N.Y. 10014

PREFACE

Basic Electronics is written for men of the U. S. Navy and Naval Reserve whose duties require them to have a knowledge of the fundamentals of electronics. Electronics concerns itself with the emission, behavior, and effect of electrons in vacuums, gases, and semiconductors. Technically speaking, electronics is a broad term extending into many fields of endeavor. Today, electronics projects itself into Navy life at every turn. It facilitates a means of rapid communications, navigates ships, helps control engineering plants, aims guns, drops bombs, and performs logistic functions. It is, therefore, important to become well informed in all areas of basic electronics, in order to be able to qualify for any of the many applicable rates or ratings.

Volume I [reprinted by Dover Publications as *Basic Electronics*] covers general information concerning naval electronics equipments, testing devices, safety procedures, basic transistor and electron tube circuits, and electronic communications. The reader should understand the concepts treated in Volume I before an attempt is made to study the more advanced subjects treated in this volume.

Volume II [here reprinted by Dover Publications as *Second-level Basic Electronics*] treats dual trace oscilloscopes, the operation and application of pulse forming and pulse shaping circuits, microwave devices such as klystrons, magnetrons, phantastrons, resonant cavities, waveguides, and duplexers. A detailed coverage of microwave receiving, transmitting, and indicating systems is also presented.

Synchro and servo systems are treated in some detail. A further dicussion of this subject is to be found in *Synchro, Servo, and Gyro Fundamentals*, NavPers 10105.

The final chapter presents a discussion of number systems and logic circuits. This material is intended as introductory, and for men of some ratings, it will be necessary to refer to manuals which present a more in-depth discussion.

As one of the Basic Navy Training Manuals, this book was prepared by the Training Publications Division of the Naval Personnel Program Support Activity, Washington, D. C., which is a field activity of the Bureau of Naval Personnel. Technical assistance was provided by the Electronics Technician (Class A) Schools at Great Lakes, Illinois and Treasure Island, California, and the Navy Training Publications Center, Memphis, Tennessee.

CONTENTS

THE UNITED STATES NAVY

GUARDIAN OF OUR COUNTRY

The United States Navy is responsible for maintaining control of the sea and is a ready force on watch at home and overseas, capable of strong action to preserve the peace or of instant offensive action to win in war.

It is upon the maintenance of this control that our country's glorious future depends; the United States Navy exists to make it so.

WE SERVE WITH HONOR

Tradition, valor, and victory are the Navy's heritage from the past. To these may be added dedication, discipline, and vigilance as the watchwords of the present and the future.

At home or on distant stations we serve with pride, confident in the respect of our country, our shipmates, and our families.

Our responsibilities sober us; our adversities strengthen us.

Service to God and Country is our special privilege. We serve with honor.

THE FUTURE OF THE NAVY

The Navy will always employ new weapons, new techniques, and greater power to protect and defend the United States on the sea, under the sea, and in the air.

Now and in the future, control of the sea gives the United States her greatest advantage for the maintenance of peace and for victory in war.

Mobility, surprise, dispersal, and offensive power are the keynotes of the new Navy. The roots of the Navy lie in a strong belief in the future, in continued dedication to our tasks, and in reflection on our heritage from the past.

Never have our opportunities and our responsibilities been greater.

CHAPTER 1

DUAL TRACE OSCILLOSCOPE

This chapter will examine the dual trace oscilloscope. Dual trace operation permits viewing of two independent signal sources as a dual display on a single cathode-ray tube (crt) screen. This operation affords an accurate means of making amplitude, phase, or time displacement comparisons and measurements between two signals.

A dual trace oscilloscope should not be confused with a dual beam oscilloscope. Dual beam oscilloscopes are those which produce two separate electron beams on a single scope, which can be individually or jointly controlled. There are multibeam scopes used for specialized applications that have more than two separate beams. Dual trace refers to a single beam in a crt that is shared by two channels.

OBTAINING THE DUAL TRACE

There are two methods by which the single beam is shared. The first method of obtaining a dual trace is called the CHOP MODE. Figure 1-1 shows a simplified block diagram of the dual trace section, using the chop mode.

The output d.c. voltage reference on each of the amplifiers is adjustable. Therefore, the beam will be deflected by different amounts on each channel, if the voltage reference is different at each amplifier output. The output voltage from each amplifier is applied to the deflection plates through the gate. The gate is actually an electronic switch; in this application, it is commonly referred to as a BEAM SWITCH. The switch is controlled by a high frequency multivibrator in the chop mode. That is, the gate selects one channel's output and then the other at a high frequency rate, which is 100 kHz in most oscilloscopes. Since the switching time is very short in a good quality oscilloscope, the resultant display is two horizontal dashed lines, as shown in figure 1-2A.

Dashed line A is the output of one channel

while line B is output of the other. The trace goes from left to right due to a sawtooth waveform applied to the horizontal plates. A more detailed analysis shows that the beam moves from 1 to 2 while the gate is connected to the output from one channel. Then when the gate samples the output of the second channel during time 3-4 (assuming channel 2 is at a different voltage reference), the beam is at a different vertical location. The beam continues in the sequence 5-6, 7-8, 9-10, 11-12 through the rest of one horizontal sweep. When the chopping frequency is much higher than the horizontal sweep frequency, the number of dashes will be very large. For example, if the chopping occurs at 100 kHz and the sweep frequency is 1 kHz, then each horizontal line would be comprised of 100 dashes. This display would then look like a series of closely spaced dots as shown in figure 1-2B. As the sweep frequency becomes lower in respect to the chopping frequency, the display will show two apparently continuous traces. Therefore, the chop mode is used at low sweep rates (low time per division settings). When signals are applied to the channel amplifiers, the outputs are changed in accordance with the input signal. The resultant pattern on the screen gives a time base presentation of the signal of each channel as shown in figure 1-3. Part A shows the chopped traces without signals; B depicts the signals into the two channels; and C is the resultant display.

The second method (ALTERNATE MODE) of obtaining a dual trace function uses the technique of gating between sweeps as shown in figure 1-4.

The gate samples one channel for one complete sweep and the other channel for the next complete sweep. The gate selection is controlled by the sweep circuitry shown in the block diagram of figure 1-4. At slow sweep speeds, one trace begins to fade while the other channel is being gated. Consequently, this mode is not used for slow sweep speeds. Because the chop mode will not operate satisfactorily at high speeds and the

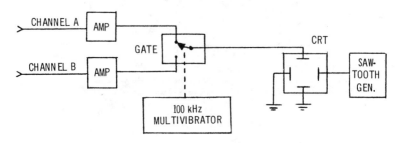

179.517
Figure 1-1.—CHOP mode.

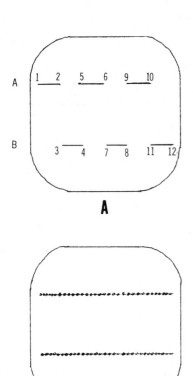

A

B

179.518
Figure 1-2.—Displaying the CHOP mode.

alternate mode is deficient at low speeds, both are used on dual trace oscilloscopes to complement each other.

CONTROLS

Because the settings of the controls on the front panel change the operation of an oscilloscope, it is imperative that the user have an understanding of the effects of these controls. (Many of these controls function identically to those on the single trace triggered oscilloscope previously discussed.) The controls in the following discussion are illustrated in figure 1-5. It should be noted that this is a drawing for clarity, and is not intended to illustrate any particular type of dual trace oscilloscope.

CRT CIRCUITRY CONTROLS

The ON-OFF switch controls the application of primary power to the oscilloscope. On many oscilloscopes, this switch is part of another control such as the INTENSITY control or GRATICULE control. (The graticule is the overlying grid on the face of the crt.) Figure 1-5 shows the ON-OFF switch as a part of the GRATICULE control. Some oscilloscopes have a lamp near the ON-OFF switch that illuminates to indicate the power on condition.

The GRATICULE control varies the graticule illumination intensity. The best setting depends on ambient light conditions and the specific use of the oscilloscope.

Adjustment of the INTENSITY or BRILLIANCE control varies the brightness of the spot or trace. Internally, this control varies the bias of the crt.

A small clear spot or trace is obtained by setting the FOCUS control, while a trace of uniform width across the entire screen is dependent on

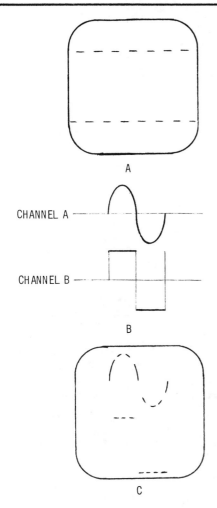

CHANNEL A

CHANNEL B

179.519
Figure 1-3.—Dual channel display in
CHOP mode.

the setting of the ASTIGMATISM (ASTIG) control. Since the INTENSITY, FOCUS, and ASTIGMA- TISM controls interact, they must be adjusted in relation to one another to obtain a uniform trace with maximum resolution at the desired intensity.

VERTICAL SECTION CONTROLS

The input coupling switch determines the type of voltage that will be applied to the vertical channel. Figure 1-6 shows the basic coupling net- works in AC, DC, and GROUND positions.

The capacitor in the AC position blocks the d.c. component of the input, so only a change in voltage is coupled to the vertical deflection circuits. The input to the vertical channel is directly coupled in the DC position. A third position (GND) is used to ground the vertical amplifier input and open the input signal path between the input jack and the amplifiers.

The VOLTS/CM SWITCH selects different settings on an input attenuator network. When the setting of the VOLTS/CM switch is changed, the amount of attenuation of the input signal is varied, and the amplitude of the display is changed. The gain of the vertical amplifiers is adjusted during calibration so that each position of the VOLT/CM switch will provide a vertical de- flection of a specified amount for a given signal strength. For example, in the 1 volt per centi- meter position (fig. 1-5), each centimeter that the display covers, vertically, represents one volt (the graticule is usually divided into divisions of one centimeter).

The VARIABLE AMPLITUDE control (gener- ally a part of the VOLTS/CM switch, see fig. 1-5) provides another means for varying the amplitude of the display. This control varies the gain of the vertical amplifiers. If the VAR- IABLE control is not fully clockwise, the verti- cal deflection will be less than what is specified by the VOLTS/CM position.

The VERTICAL POSITION (SHIFT) control affects the vertical location of the display. This is accomplished by varying the d.c. potential between the vertical deflection plates.

The BALANCE (BAL) control is normally a screwdriver adjustment which, when properly adjusted, prevents shifting of the trace as the VARIABLE AMPLITUDE control is adjusted.

The SET GAIN control is a screwdriver adjustment that varies the gain of the vertical amplifiers. When it is properly adjusted with the VOLTS/CM switch in the CAL position, the amount of vertical deflection will be correct for the other VOLTS/CM settings.

If a X10 switch is incorporated, it switches in an additional amplifier stage that has a gain of 10. This increases the amplitude of the signal applied to the vertical deflection plates by 10. The frequency response will usually be much less in this position and should not be used unless absolutely necessary.

An INVERT switch reverses the phase of the voltage applied between the vertical deflection plates. This can be used to superimpose two signals that are actually 180° out of phase.

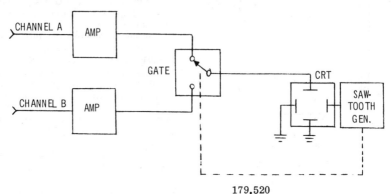

179.520
Figure 1-4.—ALTERNATE mode.

The MODE switch determines the channel that is being gated to the vertical deflection plates. Usually, the following modes are available: CHANNEL 1 ONLY (CHAN 1), CHANNEL 2 ONLY (CHAN 2), ALTERNATE (ALT), CHOP, or ADD. In the CHANNEL 1 ONLY mode, the beam switch remains connected to channel 1 and the oscilloscope behaves essentially as a single trace scope. The CHANNEL 2 ONLY mode applies the voltage on channel 2 only to the vertical deflection system. When switched to the ALTERNATE mode, the oscilloscope requires high horizontal sweep speeds for signal comparison. One channel will appear on the screen during one sweep and the other channel output will appear during the other sweep. For low sweep speeds, the CHOP mode, which presents both channels during the same sweep, is used. In the ADD mode, the signal voltage on channel 1 is algebraically added to the signal voltage on channel 2. Consequently, a single trace, which represents the sum of the two channels, is displayed.

The TRIGGER SOURCE selector control allows either channel input to trigger either sweep trace. A common example of the use of the control is the selection of channel 1 or channel 2 as the trigger or synchronizing source for the horizontal sweep. However, some oscilloscopes have two independent horizontal sweeps which can be triggered in any combination from channel 1 and channel 2, i.e., sweep A, as shown in figure 1-5, can be triggered by either channel and sweep B can be triggered by either channel. In other oscilloscopes, a composite signal of A and B can be selected to trigger the sweep circuits.

TRIGGERING AND HORIZONTAL SECTION CONTROLS

The setting of the TRIGGER SELECTION switch to the INTERNAL (INT) position makes it possible to obtain the trigger pulse internally from the vertical amplifiers. The channel that does the triggering is selected by the setting of the TRIGGER SOURCE selector as just described. By switching the TRIGGER SELECTION switch to EXTERNAL (EXT), the trigger can be obtained externally from whatever source is being applied through the external trigger jack. Some units (as shown in fig. 1-5) also have a LINE position from which can be obtained a low amplitude 60-Hz line voltage which can also be used for triggering. These positions are sometimes called TRIGGER MODES.

The TRIGGER COUPLING switch usually has three positions: DC, AC, and HIGH FREQUENCY AC (HFAC). The DC position provides direct coupling to the trigger circuits, whereas the AC position incorporates a coupling capacitor to block any d.c. component. The HIGH FREQUENCY AC position has a high pass filter which passes only those trigger signals that are above a certain frequency.

The position of the TRIGGER SLOPE switch determines whether the sawtooth sweep will be initiated on the positive going portion or on the negative going portion of the signal to be displayed. In figure 1-7, the positive going portion occurs from time A to B and the negative going portion occurs from time B to C.

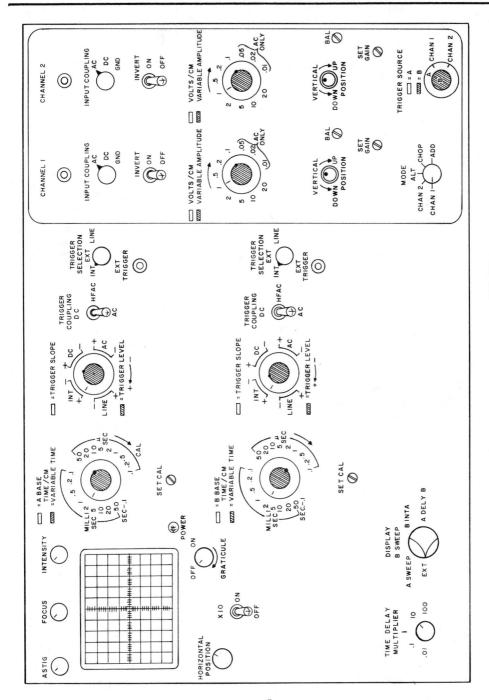

Figure 1-5.—Dual trace oscilloscope controls.

179.521

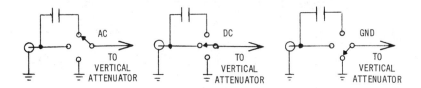

179.522
Figure 1-6.—Input coupling circuits.

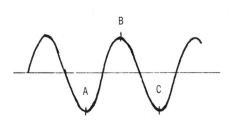

179.523
Figure 1-7.—Waveform slope.

The TRIGGER LEVEL (mounted with the TRIGGER SLOPE of fig. 1-5) determines the voltage level required to trigger the sweep. For example, in the INTERNAL TRIGGER mode, the trigger is obtained from the signal to be displayed, therefore, the setting of the TRIGGER LEVEL will determine the point of the input waveform that will be displayed at the start of the sweep.

Figure 1-8 shows some of the displays for one channel that will be obtained for different trigger levels and trigger slope settings. The level is zero and the slope is positive in diagram A, while diagram B also shows a zero level but a negative slope selection. Diagram C shows the effects of a positive trigger level setting and positive trigger slope setting, while diagram D displays a negative trigger level setting with a positive trigger slope setting. Diagrams E and F have negative slope settings. The difference is that E has a positive trigger level setting while F has a negative trigger level setting.

The automatic function of the trigger circuitry allows a free running trace without a trigger signal, but when a trigger signal is applied, the circuit reverts to the triggered mode of operation, and the sweep no longer free runs. This provides a trace when no signal is applied.

Synchronization is also used to cause a free running condition without a trigger signal. Synchronization is not synonymous with triggering. Triggering refers to a certain condition which initiates an operation. Without this condition, the operation would not occur. In the case of the triggered sweep that was just presented, the sweep will not be started until a trigger is applied and each succeeding sweep must have a trigger before a sweep commences. Synchronization means the causation of an operation or event to be brought in step with a second operation. A sweep circuit which uses synchronization instead of triggering will cause a previously free running sweep to be locked in step with the synchronizing signal. The TRIGGER LEVEL control setting must be increased until synchronization occurs; but until this time, an unstable pattern appears on the crt face.

The A TIME BASE circuitry produces a sawtooth waveform that is applied to the horizontal deflection plates on the crt in order to obtain a sweep. The rate of the sweep is changed by the TIME/CM switch. For a given setting of the TIME/CM switch, and with the VARIABLE TIME control set to the CAL position, the movement of the beam from left to right across one centimeter (the graticule is divided into centimeters) occurs in the time specified by the TIME/CM switch setting. The VARIABLE TIME control varies the rate of the sweep so that any sweep speed in between those set by the TIME/CM switch can be obtained. Most oscilloscopes have a B TIME BASE circuitry which has a TIME/CM and VARIABLE TIME control which is similar to the A TIME BASE circuitry. The HORIZONTAL POSITION or X SHIFT control adjusts the horizontal position of the trace in both the A and B TIME BASE.

There is normally a screwdriver adjustment called SET CAL which varies the sweep rate.

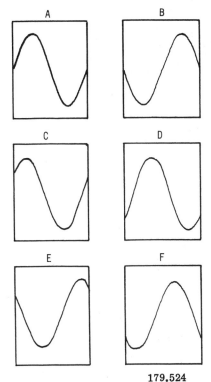

179.524

Figure 1-8.—Effects of trigger controls.

This is adjusted to obtain the exact time per centimeter sweep rate that is indicated by the TIME/CM setting.

Many oscilloscopes have a X10 switch in the horizontal amplifier section. Just as in the vertical section, this switch adds another amplifier. In the sweep mode, the sawtooth voltage change will be 10 times greater than that required for a complete left to right deflection. As a result only one-tenth of the entire sweep will be visable on the screen.

The heart of sweep circuit control is the DISPLAY switch. In the EXTERNAL (EXT) position of this switch, an external signal must be applied to the external horizontal input jack in order to obtain horizontal deflection. The relationship between the type of waveform applied to the external jack and the Y input jack will determine the resultant display. Two examples are time base presentation, like those obtained when using internal sweeping, and lissajous patterns.

In the A SWEEP position of the DISPLAY switch, internal sweeping occurs as determined by the A TIME/CM controls.

Some oscilloscopes have a B SWEEP position on the DISPLAY switch so that either the A sweep or the B sweep can be used. In many oscilloscopes, the B sweep is used only for the functions of B INTENSIFIED BY A and A DELAYED BY B.

In the B INTENSIFIED BY A (B INT A) position of the DISPLAY switch, the B sweep is applied to the horizontal deflection plates and a rectangular pulse equal to the rise time of the A sweep is added to the unblanking pulse from the B sweep. This will cause the beam to be deflected at the B sweep rate and intensified a greater amount during a time equal to the A sweep. The A sweep time should always be shorter than the B sweep time. The segment of the B sweep which will be intensified will be determined by the setting on the TIME DELAY MULTIPLIER. The settings on the TIME DELAY MULTIPLIER multiplied by the B sweep TIME/CM setting determines the amount of time that will elapse between the initiation of the B sweep and the initiation of the A sweep. An example is shown in figure 1-9.

Waveform A represents the A sweep, while waveform B represents the B sweep. The input signal to one of the vertical channels is represented by waveform C. Waveform D is the composite unblanking pulse, which is the addition of a pulse of a time duration equal to the rise time of waveform B and a pulse of a time duration equal to the rise time of waveform A. As was stated in the preceeding paragraph, the start of waveform A is determined by the setting of the TIME DELAY MULTIPLIER. Varying the setting of this control will change the portion of the input waveform that is intensified. Figure 1-9E shows the resultant display.

The A DELAYED BY B (A DELY B) mode display occurs only during the duration of the A sweep. Again the B sweep must be longer than the A sweep and the start of the A sweep in relation to the B sweep is determined by the TIME DELAY MULTIPLIER. In this case the beam is swept by the A sweep and only that segment of the input signal which occurs during the time of the A sweep will be presented on the crt, since the crt will be unblanked once during the A sweep. Figure 1-10 shows the typical waveforms when operating in the A DELAYED BY B mode.

Waveform A is the A sweep and waveform B is the B sweep. The input signal applied to a vertical channel is represented by waveform C. The unblanking pulse is shown by waveform D.

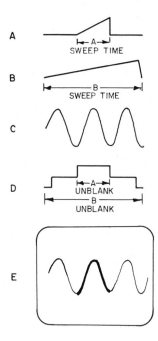

179.525
Figure 1-9.—B INTENSIFIED BY A mode.

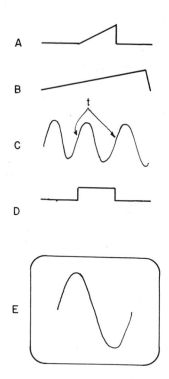

179.526
Figure 1-10.—A DELAYED BY B mode.

The resultant display (waveform E) is that portion of the input waveform that occurs during the A sweep as represented by time (t) in the diagram. This mode of operation is especially useful in displaying an extremely small portion of a complex waveform or for viewing only the rise time of a rectangular wave.

When in dual trace operation and operating in the B INTENSIFIED BY A mode, the waveforms of both inputs will be intensified. In the A DELAYED BY B mode, a segment of both channels will be displayed. These conditions are shown in figure 1-11.

OPERATION

Applications of triggered oscilloscopes will be covered in this section as well as the basic characteristics and operation of dual trace oscilloscopes.

Figure 1-12 is the block diagram of a typical dual trace oscilloscope without the power supplies. The circled letters in the block refer to the waveform designations to the right of the diagram.

The waveform to be observed (waveform A) is fed into the channel A vertical amplifier section. The VOLTS/DIV control sets the gain of the amplifier. In a similar manner waveform B is applied to the channel B vertical amplifier section. The beam switch determines the channel as described previously under "OBTAINING THE DUAL TRACE". Assuming that the CHOP mode is selected, the output of the beam switch is shown by waveform C. The waveform is fed through the delay line and vertical output amplifier to the vertical deflection plates of the cathode-ray tube. The purpose of the delay line will be explained presently. But, first, the time-base generator will be described.

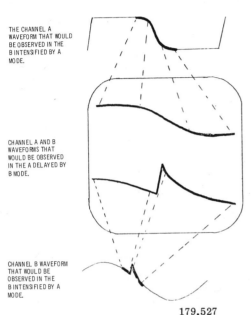

THE CHANNEL A WAVEFORM THAT WOULD BE OBSERVED IN THE B INTENSIFIED BY A MODE.

CHANNEL A AND B WAVEFORMS THAT WOULD BE OBSERVED IN THE A DELAYED BY B MODE.

CHANNEL B WAVEFORM THAT WOULD BE OBSERVED IN THE B INTENSIFIED BY A MODE.

179.527

Figure 1-11.—Dual trace A DELAYED BY B mode.

TIME-BASE GENERATOR

Just as in the single trace oscilloscope, the time-base generator or sweep generator develops a sawtooth wave (waveform D) that provides the horizontal deflection voltage. The rising or positive going part of this sawtooth is linear. That is, the waveform rises through a given number of volts during each unit of time. This rate of rise is set by the TIME/DIV control. The sawtooth voltage is fed to the time-base amplifier. This amplifier includes a phase inverter so that the amplifier supplies two output sawtooth waveforms simultaneously. One of them is positive going (waveform E), while the other is negative going (waveform F). The positive going sawtooth is applied to the right-hand horizontal deflection plate and the negative going sawtooth goes to the left-hand plate. As a result, the cathode-ray beam is swept horizontally to the right through a given number of graticule divisions during each unit of time. The sweep rate is controlled by the TIME /DIV control.

DELAY LINE

In order to maintain a stable display on the screen, each sweep must start at the same point on the waveform being displayed. This is accomplished by feeding a sample of the displayed waveform to a trigger circuit that produces a negative output voltage spike (waveform G) at a selected point on the displayed waveform. This triggering spike is used to start the run-up portion of the horizontal sweep. Since the leading edge of the waveform to be displayed is used to actuate the trigger circuit, and since the triggering and unblanking operations require a measurable time, the actual start of the trace on the screen is lagging the start of the waveform to be displayed. This difference is approximately .14 usec. in many oscilloscopes. Time interval t represents the difference in figure 1-12. In order to display the leading edge of the input waveforms, a delay, Q, is introduced by the delay line in the vertical deflection channel, after the point where the trigger is obtained. The delayed vertical signal is shown by waveform H and a push-pull version of waveform H comes from the vertical output amplifier. To reemphasize the purpose of the delay line; it is to retard the application of the observed waveform to the vertical deflection plates until the trigger and time-base circuits have had an opportunity to initiate the unblanking and horizontal sweep operations. In this way the entire waveform can be observed, even though the leading edge of the waveform was used to trigger the horizontal sweep.

UNBLANKING

The unblanking operation alluded to previously is the application of a rectangular unblanking wave (waveform I in fig. 1-12) to the grid of the crt. The duration of the positive part of this rectangular wave corresponds to the duration of the positive going part of the time-base output (waveform D), so that the beam is switched on during its left-to-right excursion and is switched off during its right-to-left retrace.

For the input signals shown in figure 1-12, the waveform at many points will be different when operating in the alternate mode. Figure 1-13 depicts the waveforms in the alternate mode. Two sweeps are shown since it requires two sweeps to display the information of both channels in this mode. Notice that the unblanking voltage is removed in waveform I during the retrace. The trigger pulse (waveform G) is obtained from the same channel regardless of which input is being

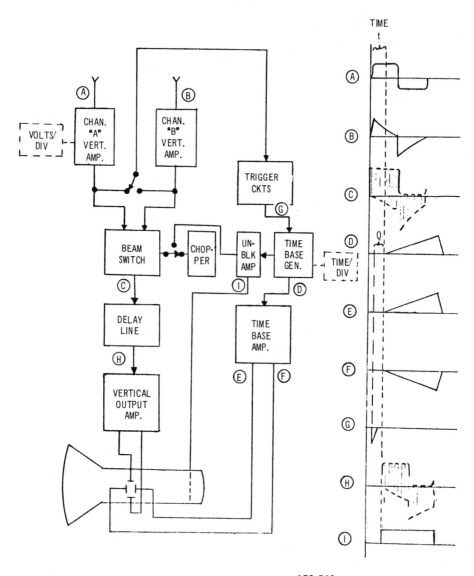

179.528
Figure 1-12.—Basic dual trace oscilloscope.

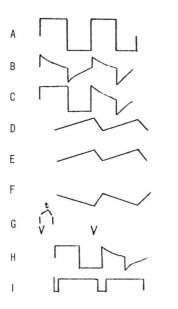

A

B

C

D

E

F

G

H

I

179.529

Figure 1-13.—ALTERNATE mode waveforms
as applied to figure 1-12.

sampled. The input channel that triggers the sweep depends on the front panel setting of the TRIGGER selector. It can be seen from waveform C that during the first entire sweep one channel is sampled and during the next sweep the other channel is sampled as was described in the section on methods of obtaining a dual trace.

The presentation of a basic dual trace oscilloscope as described here is close to actual operation. A few operations have been simplified because of the complex circuitry involved. As a result there are some slight deviations from what will be found in actual oscilloscopes.

ACCESSORIES

The basic dual trace oscilloscope has one gun assembly and two vertical channels. However there are many variations. The horizontal sweep channels vary somewhat from equipment to equipment. Some have one time base circuit while others have two, and these two are interdependent in some oscilloscopes while others are independently controlled. Also, most modern general purpose oscilloscopes are modular contructed.

That is, most of the vertical circuitry is contained in a removable plug-in unit and most of the horizontal circuitry is contained in another plug-in unit. The main frame of the oscilloscope can then be adapted for many applications by designing a variety of plug-in assemblies. This modular feature provides much greater versatility in a single oscilloscope. For instance, the dual trace plug-in module can be replaced with a semiconductor curve tracer plug-in module if it is desired to analyze transistor characteristics. Other plug-in modules available with some oscilloscopes are high-gain wide bandwidth amplifiers, differential amplifiers, spectrum analyzers, physiological monitors, and other specialized units. Therefore, the dual trace capability is a function of the type of plug-in unit that is used with some oscilloscopes.

In order to derive maximum usefulness from an oscilloscope there must be a means of connecting the desired signal to the oscilloscope input. Aside from cable connections between an equipment output and the oscilloscope input there are a variety of probes available which facilitate monitoring of signals at any point desired in a circuit. The more common types include; 1:1 probes, attenuation probes, and current probes. Each of these probes may be supplied with several different tips to allow measurement of signals on any type of test point. Figure 1-14 shows some of the more common probe tips.

In choosing the probe to use for a particular measurement, one must consider such factors as circuit loading, signal amplitude, and scope sensitivity.

The 1:1 probe offers little or no attenuation of the signal under test and is therefore useful for the measurement of low level signals. However, circuit loading with a 1:1 probe may be detrimental: the impedance at the probe tip is the same as the input impedance of the oscilloscope.

An attenuator probe has an internal high value resistor in series with the probe tip. This gives the probe a higher input impedance than that of the oscilloscope, providing the capability of measuring high amplitude signals that would overdrive the vertical amplifier if connected directly to the oscilloscope. Figure 1-15 shows a schematic representation of a basic attenuation probe. The 9-megohm resistor in the probe and the 1-megohm input resistor of the oscilloscope form a 10:1 voltage divider.

11

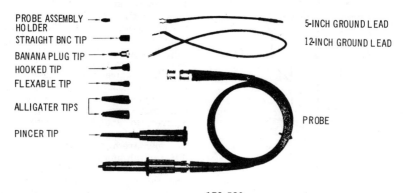

179.530
Figure 1-14.—Probe tips.

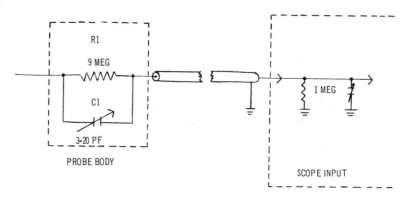

179.531
Figure 1-15.—Basic attenuation probe.

Since the probe resistor is in series, the oscilloscope input resistance when using the probe is 10 megohms. Thus, using the attenuator probe with the oscilloscope will cause less circuit loading than using a 1:1 probe.

Before using an attenuator probe for measurement of high frequency signals or for fast rising waveforms the probe compensating capacitor (C1 in fig. 1-15) must be adjusted according to instructions in the applicable technical manual. Some probes will have an impedance equalizer in the end of the cable that attaches to the oscilloscope. The impedance equalizer, when adjusted as per manufacturer's instructions, assures proper impedance matching between probe and oscilloscope. An improperly adjusted impedance equalizer will result in erroneous measurements especially when measuring high frequencies or fast rising signals.

Current probes utilize the electromagnetic fields produced by a current. The probe is designed to be clamped around a conductor without having to disconnect the conductor. The current probe is electrically insulated from the conductor, but the magnetic fields about the conductor induce a potential in the current probe proportional to the current through the conductor. Thus the vertical deflection of the oscilloscope display will be directly proportional to the current through the conductor.

CHAPTER 2

PULSE FORMING AND PULSE SHAPING CIRCUITS—PART I

Pulse circuits in electronic equipments accomplish timing functions by producing a variety of voltage waveforms such as square waves, trapezoidal waves, sawtooth waves, rectangular waves and sharp peaks. Although all these circuits are generally classified as timing circuits, the specific function performed may be one of timing, waveshaping, or wave generating.

TRANSIENT AND NONSINUSOIDAL VOLTAGES

Waveforms associated with timing circuits are produced by combinations of active devices such as diodes, transistors, or electron tubes and passive devices such as capacitors, resistors, or inductors. The exponential charge and discharge characteristics of capacitors and inductors make these components extremely useful in timing circuits because of their effects on transient and nonsinusoidal voltages.

Any momentary change in voltage must be considered a transient voltage, whether it be a single change or one which occurs at fixed or indeterminate intervals of time. Voltage waveforms such as sine waves, square waves, trapezoidal waves, sawtooth waves, rectangular waves, and peaked waves are all periodic in nature since the change occurs at fixed intervals. A sine wave, however, is not considered to be a transient voltage since the voltage is changing continuously.

For the purpose of circuit explanation, there are two ways of analyzing transient and non-sinusoidal voltages. One is to consider that the transient waveshape is a momentary voltage, which is followed after a certain interval by another similar change. The other is to assume that the waveshape is the algebraic sum of many sine waves having different frequencies and amplitudes. The second method is more useful in most cases. For example, in the design of an amplifier, this type of analysis is necessary because the amplifier cannot handle a transient

without causing distortion unless it is capable of passing all the sine wave frequencies contained within the transient.

Figure 2-1A illustrates a square wave with the voltage plotted against time. In terms of rapid voltage change analysis the square wave illustrated can be considered as a voltage that remains unchanged at +50 volts until t_1, when it suddenly drops to a -50 volts. It remains at this value until t_2, when it abruptly increases back to +50 volts and remains at this value until t_3, etc. (As can be seen, this type of analysis has a rather limited usage.)

Using the algebraic sum method, the waveform can be analyzed by determining what sine waves are required to produce it. The sine wave that has the same frequency as the complex periodic wave is called the fundamental frequency. The type and number of harmonics (multiples of the fundamental frequency) included in the complex waveform are dependent on the shape of the wave, in this case a square wave. A perfect square wave consists of the fundamental frequency plus an infinite number of odd harmonics (3rd, 5th, etc.) which cross the zero reference line in phase with the fundamental.

Figure 2-1B graphically illustrates a fundamental frequency (A) combined with its 3rd harmonic (B). The resultant waveform (C) slightly resembles a square wave. Figure 2-1C shows the results of algebraically adding the 5th harmonic (D) to the resultant wave (C). Notice that the sum of waves C and D result in wave E which has steeper sides and a flatter top. With the addition of the 7th harmonic (F), as shown in figure 2-1D, the shape of the composite waveform more nearly resembles that of a square wave. Thus, as more and more odd harmonics are added the shape of the composite waveform more nearly approaches that of a perfect square wave.

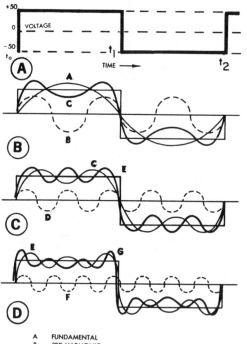

A FUNDAMENTAL
B 3RD HARMONIC
C FUNDAMENTAL PLUS 3RD HARMONIC
D 5TH HARMONIC
E FUNDAMENTAL PLUS 3RD AND 5TH HARMONIC
F 7TH HARMONIC
G FUNDAMENTAL PLUS 3RD, 5TH AND 7TH HARMONIC

20.295(179)
Figure 2-1.—Analysis of a square wave.

Sawtooth waves like square waves can be constructed from a series of sine waves. A sawtooth contains both odd and even harmonics, with the odd order harmonics (3, 5, 7, etc.) crossing the reference line in phase with the fundamental and the even harmonics (2, 4, 6, etc.) crossing 180° out of phase with the fundamental.

Figure 2-2A, B, C, and D show the results of algebraically adding harmonics to a fundamental in order to produce a sawtooth waveform. Figure 2-2D depicts the resultant wave of all harmonics up to the seventh algebraically added. It closely resembles the superimposed sawtooth waveform.

A peaked or triangular wave can also be constructed from a number of sine waves. Figure 2-3 illustrates the harmonic composition of a peaked wave. Notice that it consists of odd order harmonics only. The phase relationship between the harmonics and the fundamental in the peaked wave is different than in the square wave. All odd order harmonics in the square wave crossed the reference line in phase with the fundamental. In the peaked wave, the 3rd, 7th, 11th, etc. harmonics cross the zero reference line 180° out of phase with the fundamental, while the 5th, 9th, 13th, etc. harmonics cross the reference line in phase with the fundamental. As can be seen in figure 2-3C, the addition of each harmonic produces a composite wave with higher peaks and steeper sides.

RC AND RL SHAPING CIRCUITS

Shaping circuits are used to change the shape of applied nonsinusoidal waveforms. Shaping circuits may be either series RC or RL circuits.

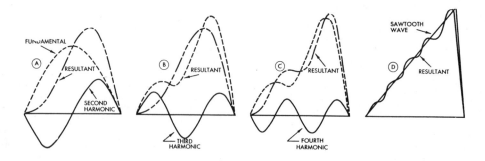

20.296(179)
Figure 2-2.—Analysis of a sawtooth wave.

14

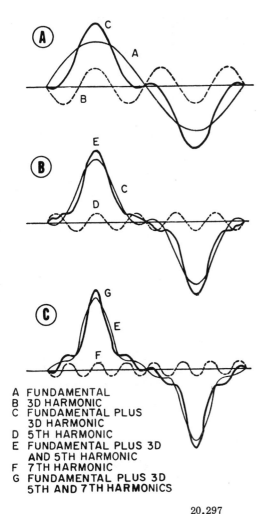

A FUNDAMENTAL
B 3D HARMONIC
C FUNDAMENTAL PLUS
   3D HARMONIC
D 5TH HARMONIC
E FUNDAMENTAL PLUS 3D
   AND 5TH HARMONIC
F 7TH HARMONIC
G FUNDAMENTAL PLUS 3D
   5TH AND 7TH HARMONICS

20.297

Figure 2-3.—Analysis of a peaked wave.

These circuits electrically perform the mathematical operations of integration and differentiation and are called INTEGRATORS and DIFFERENTIATORS.

To understand the principles of integration and differentiation, it is helpful to review the operation of low pass and high pass filters. Figure 2-4 illustrates an RC circuit with a

100-volt, 1-kHz applied sine wave. With a capacitance of 0.0318 microfarads and a frequency of 1 kHz, the capacitive reactance will be 5 kohms. This means that at a frequency of 1 kHz, there will be equal voltage drops across the resistor and capacitor of 70.7 volts. The phase angle between the reference (I) and the applied voltage will be 45° as shown in figure 2-5. In terms of discrimination, 1 kHz will be the cutoff frequency. The cutoff frequency may be defined as that frequency where the phase angle is 45°, or that frequency where the voltage drops across the reactive and resistive components are equal.

Now assume that the applied frequency is variable; when the frequency is increased to 5 kHz, the capacitive reactance will decrease to a value of 1 kohm and the phase angle will now be 11.3°. Since the phase angle dropped below 45° to 11.3°, the voltage across the capacitor decreases to 19.6 volts and increases to 98 volts across the resistor.

Conversely, if the frequency applied to the circuit decreases to 500 Hz, the reactance offered by the capacitor would increase to 10 kohms and the phase angle would be 63.5°. The voltage across the capacitor now increases to 89.5 volts and the voltage across the resistor decreases to 44.6 volts. Thus, high frequencies are being attenuated by the circuit or it can be said that the circuit is discriminating against high frequencies.

It can be seen from the foregoing discussion that if the output were taken across the resistor, then the reverse would be true. That is, the circuit would now attenuate or discriminate against the low frequencies and pass the high

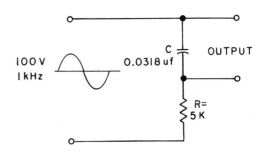

1.37(179)

Figure 2-4.—Low pass filter.

15

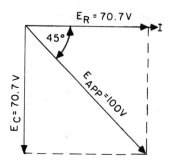

179.532
Figure 2-5.—Vector diagram.

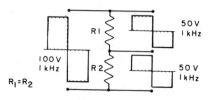

179.533
Figure 2-6.—Application of square waves to
series resistive circuit.

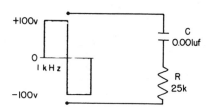

179.534
Figure 2-7.—Application of a square wave to
series RC circuit.

frequencies, in other words, act as a high pass filter. In either case, 1 MHz would be the cut-off frequency.

INTEGRATION

Figure 2-6 shows a pure square wave applied to a series resistive circuit. If the values of the resistors are equal, the voltage drops across each each resistor will be equal. From the one pure square wave input, two pure square waves of equal amplitude will be produced. The resistance of the resistors will not affect the phase or amplitude relationships of the harmonics contained in the square wave. However, if the same square wave is applied to a series RC circuit as shown in figure 2-7, the action is not the same.

RC Integrator

The RC integrator is used as a waveshaping network in radio, television, radar, and computers, as well as many other special electronic applications.

Since the harmonic content of the square wave is odd multiples of the fundamental frequency, there will be significant harmonics as high as 50 or 60 times the frequency of the fundamental and the capacitor will offer a reactance of different magnitude to each harmonic. This means that the voltage drop across the capacitor for each harmonic frequency present will not be the same. To low frequencies, the capacitor will offer a large opposition providing a large drop across the capacitor. To high frequencies, the reactance of the capacitor will be extremely small causing a small voltage drop

across the capacitor. If the voltage component of the harmonic is not developed across the reactance of the capacitor, then it must be developed across the resistor, (Kirchoff's voltage law must be observed).

It must be remembered that the reactance offered to each harmonic frequency will not only cause a change in the amplitude of the harmonics, but will also cause a change in the phase of each individual harmonic frequency with respect to the current reference. The amount of phase and amplitude change taking place across the capacitor is dependent upon the capacitive reactance, which is a function of the capacitance and the frequency. The value of the resistance offered by the resistor must also be considered because it controls the ratio of the voltage drops across itself and the capacitor.

Since the amplitude and phase angle of each harmonic is changed, the output when taken across the capacitor will look quite different from the input. The square wave applied to the circuit is 100-volts peak at a frequency of 1 kHz. The odd harmonics will be 3 kHz, 5 kHz, 7 kHz, etc. Table 2-1 shows the value of the reactance offered to several of the harmonics and indicates

Table 2-1.—CHANGE IN X_C WITH EACH
HARMONIC.

HARMONIC	X_C	R
Fund.	159k	25k
3rd	53k	25k
5th	31.8k	25k
7th	22.7k	25k
9th	17.7k	25k
11th	14.5k	25k

179.535

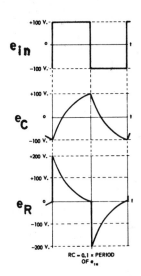

179.536
Figure 2-8.—Partial integration.

the approximate value of the cutoff frequency. It can be seen from the table that the cutoff frequency lies between the fifth and seventh harmonics. That is, at some point between these two values, the capacitive reactance will equal the resistance. Therefore, all of the harmonic frequencies above the fifth will not be effectively dropped across the output capacitor, and the absence of the higher order harmonics will cause the leading edge of the waveform developed across the capacitor to be rounded. An example of this effect is shown in figure 2-8.

To satisfy Kirchoff's voltage law, the harmonics not effectively developed across the capacitor must be developed across the resistor. If the waveform across both the resistor and the capacitor were added graphically, the resultant would be an exact duplication of the input square wave. Note the pattern of the voltage waveform across the resistor.

Increasing the capacitance (decreasing the capacitive reactance) will change the output waveforms as shown in figure 2-9. NOTE: The same effect could be obtained by increasing the resistance. Thus, when the output is taken across the capacitor, the circuit acts as an integrator and the degree of integration that takes place will be dependent on the RC time of the circuit. If the RC time is short in comparison to the time duration of the pulse, little or no integration takes place and the output waveform will resemble the input waveform. However, as the RC time is increased (made longer in comparison to the time duration of the pulse) the degree of integration increases as can be seen by studying figures 2-8 and 2-9. Eventually, a point will be

reached where there will be little or no output (complete integration). This usually occurs when the RC time is increased to a value of 10 times that of the time duration of the pulse.

INTEGRATOR WAVEFORM ANALYSIS.—It is not necessary to actually compute and graph the waveforms that would result from a long time constant (10 times the pulse duration), a short time constant (one-tenth of the pulse duration), and an intermediate time constant (some time constant between the long and the short). Instead, the capacitor output voltage will be plotted by using the universal time constant chart shown in figure 2-10.

Capacitor charge will follow curve A in figure 2-10, while resistor voltage change follows curve B. On discharge, the reverse will be true, capacitor voltage follows curve B while resistor voltage follows curve A. Thus, knowing the RC time of the circuit and the amplitude of the applied voltage, E_R and E_C may be accurately plotted for any given instant of time. In figure 2-11, a pulse of 100-microseconds duration at an amplitude of 100 volts will be applied to the circuit composed of the 0.01-μf capacitor and the variable resistor, R. The square wave applied is a symmetrical square wave. The resistance of the variable resistor will be set at a value of 1,000 ohms. The time constant (TC) of the circuit is given by:

17

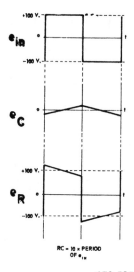

RC = 10 x PERIOD OF e_{IN}

179.537
Figure 2-9.—Integration.

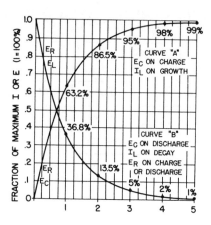

179.538
Figure 2-10.—Number of time constants =
$$\frac{t}{RC}, \frac{Rt}{L} \;.$$

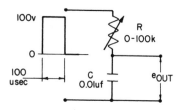

179.539
Figure 2-11.—RC circuit used to plot curves in figures 2-12, 13, and 14.

TC = RC

Substituting values:

TC = 1 x 10^3 x 1 x 10^{-8}

TC = 10 microseconds

Since the time constant of the circuit is 10 microseconds, and the pulse duration is 100 microseconds, the time constant is short in comparison to the pulse duration. The capacitor will charge exponentially through the resistor. In five time constants, the capactor will be, for all practical purposes, completely charged. At the end of the first time constant the capacitor will be charged to 63.2 volts; at the end of the second, 86.5 volts; at the end of the third, 95 volts; at the end of the fourth, 98 volts; and finally at the end of the fifth time constant (50 microseconds) the capacitor is fully charged. At 100 microseconds, the input voltage drops to zero and the capacitor starts to discharge. The capacitor discharge path is through the resistor. Thus, the capacitor will follow the same type of exponential curve on discharge as it followed on charge, and it is considered to be fully discharged after 5 TC. Therefore, at the end of 200 microseconds, the input voltage again rises to 100 volts and the whole process will repeat itself. This is shown graphically in figure 2-12.

To change the time constant, the variable resistor in figure 2-11 will be increased to a value of 10,000 ohms. The time constant will now be equal to 100 microseconds.

A graph of the input (e_{in}) and output (e_C) waveforms is shown in figure 2-13. The long sloping rise and fall of voltage is because of the capacitors inability to charge and discharge rapidly through the 10,000-ohm series resistance.

At the first instant of time, one hundred volts is applied to the intermediate time constant circuit. One time constant is exactly equal, in this circuit, to the duration of the input pulse. After one time constant, the capacitor will charge to 63.2% of the input voltage (100 volts). Therefore, at the end of one time constant (100 microseconds) the voltage across the capacitor is equal

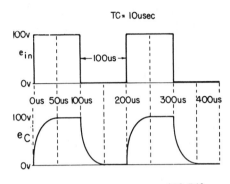

TC = 10usec

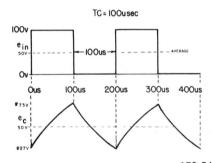

179.540

Figure 2-12.—Waveshapes when TC =
10 μsec (short TC).

TC = 100usec

179.541

Figure 2-13.—Waveshape when TC = 100 μsec.

to 63.2 volts. However, as soon as 100 microseconds has elapsed, and the initial charge on the capacitor has risen to 63.2 volts; the input voltage suddenly drops to zero, where it remains for 100 microseconds. Since the discharge time is 100 microseconds (one time constant), the capacitor will discharge 63.2% of its total 63.2 volt charge—to a value of 23.3 volts. During the next 100 microseconds, the input voltage will increase from zero to 100 volts very rapidly. The capacitor will now charge for 100 microseconds (one time constant). The voltage available for this charge is the difference between the voltage applied (100 volts) and the charge on the capacitor (23.3 volts), or 76.7 volts. Since the capacitor will only be able to charge for one time

constant, it will charge to 63.2% of the 76.7 volts, or 48.4 volts. The total charge on the capacitor at the end of 300 microseconds will be 23.3 volts plus 48.4 volts or 71.7 volts.

Notice that the capacitor voltage at the end of 300 microseconds is greater than the capacitor voltage at the end of 100 microseconds. The voltage at the end of 100 microseconds is 63.2 volts, and the capacitor voltage at the end of 300 microseconds is 71.7 volts—an increase of 8.5 volts.

The output waveform in this graph (fig. 2-13) is the waveform realized after many cycles of input signal to the integrator. The capacitor charges and discharges in a step-by-step manner until, finally, the capacitor will charge and discharge above and below a fifty volt level as shown in figure 2-13. The fifty volt level is governed by the amplitude of the symmetrical input pulse.

If the resistance in the circuit of figure 2-11 is increased to 100,000 ohms, the time constant of the circuit will be 1,000 microseconds. This time constant is ten times the pulse duration of the input pulse. It is, therefore, a long time constant circuit.

The shape of the output waveform across the capacitor is shown in figure 2-14. The shape of the output waveform is characterized by a long sloping rise and fall of capacitor voltage.

The universal time constant chart must be consulted to determine the value of charge on the capacitor at the end of the first 100 microseconds of the input signal. On the time constant chart, the percentage of voltage corresponding to 1/10 (100 microseconds/1000 microseconds) of a time constant is found to be 9.5 percent.

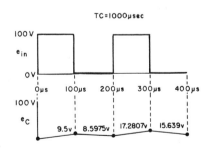

179.542

Figure 2-14.—Waveshape when TC = 1000 μsec.

19

This is accomplished by projecting a line upward from the 1/10 of a time constant point on the base to where it intersects with the capacitor charge curve. From this point, another line is drawn at right angles to the first and the percent of charge may be read from the scale at the left of the chart.

Since the applied voltage is 100 volts, the charge on the capacitor at the end of the first 100 microseconds will be 9.5 volts. At the end of the first 100 microseconds, the input signal will fall suddenly to zero; and the capacitor will discharge. It will be able to discharge for 100 microseconds. Therefore, the capacitor will discharge 9.5% of its accumulated 9.5 volts, or 0.9025 volts. The loss of the 0.9025 volts will result in a remaining charge on the capacitor of 8.5975 volts. At the end of 200 microseconds, the input signal will again suddenly rise to a value of 100 volts. The capacitor will be able to charge to 9.5% of the 91.4025 volt difference (100 volts-8.5975 volts = 91.4025 volts), or to a value of 8.6832 volts plus the initial 8.5975 volts. This will result in a total charge on the capacitor at the end of the first 300 microseconds of 17.2807 volts (8.6832 volts plus 8.5975 volts).

Notice that the capacitor voltage at the end of the first 300 microseconds is greater than the capacitor voltage at the end of the first 100 microseconds. The voltage at the end of the first 100 microseconds is 9.5 volts, and the capacitor voltage at the end of the first 300 microseconds is 17.2807 volts — an increase of 7.7807 volts.

The capacitor charges and discharges in a step-by-step manner until, finally, the capacitor will charge and discharge above and below a fifty volt level.

RL Integrator

The RL integrator is used as a waveshaping network in various types of electronic equipments such as radio, radar, television and in other special electronic applications. It is also used as an analog in performing the mathematical function of integration in computers.

The RL circuit, shown in figure 2-15 may also be used as an integrating circuit. To obtain an integrated waveform from the series RL circuit, the output must be taken across the resistor. The characteristics of the inductor are such that at the first instant of time in which voltage is applied, the current flow through the inductor is minimum; and the voltage drop across it is maximum. Therefore, the value of the voltage drop across the series resistor at the

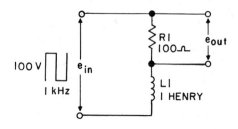

179.543

Figure 2-15.—RL integrating circuit.

same instant (first instant) of time must be negligible because there is negligible current flow through it. However, as time passes, current begins to flow through the circuit, and the voltage developed across the resistor begins to increase.

Thus, the circuit is unable to respond to the sudden changes in input voltage and the output waveform will be an integrated version of the input waveform.

DIFFERENTIATION

The RC differentiator is used to produce a pip or peaked waveform, for timing or synchronizing purposes, from a square (or rectangular) shaped input signal; to perform the electrical analog of differentiation for computer applications; and to produce specifically distorted waveshapes for special applications, such as trigger and marker pulses.

Differentiation is the direct opposite of integration. In the RC integrator, the output is taken from the capacitor. In the RC differentiator, the output is taken across the resistor. This, of course, means that when the RL circuit is used as a differentiator, the differentiated output is taken across the inductor.

The RL differentiator is used to distort an applied waveform (such as a square wave) into a peaked wave for the purpose of providing trigger and marker pulses. It is also used to electronically perform the mathematical function of differentiation in computers, and for separating the horizontal sync in television receivers.

An application of Kirchhoff's law shows the relatiohship between the waveforms across the resistor and capacitor in a series network. Since the sum of the voltage drops in a closed loop must equal to applied voltage, the graphical sum of the voltage waveforms in a closed loop must equal the applied waveform. Figure 2-16 shows the output taken across the variable resistor.

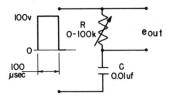

179.544

Figure 2-16.—RC circuit as differentiator.

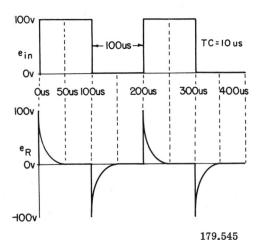

179.545

Figure 2-17.—Waveform when TC = 10 μsec.

With the variable resistor set at 1 kohm and a capacitance value of 0.01 microfarad, the time constant of the circuit will be 10 microseconds. Since the input waveform has a duration of 100 microseconds, the circuit is a short time constant circuit.

In the short time constant circuit at the first instant of time, the voltage across the capacitor is zero; and the current flow through the resistor will cause a maximum voltage to be developed across it. This is shown at the first instant of time in the graph of figure 2-17.

As the capacitor begins assuming a charge, the voltage drop across the resistor will begin to decrease. At the end of the first time constant, the voltage drop across the resistor will have decreased by a value equal to 63.2% of the applied voltage. Since there are 100 volts applied, the voltage across the resistor after one time constant will be equal to 36.8 volts. After the second time constant, the voltage across the resistor will be down to 13.5 volts. At the end of the third time constant, e_R will be 5 volts, and at the end of the fourth time constant, 2 volts. At the end of the fifth time constant, the voltage across the resistor will be very close to zero volts. Since the time constant is equal to 10 microseconds, it will take a total of 50 microseconds for the capacitor to be considered fully charged.

As shown in figure 2-17, the slope of the charge curve will be very sharp and after 5 time constants the voltage across the resistor will remain at zero volts until the end of 100 microseconds. At that time, the applied voltage suddenly drops to zero, and the capacitor will discharge through the resistor. At this time, the discharge current will be maximum, causing a large discharge voltage drop across the resistor. This is shown as the negative spike in figure 2-17. Since the current flow from the capacitor, which now acts like a source, is decreasing exponentially, the voltage across the resistor will

also decrease. The resistor voltage will decrease exponentially to zero volts in five time constants. All of this discharge action will take a total of 50 microseconds.

After 200 microseconds, the action begins again. The output waveform taken across the resistor in this short time constant circuit is an example of differentiation. With the square wave applied, the output is composed of positive and negative spikes. These spikes approximate the rate of charge of the input capacitor.

The output across the resistor in an RC circuit of intermediate time constant is shown in figure 2-18. The value of the variable resistor has been increased to 10 kohms. This means that the time constant of the circuit is equal to the duration of the input pulse (100 microseconds). For clarity, the voltage waveforms developed across both the resistor and the capacitors are shown. At all times, the sum of the voltages across the resistor and capacitor must be equal to the applied voltage.

When a pulse of 100 volts in amplitude is applied for a duration of 100 microseconds, the capacitor cannot respond quickly to the change in voltage, and all of the applied voltage is felt across the resistor. Figure 2-18 shows the voltage across the resistor, e_R, to be 100 volts and the voltage across the capacitor, e_C, to be zero volts at this time. As time progresses, the

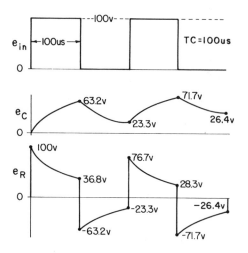

Figure 2-18.—Waveform when TC = 100 μsec.

179.546

capacitor will charge. As the capacitor voltage increases, the resistor voltage will decrease. Since the time that the capacitor is permitted to charge is 100 microseconds (equal to one time constant in this circuit), the capacitor will charge to 63.2% of the applied voltage at the end of one time constant, or 63.2 volts. Because Kirchhoff's law must be adhered to at all times, the voltage across the resistor must be equal to the difference between the applied voltage (100 volts) and the charge on the capacitor (63.2 volts), or 36.8 volts.

At the end of the first 100 microseconds, the input voltage suddenly drops to zero volts, a change of 100 volts. Since the capacitor is not able to respond to so rapid a voltage change, the 100-volt change must occur across the resistor. The voltage across the resistor must, therefore, reverse polarity and attain a magnitude of -63.2 volts. The capacitor now acts as a source and the sum of the voltage across the two components is now zero.

During the next 100 microseconds, the capacitor discharges. To maintain the total voltage at zero, the voltage across the resistor must decrease at exactly the same rate. This exponential decrease in resistor voltage is shown in figure 2-18. Since the capacitor will discharge 63.2% of its charge (to a value of +23.3 volts) at the end

of the second 100 microseconds, the resistor voltage must decrease to a value of -23.3 volts in order to maintain the total voltage at zero volts.

At the end of 200 microseconds, the input voltage again rises suddenly to +100 volts. Since the capacitor cannot respond to the 100-volt increase instantaneously, the 100-volt change takes place across the resistor. The voltage across the resistor suddenly rises from -23.3 volts to +76.7 volts. The capacitor will now begin to charge for 100 microseconds, thus decreasing the voltage across the resistor. This charge and discharge action will continue for many cycles. Finally, the voltage across the capacitor will rise and fall, by equal amounts, about a positive fifty volt level. The resistor voltage will also rise and fall, by equal amounts, about a zero volt level.

If the time constant for the circuit in figure 2-16 is increased to make it a long time constant circuit, the differentiator output will appear more like the input. The time constant for the circuit can be changed by either increasing the value of capacitance or resistance. In this circuit the time constant will be increased by increasing the value of resistance from 10 kohms to 100 kohms. This will result in a time constant of 1000 microseconds. This time constant is ten times the duration of the input pulse. The output of the long time constant circuit is shown in figure 2-19.

When a pulse of 100-volts amplitude is applied for a duration of 100 microseconds, the capacitor cannot respond instantaneously to the change in voltage and all of the applied voltage is felt across the resistor. As time progresses, the capacitor will charge and the voltage across the resistor will be reduced. Since the time that the capacitor is permitted to charge is 100 microseconds, the capacitor will charge for only 1/10 of one time constant, or to 9.5% of the applied voltage (as found using the universal time constant chart). Because Kirchhoff's law must be observed, the voltage across the resistor must be equal to the difference between the applied voltage and the charge on the capacitor (100 volts - 9.5 volts), or 90.5 volts.

At the end of the first 100 microseconds of input, the applied voltage suddenly drops to zero volts, a change of 100 volts. Since the capacitor is not able to respond to so rapid a voltage change, the 100-volt change must occur across the resistor. The voltage across the resistor must therefore reverse polarity and attain a magnitude of -9.5 volts. The sum of the voltage across the two components is now zero.

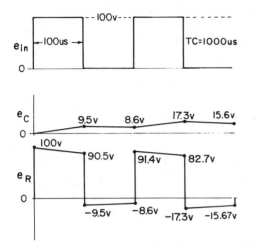

179.547

Figure 2-19.—Waveform when TC = 1000 μsec.

During the next 100 microseconds, the capacitor discharges. To maintain the total voltage at zero, the voltage across the resistor must decrease at exactly the same rate as the capacitor discharge. This exponential decrease in resistor voltage is shown in figure 2-17 during the second 100 microseconds of operation. Since the capacitor will now discharge 9.5% of its charge (to a value of +8.6 volts) at the end of the second 100 microseconds, the resistor voltage must rise, in a positive direction, to a value of -8.6 volts in order to maintain the total voltage at zero volts.

At the end of 200 microseconds, the input voltage again suddenly rises to +100 volts. Since the capacitor cannot respond to the 100-volt change instantaneously, this step-by-step action will continue until stabilization. After many cycles have passed, the capacitor voltage varies above and below the positive fifty volt level by equal amounts and the resistor voltage varies above and below the zero volt level by equal amounts.

The RC networks which have been discussed in this chapter may also be used as coupling networks. Normally, when an RC circuit is used as a coupling circuit, the output is taken from across the resistor and a long time constant circuit is used. Thus, the output waveform will closely resemble the input waveform as shown in figure 2-19.

If a pure sine wave is applied to a long time constant RC circuit (R is much greater than X_C), a large percentage of the applied voltage will be dropped across the resistor, and a small amount of voltage will be dropped across the capacitor.

RC OSCILLATORS

An RC oscillator uses a circuit consisting of resistance and capacitance to control the frequency of oscillations. This type of oscillator is used in the audio and low radio frequency range where tuned LC circuits are relatively unstable, difficult to construct, and economically unfeasible. It also offers a greater tuning range than the LC type for a specific capacitance range, so that fewer parts are needed to cover a given range of frequencies. There are two classes of RC oscillators which will produce a sine wave output, the BRIDGE and the PHASE-SHIFT. There are a number of circuit variations in each of these classes; however, the basic principles of operation are the same.

The bridge circuit oscillator uses two amplifying devices to shift the phase 360 degrees (from input to output) and a bridge network to control the frequency of operation. Normally, the bridge circuit oscillator will also incorporate an amplitude control circuit which is used to control the linearity and stability of the output signal.

The phase-shift oscillator usually consists of a single amplifying device and a series of phase-shift networks composed of resistive and capacitive elements. The amplifying device produces an initial 180° phase shift in the signal and then the phase-shift networks are used to produce an additional 180° phase shift to produce an output which has been shifted 360° with respect to the input and thus, is in phase with the input. A portion of this output is then applied to the input in order to sustain oscillations (regenerative feedback).

To provide a sinusoidal output signal, the RC oscillator must operate as a class A linear amplifier with regenerative feedback. Thus, the overall efficiency is low. As a result, this type of oscillator is generally used for laboratory and test equipments, rather than as a power oscillator.

Stability of the RC oscillator in the audio range is generally much better than that of a comparable LC circuit because the LC circuit requires a large inductor which is susceptible to disturbance from stray fields and is difficult to shield adequately for maximum stability.

23

THE WIEN BRIDGE OSCILLATOR

When the output of a linear amplifier is applied to its input, a feedback loop is produced. If the feedback is out of phase with the input (negative feedback) the amplifier output will be reduced. If the feedback is in phase (positive feedback) the amplifier can oscillate. The frequency of osciallation for positive feedback can be controlled by using a frequency selective network in the feedback loop, such as the Wien bridge. When negative feedback is applied to the emitter or cathode circuit of an amplifier, it produces a degenerative effect which reduces the output and improves the amplifier response (this is called INVERSE FEEDBACK). By use of the impedance bridge circuit a differential input can be used to provide oscillation at the desired frequency, with amplitude and waveform control.

The Wien bridge oscillator is used as a variable frequency oscillator for test equipment and laboratory equipment to supply a sinusoidal output of practically constant amplitude and exceptional stability over the audio frequency and low radio frequency ranges.

Circuit Operation (General)

The bridge circuit and feedback loop are shown in the simplified schematic (fig. 2-20). The operation of the circuit is essentially the same regardless of whether transistor or electron tube amplifier circuits are used to provide the necessary amplification; therefore, the general symbol for an amplifier (triangular symbol) is used to represent each stage of amplification in the schematic.

In the actual circuit, R4 is a small incandescent lamp with a tungsten filament, and is normally operated at a temperature that gives automatic control of amplitude (thermistors and varistors are also used). Resistors R1 and R2 are of equal value (as are capacitors C1 and C2), with R3 having twice the resistance of lamp R4 at the operating temperature. The bridge is balanced and the circuit oscillates at the operating frequency (f_O), as determined by the equation:

$$f_O = \frac{1}{2\pi\ R1C1}$$

It can be seen by inspection that resistors R3 and R4 form a resistive voltage divider across which the output voltage of A2 is applied.

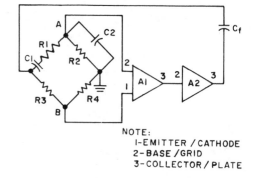

NOTE:
1-EMITTER / CATHODE
2-BASE /GRID
3-COLLECTOR / PLATE

20.160(179(A)

Figure 2-20.—Simplified Wien-bridge oscillator.

Since these resistors are not frequency responsive, the voltage at any instant from point B (emitter/cathode of A1) of the bridge with respect to ground is dependent upon the ratio of R3 to R4 for any frequency which the amplifier produces at its output. Components R1, C1 and R2, C2 form a frequency-responsive reactive voltage divider between the output of A2, the base/grid of A1 (point A) and ground. Thus the voltage across R2 is applied to the input of the amplifier (between the base/grid and ground). When the voltage between point A and ground is in phase with the output voltage of A2, maximum voltage will appear between the base/grid and ground; therefore, maximum amplification occurs and a large output voltage is produced by A2. Two amplifier stages, A1 and A2, are used to produce a total phase shift of 360 degrees (180 degrees in each stage) to ensure that the voltage at the output is in phase with the input. Thus reactive networks R1, C1 and R2, C2 are not required to shift the phase to produce oscillation, but are used to control the frequency at which oscillation takes place.

The manner which these various feedback voltages vary amplitude and phase are best shown by graphic representation (fig. 2-21). The center (dotted) vertical line (ordinate) represents the frequency at which the oscillator operates. The $-f_O$ and $+f_O$ vertical lines represent, respectively, frequencies much lower than and much higher than the operating frequency.

Curve A represents the negative feedback between point B of the simplified schematic and

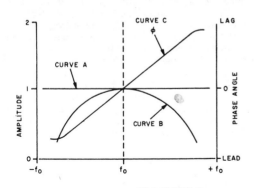

20.160(179)C
Figure 2-21.— Phasing diagram.

ground. Since it is the same at all frequencies, it is represented by the horizontal line across the middle of the graph. Curve B represents the positive feedback voltage existing between the base/grid (point A) of A1 and ground, or the voltage across R2. At frequencies below the operating frequency, the series reactance of C1 is large, and the voltage across R2 is reduced. As the operating frequency is approached, the reactance diminishes, and the voltage across R2 reaches a maximum at f_0. As the frequency is increased above f_0, the parallel reactance of C2 shunts R2, effectively reducing the voltage across R2. Thus, the voltage across R2 is reduced both above and below f_0 and is maximum only at the operating frequency. At this frequency, the positive feedback voltage (at the base/grid) is exactly equal to (or slightly greater than) the negative feedback voltage (at the emitter/cathode), and amplification is at a maximum for A1 and A2. Now consider the phase of the output voltage which is fed back to the input of A1, as shown by curve C. Because of the phase shift produced by R1, C1 a phase shift occurs above or below the operating frequency. At the operating frequency however, the phase change is zero, and the output of A2 is exactly 360 degrees from the input voltage because of the phase inversion through the two stages of amplification. Thus, below f_0 the phase angle leads and above f_0 it lags. The out of phase voltage above and below f_0, together with the decrease in the regenerative feedback voltage applied to the base/grid as compared with the degenerative feedback applied to the emitter/cathode of A1 effectively stops oscillation at all frequencies

except the operating frequency, where R1C1 equals R2C2.

Semiconductor Wien
Bridge Oscillator

The semiconductor Wien bridge oscillator is shown schematically in figure 2-22. The operation of this circuit is similar to that of the basic circuit described above, with the oscillation frequency being at the frequency where R1C1 equals R2C2 and with waveform linearity retained by inverse feedback through R3 and R4. The bridge arrangement can be seen by comparing the components inside the dashed line with the bridge of figure 2-20.

Voltage divider base bias is used, with R2 and R5 biasing Q1, and R7 and R8 biasing Q2. Temperature stabilization is provided by DS1(R4).

The resistance of R3 and DS1 (R4) form the resistive arm of the bridge across which the output of Q2 is applied; a portion of this voltage appears across DS1 (R4) as a negative feedback, being in phase with the emitter voltage. DS1 (R4) is an incandescent lamp. However, it could be a thermistor with a positive temperature coefficient. When a lamp is used, it is operated at a current which produces a temperature sensitive point (where resistance varies rapidly with

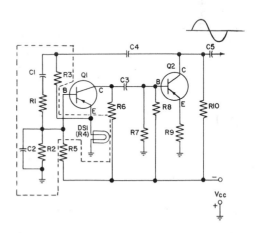

20.444
Figure 2-22.—Wien bridge oscillator using
PNP transistor.

temperature); when a thermistor is used, it is selected to have the desired temperature-current characteristic. In either case, the bias developed across this resistance is in opposition to the normal (forward) bias, and produces a degenerative effect. The feedback voltage is of the same polarity as the degenerative bias and increases the degeneration. Since the output of the voltage divider is not frequency sensitive, the feedback voltage is always constant regardless of the frequency of operation. At frequencies other than the frequency of operation the degenerative feedback predominates and prevents oscillation. At the frequency of operation, which is controlled by the bridge reactive arms consisting of R1, C1 and R2, C2 the positive feedback is a maximum. This in-phase feedback signal is applied to the base and is slightly greater than the negative feedback at the balance point or frequency of operation.

The amplified output of Q1 is developed across collector resistor R6, and it is applied by capacitor C3 and base resistor R7, which form a conventional resistance-coupling network (designed for minimum phase shift), to the input (base) of transistor Q2. The signal is further amplified by Q2, and the voltage developed across collector resistor R10 is supplied as an output through capacitor C5, and, as a positive feedback, through C4 to the bridge network. Note that the Q2 emitter resistor, R9, is not bypassed and that the circuit of Q2 is therefore degenerative. Note also that R10, C4 and C5 are designed to provide a minimum amount of phase shift. Thus, with a highly degenerative two-stage amplifier and class A bias, the output signal is essentially a pure sine wave.

Since the coupling networks are arranged for minimum phase shift, the phase shift required for regeneration is obtained from the inverting action of the common-emitter configuration, with each amplifier stage providing a 180-degree shift. The feedback input signal is thus shifted 360 degrees in phase to produce a regenerative feedback independent of circuit parameters. The reactive portion of the bridge (C1, C2 and R1, R2) determines the frequency at which maximum amplification (and feedback) occurs. DS1 (R4) controls the degenerative feedback and also the output amplitude; that is, when the input signal to Q1 increases, more emitter current flows through DS1 (R4) and the lamp or thermistor resistance increases, producing a degenerative voltage which opposes the input signal, and tends to restore it to the original operating value by reducing the amount of amplification through the

feedback loop. This oscillator then, with amplitude stability, temperature stabilization, and degenerative feedback to control the waveform, and with an RC frequency-selective circuit to determine the frequency of operation provides a signal of excellent stability and pure waveshape for test applications. In order to control the frequency, either resistors R1 and R2 or capacitors C1 and C2 are changed in value or made variable. With a two-gang variable capacitor and a selector switch (or fixed and variable resistors), a continuous range of frequencies over a number of bands may be obtained.

Electron Tube Wien
Bridge Oscillator

The schematic of an electron tube Wien bridge oscillator is shown in figure 2-23. The bridge comparison with figure 2-20 and the components inside the dashed line can be made, as with the transistor circuit. Except for bias arrangement, this circuit is practically identical to the transistor Wien bridge oscillator just discussed. Biasing is a combination of cathode and contact bias, with a large amount of degeneration (inverse feedback) being provided by the unbypassed cathode bias circuits; the output waveform is extremely linear. The output amplitude is small because of the large amount of degeneration employed, and circuit stability is excellent with a minimum of phase shift or frequency variation.

THE RC PHASE-SHIFT
OSCILLATOR

The RC phase-shift oscillator is used to provide a sine wave output of relatively constant amplitude and frequency in the audio and low radio frequency range where the use of LC circuits is impractical due to design considerations.

The basic RC phase-shift oscillator is actually a modification of a conventional RC coupled amplifier (single stage), as shown in figure 2-24. Since in the conventional amplifier, the output is shifted in phase 180° with relation to the input it will be necessary to shift the output signal an additional 180° in order to provide a feedback signal which is in phase with the input (regenerative). This may be accomplished through the use of three or more inverted-L type RC sections as shown in figure 2-24.

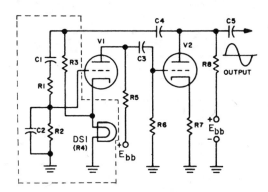

20.160(179)B
Figure 2-23.—Wien bridge oscillator.

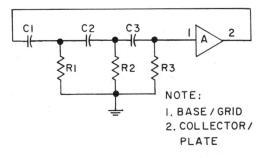

NOTE:
1. BASE / GRID
2. COLLECTOR/
PLATE

33.218(179)
Figure 2-24.—RC phase-shift oscillator
(basic).

Circuit Operation (General)

The current in a series circuit comprised of resistance and capacitance is determined by the applied voltage divided by the series impedance of the components $(I = \frac{E}{Z})$. Since a series RC circuit exhibits capacitive reactance, the current leads the applied voltage by a specific phase angle. The phase angle is determined by the relationship of resistance and capacitance. The voltage drop produced across the resistance is determined by the current through the resistance, and therefore, leads the applied voltage by a given phase angle.

A series resistance-capacitance circuit is shown in figure 2-25, together with its vector diagram. Assuming that this circuit represents the first section of the filter network (fig. 2-24), it is across this section that the amplifier output is impressed. The input voltage to this circuit (amplifier output voltage) is designated E_a in figure 2-25.

The values of capacitive reactance and resistance for the circuit are chosen so that at the operating frequency the ratio of capacitive reactance (X_C) to resistance (R) is such that the current in the circuit will lead the applied voltage by 60 degrees. Since the voltage drop across a resistance is in phase with the current flow through the resistance, E_R (the circuit output voltage) leads E_a (the applied voltage) by 60 degrees. If all sections of the filter are alike, then E_R, when applied to the second section of the filter, will be shifted an additional 60 degrees in phase and the output of the second section of the filter will have been shifted in phase a total of 120 degrees with relation to E_a. The output of the second section of the filter is then applied to the third section of the filter where it will be shifted in phase an additional 60 degrees for a total phase shift through the filter of 180 degrees. Thus, the output of the filter when applied to the input of the amplifier will be of the correct phase and of sufficient amplitude (assuming circuit losses are minimal) to maintain oscillations.

Since X_C varies with frequency, it is apparent from the vector diagram (fig. 2-25) that the phase shift through each section of the filter will vary and the total phase shift through the filter will be something other than 180 degrees at frequencies other than the one for which the circuit was designed. Thus, the circuit will not oscillate at a frequency other than its design frequency

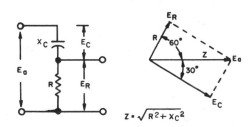

179.548
Figure 2-25.—Vector analysis of RC circuit.

27

since the feedback voltage will not be of the correct phase to sustain oscillations.

The RC phase-shift oscillator is normally fixed in frequency, but the output frequency can be made variable over a range of frequencies by providing ganged variable capacitors or resistors in the phase-shift network. An increase in the value of either R or C will produce a decrease in the output frequency; conversely, a decrease in the value of either R or C will produce an increase in the output frequency.

By increasing the number of phase-shift sections comprising the network, the losses of the total network can be decreased; this means that the additional sections will each be required to have a lesser degree of phase shift per section so that the overall phase shift of the network remains at 180 degrees for the desired frequency of oscillation. Since the loss per section is decreased as the amount of phase shift (per section) is reduced, many oscillators use networks consisting of four, five, and six sections. Assuming that the values of R and C are equal for each section, the individual sections are designed to produce phase shifts per section of 45, 36, and 30 degrees, respectively.

Semiconductor RC
Phase-Shift Oscillator

The RC phase-shift oscillator circuit (fig. 2-26) uses a PNP transistor connected in the common-emitter configuration to provide the necessary amplification. Resistors R1, R2, and R3 and capacitors C1, C2, and C3 comprise the feedback and phase-shift network. Resistors R3 and R4 establish base bias for the PNP transistor. Resistor R5 is the emitter swamping resistor, which prevents large increases in emitter current and causes the variation of emitter-base junction resistance to be a small percentage of the total emitter circuit resistance. Capacitor C4 bypasses the emitter swamping resistor, R5, and effectively places the emitter at signal ground potential. Resistor R6 is the collector load resistance across which the output signal is developed. Capacitor C5 is the output coupling capacitor.

Oscillations are started when input power is first applied to the circuit. A change in the base current results in an amplified change in collector current which is shifted in phase 180 degrees. The output signal developed across the collector load resistance, R6, is returned to the transistor base as an input signal inverted 180 degrees by the action of the feedback and phase-shift network, making the circuit regenerative.

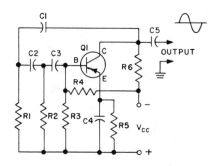

20.443
Figure 2-26.—RC phase-shift oscillator using PNP transistor.

The output waveform is essentially a sine wave at a fixed frequency. When fixed values of resistance and capacitance are used for the feedback network, the 180-degree phase shift occurs at only one frequency. At all other frequencies, the capacitive reactance either increases or decreases, causing a variation in phase relationship; thus, the feedback is no longer in phase. Note, however, that if the components comprising the phase-shift network should change value, the frequency of oscillation will change to the frequency at which a phase shift of 180 degrees will occur to sustain oscillations.

Electron Tube RC
Phase-Shift Oscillator

In the circuit shown in figure 2-27, a sharp cutoff, pentode tube is used as the amplifier tube; however, a triode tube can be employed in a similar circuit. Bias voltage is developed across cathode resistor R4. Cathode bypass capacitor C4, by virtue of its filtering action, keeps the bias voltage relatively constant and places the cathode at signal ground potential. The sine wave voltage is developed across plate load resistor R6; capacitor C6 is the output coupling capacitor. Any variation in plate current will cause a corresponding change in plate voltage. These plate voltage variations will also be present at the grid of the tube, since the plate is coupled to the grid through the phase-shift network.

Oscillations are initially started in this circuit by small changes in E_{bb} or by random noise. It it were not for the action of the phase-shift network, the voltage variations fed from the plate

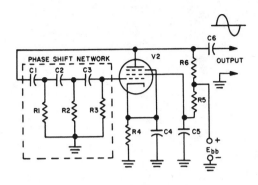

179.549

Figure 2-27.—Electron tube RC oscillator
circuit.

variable resistors or capacitors in the phase-shift network. An increase in the value of resistance or capacitance will decrease the operating frequency; conversely, a decrease in the value of resistance or capacitance will increase the operating frequency. In several practical applications of this circuit, three or more fixed RC sections are employed together with a variable section to provide a limited range of output frequencies which are determined by the setting of a variable capacitor. In this circuit variation, the fixed RC sections use values of R and C which will provide a phase shift somewhat less than 180 degrees at the operating frequency desired, and the last (variable) RC section completes the required phase shift to exactly 180 degrees. The operating frequency is then determined by the setting of the variable capacitor.

back to the grid of the tube would cancel the plate current variations, since the tube introduces a polarity inversion between the grid and plate signals. For example, if the plate voltage variation at any instant of time was positive, the positive variation would be present on the grid. This positive going grid voltage would then cause the plate current to increase, in turn causing the plate voltage variation to go negative and, thus, cancelling out the original grid voltage variation.

Assuming that the plate voltage variations are applied to the grid 180 degrees out of phase with respect to the initial grid signal voltage, maximum degeneration (or cancellation) will occur. However, if the plate voltage variations fed back to the grid approach zero-degree phase difference, minimum degeneration will occur. Therefore, if the phase difference between the plate voltage variations and the initial grid signal voltage is exactly zero (in phase), the plate voltage variations will reinforce the grid signal voltage at any instant of time, causing regeneration; furthermore, these variations will be amplified by the tube and reapplied to the grid, amplified again, and so on, until a point of stage equilibrium is reached and no further amplification takes place. The phase-shift network provides the required phase shift of 180 degrees to bring the voltage fed back to the grid in phase with the initial grid-signal voltage and cause regeneration. The circuit then oscillates under these conditions with relatively constant amplitude.

The phase-shift oscillator is designed primarily for fixed frequency operation; however, the operating frequency can be made variable by using

LIMITERS

Limiting refers to the detachment (i.e., elimination) of one or both extremities of a waveform or parts thereof. Clipper is another name for a limiter because the output waveform presents a clipped appearance on the wave peaks and the two terms may be used interchangeably. A wide variety of applications require utilization of limiters such as: modulation compression in transmitters, which allows a higher average percent of modulation; noise limiting in communications and FM receivers to improve fidelity; and wave shaping in radar timing circuitry. An example of waveshaping is limiting or clipping both portions of a sinusoidal waveform to produce a close approximation of either a square wave or a rectangular wave.

Clipping may be accomplished with diodes or amplifying devices (transistors or tubes). Diode limiters or clippers may be classified according to the way they are connected (series or parallel). Clipper circuits are also categorized as positive or negative lobe limiters. A positive lobe circuit abolishes either part or all of that portion of a waveform which is positive in respect to some reference level. Conversely, the negative lobe limiter affects a waveform's negative portion.

Circuits which implement the limiting operation will be explained in the following sequence: series positive and negative (lobe) limiters, parallel positive and negative (lobe) limiters, dual diode limiters, grid limiters, saturation limiters, and cutoff limiters.

29

SERIES DIODE LIMITERS

The series diode limiter consists of a diode in series with a load resistor. The polarity of the diode with respect to ground determines which input polarity is clipped. Figure 2-28 is the schematic diagram of a positive lobe series limiter, which is used when it is necessary to square off part, or all, of the positive portion of an input waveform and allow the negative portion to pass without modification.

During the positive alternation, D1 will be reverse biased (cathode positive with respect to plate or anode) and little or no current will flow, since the impedance of the diode (termed the reverse resistance) will be quite high. Normally, the load resistance will be made small in comparison to the reverse resistance of the diode. Thus, most of the input voltage is dropped across D1 and little or no voltage appears in the output (E_{RL} will be negligible).

During the negative alternation, D1 will be forwarded biased (cathode negative with respect to plate or anode) and its impedance (now termed the forward resistance) will be quite low in comparison to R_L. Under these conditions, considerable current will flow (limited primarily by RL) and most of the input voltage will appear in the output ($E_{RL} = E_{in} - E_{D1}$).

The larger R_L is made in comparison to the forward resistance of the diode, the more efficient the circuit becomes. However, if the value of R_L should approach or exceed the reverse resistance of the diode, effective limiting action will no longer take place. That is, as R_L is made larger in comparison to the reverse resistance of the diode, more and more of the positive alternation will appear in the output.

Figure 2-29 depicts the schematic diagram of a series negative lobe limiter. The diode conducts on the positive alternation. This pro-

duces a positive output, thus clipping the negative alternation or lobe.

Many applications do not require complete elimination of a lobe. Partial clipping may be accomplished by adding a bias voltage. Figure 2-30 shows a battery bias network connected in negative lobe limiter circuits. In figure 2-30A, the diode will no longer conduct when the input voltage level is less than the bias voltage and the bias voltage continues to maintain an output level when the diode is not conducting. In figure 2-30B, the diode will continue to conduct until the input is more negative than the bias voltage.

Figure 2-31 illustrates a biased positive lobe diode limiter. This circuit uses a voltage divider network to establish a bias voltage level. The output also shows the results of clipping three different waveshapes. Decreasing the bias level in this circuit will increase the amount of clipping. If the polarity of the bias is reversed, clipping will be even more severe.

Parallel Diode Limiters

Parallel diode limiters offer an alternate method of employing diodes in clipping circuits. In this configuration, the diode again forms a series network with a resistor across the source terminals. However, the output is taken across the diode so that the output current flows in parallel with the diode. Since the diode shunts the load, an appreciable output will be obtained only when the diode is cut off.

A positive lobe parallel diode limiter is shown in figure 2-32. The positive input alternation causes D1 to conduct. Since the series dropping resistance is large compared to the conduction resistance of D1, most of the voltage is dropped across R_S during this portion of the input signal. Consequently, the output voltage is close to zero. The exact output voltage is determined by the ratio of the conduction resistance of D1 to the value of R_S. Ideally, the diode would be a short

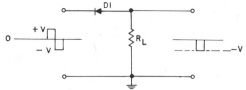

179.550
Figure 2-28.—Series positive lobe diode limiter.

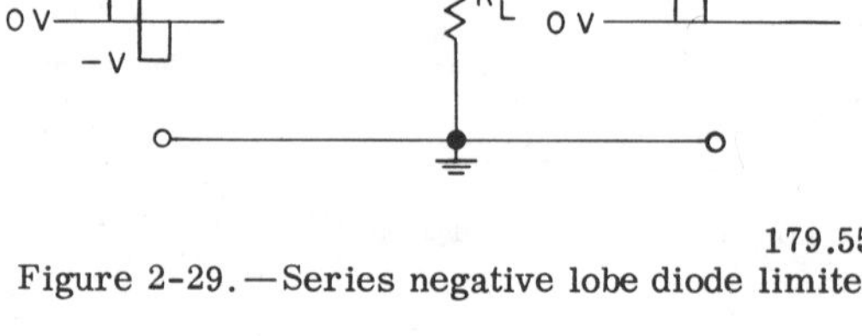

179.551
Figure 2-29.—Series negative lobe diode limiter.

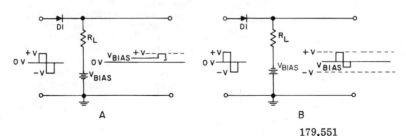

179.551

Figure 2-30.—Biased series negative lobe diode limiter.

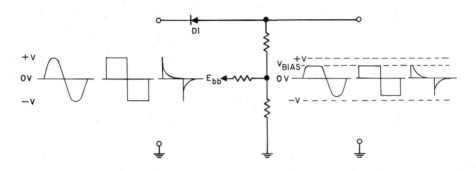

179.553

Figure 2-31.—Biased series positive lobe limiter.

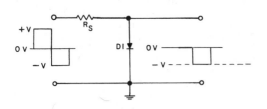

179.554

Figure 2-32.—Positive lobe parallel diode
limiter.

circuit when it conducts, resulting in a zero output voltage during the positive portion of the input. During the negative alternation, D1 is cut off and current is permitted to flow through the load.

Reversing D1 results in negative lobe clipping. The input and output waveforms, as well as the schematic diagrams for the parallel negative lobe diode clipper, are contained in figure 2-33.

As in series limiters, biasing the circuit results in clipping at levels other than at the zero reference. An input voltage may be limited to any desired positive or negative value by holding the proper diode electrode at that voltage by means of a battery or a biasing network. In figure 2-34, two such circuits are shown.

The cathode of the diode in figure 2-34A is more positive than the plate by the value of V_{bias} when no signal is applied at the input. As long as the input voltage remains less positive than V_{bias} the diode acts as an open switch and the output equals the input minus the drop across R. If the input increases to a value greater than V_{bias}, the diode conducts and behaves as a closed switch which effectively connects the upper right output terminal to the positive terminal of the battery.

31

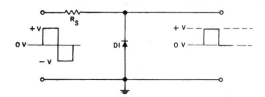

179.555

Figure 2-33.—Parallel negative lobe diode clipper.

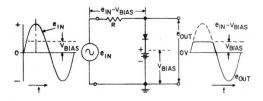

(A) POSITIVE LIMITING

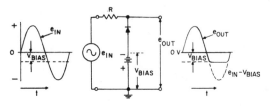

(B) NEGATIVE LIMITING

Figure 2-34.— Parallel diodes limiting above and below ground potential. 179.556

Thus, during the portion of the cycle that the input voltage equals or exceeds V_{bias}, the difference between e_{in} and V_{bias} (neglecting e_p) appears as an i_R drop across the resistor (R) and E_{out} will equal V_{bias}.

The anode of the diode in figure 2-34B is negative by the value of battery voltage, V_{bias}. Thus, as long as the input is positive or is less negative than V_{bias} the diode will not conduct and the output voltage, e_{out}, is equal to the input voltage. When the input becomes more negative than V_{bias}, the diode conducts and effectively connects the junction of the resistor and diode

to the negative terminal of the battery. During this portion of the input cycle, e_{out} equals V_{bias} and the difference between e_{in} and V_{bias} appears as an i_R drop across R.

It is sometimes desirable to pass only the positive or negative extremity of a waveform on to a succeeding stage. To accomplish this, the parallel diode limiters shown in figure 2-35 can be employed. In figure 2-35A, the entire portion of the input waveform above the negative potential, V_{bias}, causes the diode to conduct, thus producing an output voltage which varies between the negative level of V_{bias} and the negative extremity of the input. In figure 2-35B, the diode conducts during the entire portion of the input waveform which is below the positive potential of V_{bias}. The output voltage then varies between the positive level of V_{bias} and the positive extremity of the input waveform. In either case, the difference between the value of V_{bias} and e_{in}, during the time the diode conducts, is represented by the i_R drop across the series resistor, R. In actual circuitry, the batteries would be replaced by voltage divider networks and power supplies.

Dual Diode Limiters

Dual diode limiters are used to limit both positive and negative amplitude extremities. With the elimination of both peaks, the remaining signal is generally of square wave shape; therefore, this circuit is often used as a square wave generator. A circuit connected to provide this type of limiting is illustrated in figure 2-36.

Diode D1 conducts whenever the positive portion of the input signal exceeds E_1, thus limiting the positive output to the value of E_1. This results from the fact that D1, in effect, connects the output terminals across E_1. D2 is nonconducting, or an open circuit, during this time. The difference between e_{in} and E_1 appears as an i_R drop across R.

Diode D2 conducts whenever the negative portion of the input signal exceeds E_2, thus limiting the output to the value of E_2 during the negative half cycle. During this time, the output terminals are connected, in effect, across E_2, and D1 is an open circuit. By increasing the severity of the clipping, the circuit may produce a close approximation of a rectangular or square wave.

An alternate method of clipping both extremities is to connect two zener diodes back to back across the input terminals as shown in figure 2-37. During the positive portion of the input signal, D1 is in the forward biased direction and offers little impedance to current. However,

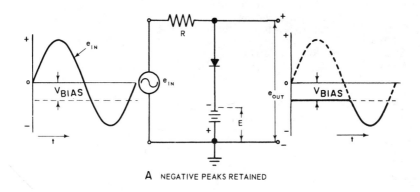

A   NEGATIVE PEAKS RETAINED

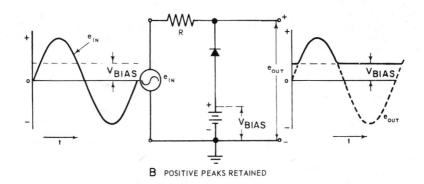

B   POSITIVE PEAKS RETAINED

179.557
Figure 2-35.—Parallel diode limiters that pass peaks only.

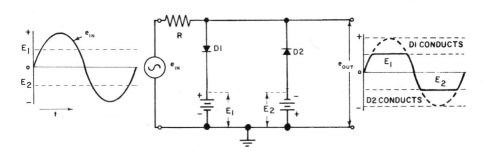

179.558
Figure 2-36.—Double-diode limiter.

33

the impedence of D2 compared to R_S is very large until the zener breakdown potential is reached. When the zener breakdown level occurs, the output voltage will no longer increase and therefore will be limited to V_{Z1}. The negative input alternation causes D1 to limit the output excursion to V_{Z2}.

Diode limiters discussed prior to this point required an input signal strength appreciably greater than a few tenths of a volt. This is because a diode has an appreciable forward resistance with low values of forward voltage. Figure 2-38 illustrates a graph of forward voltage to forward current relationships for a diode.

This high forward resistance characteristic of a diode may be used to an advantage for limiting small signal amplitudes. Figure 2-39 shows such a limiting circuit. The diodes are chosen for their high resistance with low forward voltages and for a sharp decrease in resistance past that level. For purposes of illustration, let it be assumed that a silicon diode requires five-tenths of a volt to cause appreciable conduction. During the positive alternation until the .5-volt level is reached the forward resistance of D1 is high compared to R_S so that the input voltage is dropped primarily across D1. As the input voltage is made more positive than 0.5 volts, there will be a sharp increase in the current flow in the circuit when D1's resistance starts to drop. The change is such that the drop across D1 now remains essentially constant at, or near, 0.5 volts while the drop across R_S increases with the applied voltage. Thus, the output is limited to approximately 0.5 volts since, at this time, D2 is reverse biased and acts as an open circuit (has no effect). During the negative alternation, D2 conducts and the action for D2 is the same as for D1 with the exception that the output voltage will now be limited to a -0.5 volts.

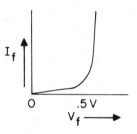

179.560
Figure 2-38.—Diode characteristic curve.

Grid Limiting

Three element devices may produce clipping or limiting action by being biased at or near saturation or cut off. When used as a grid limiter, the action is similar to that of the parallel diode limiter. Figure 2-40 is an illustration of a grid limiter circuit.

The grid and cathode act as a diode. That is, an output taken from grid to ground would be identical to that of a parallel diode clipper. In certain applications, an output may be taken from the grid with respect to ground. However, in most applications the output is taken from the plate with respect to ground and is, of course, reversed in phase. The circuit parameters are such that when the positive alternation of a signal is applied to the limiter, the grid will draw current. For all practical purposes the positive alternation is effectively dropped across R_S, since the size of R_S is large compared to the conduction resistance of the grid to cathode. With no appreciable change in voltage on the grid, plate voltage also remains constant. As the negative input alternation is applied the grid is no longer able to draw grid current, and the entire negative input alternation is developed across the extremely high grid to

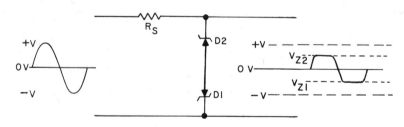

179.559
Figure 2-37.—Zener diode limiter.

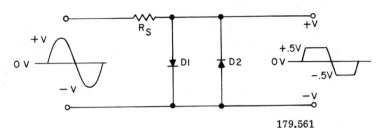

179.561

Figure 2-39.—Limiting using forward biased diodes.

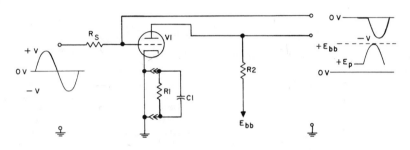

179.562

Figure 2-40.—Grid limiter.

cathode resistance. This negative grid voltage controls the conduction of the tube. It decreases plate current and increases plate voltage, therefore causing the grid and plate signals to be of opposite phase. Adding R1 and C1 in the cathode circuit will cause the operating point to shift (and provide protective bias for the tube). Current flow from ground through R1 to the plate establishes a voltage drop across R1. Now the grid cannot draw current until the positive input is above that bias potential. Input levels above the bias level are dropped across R_S due to grid current. By changing the value of R1, any desired level of positive clipping may be achieved. If an output is obtained from the plate, the negative alternation would be clipped because of the phase reversal characteristic.

With a large enough negative input alternation to the grid limiter, the tube will be driven into cutoff. The output from the plate would thus be clipped or limited on both extremities.

For any amplifier, there are limits as to how large an input signal it can handle without clipping occurring in the output. Figure 2-41 shows typical characteristic curves for solid-state and electron tube amplifiers.

Methods of establishing the point of operation of an amplifier are discussed in Volume 1, chapters 3, 6, 7, 13, and 14. Biasing at point C will clip or limit one alternation, while biasing at point A will clip or limit the other. Biasing away from either extremity by a small amount, produces partial limiting if the input signal is sufficiently large.

Circuits biased near point C are called SATURATION LIMITERS. It is to be remembered that saturation is achieved with smaller inputs by reducing source voltage and increasing R_L. Therefore, saturation limiters are normally operated with low plate voltages. Circuits operated near point A, with an input signal such that clipping occurs, are called CUTOFF LIMITERS.

If a circuit is biased at point B and a large input signal is applied, the amplifying device reaches both cutoff and saturation. Figure 2-42 shows how a square wave may be produced from a sinusoidal wave using a saturation and cutoff limiter. This circuit is commonly called an OVERDRIVEN AMPLIFIER.

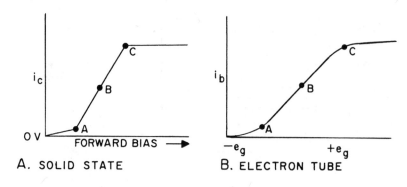

179.563
Figure 2-41.—Amplifier characteristic curves.

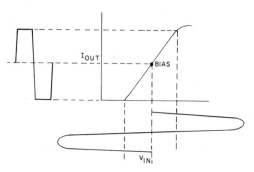

179.564
Figure 2-42.—Output plotted against input
for an overdriven amplifier.

CLAMPERS

A circuit that introduces a d.c. reference level into a pulsed or a.c. waveform is called a CLAMPING CIRCUIT, or D.C. RESTORER.

Before discussing clamping circuits, it will be helpful to briefly review the action of RC networks when used for interstage coupling in amplifier circuits. The primary purpose of the coupling capacitor (sometimes referred to as a blocking capacitor) in such a circuit is to prevent (or block) the d.c. potential applied to the collector or anode of one stage of the amplifier from appearing at the input to the next stage. Thus, only the varying component (signal component) of the collector or anode voltage of the first stage is allowed to appear at the input of the next stage where it will be referenced to some predetermined level (usually ground or zero potential). That is, if the lower end of the base or grid resistor is grounded, then the reference level is ground and the signal varies above and below ground as shown in figures 2-43A and 2-43B. Thus, the input of an ordinary RC coupling network is alternating in character about the average voltage level of the applied waveform. After the coupling capacitor charges to the average collector or anode voltage, any decrease in the collector or anode voltage causes the output voltage of the RC network to swing negative and any increase above the average causes the output voltage to swing positive.

If a biasing potential is employed, the signal applied to the base or grid will vary above and below this d.c. bias voltage as shown in figure 2-44.

For a class A amplifier, the biasing potential is adjusted to the center of the operating range and the signal is kept within the limits of this range.

In other circuits, however, it may be desired that the entire voltage excursion be above or below a fixed reference level. In such applications, a clamping circuit is used to maintain either extremity (positive or negative) of the applied waveform at some fixed level as shown in figure 2-45. The fixed level or reference level is determined by the design of the clamper.

To obtain positive clamping, the maximum negative point of a waveform is positioned on some value of d.c. reference voltage in such a

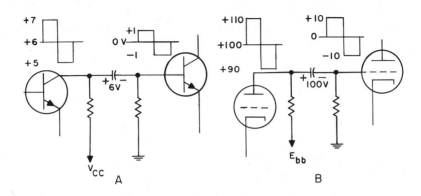

179.565
Figure 2-43.—RC coupling.

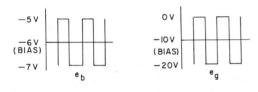

179.566

Figure 2-44.—Base/grid waveforms of figure
2-43 if biasing is employed.

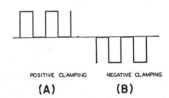

179.567

Figure 2-45.—Positive and negative clamping.

manner as to cause the entire waveform to lie in a more positive area than the reference voltage. This is shown in figure 2-45A.

To obtain negative clamping, the maximum positive point of the waveform is positioned on some value of d.c. reference voltage in such a manner as to cause the entire waveform to lie in a more negative area than the reference voltage as shown in figure 2-45B.

Clampers must not appreciably change the shape of a waveform in any manner. Any significant distortion which results from clamping is an indication of clamping circuit deficiencies in either component types or values.

POSITIVE DIODE
CLAMPER (UNBIASED)

Perhaps the simplest type of clamper is the diode clamper which utilizes a diode in conjunction with a series RC network. In figure 2-46, two versions of a positive clamper of this type

are shown. The circuit in figure 2-46A utilizes a semiconductor device while the circuit in figure 2-46B utilizes an electron tube. The operation of either circuit is essentially the same.

In general, whenever the input voltage is such as to cause the diode to conduct, the circuit will have a very short time constant since the forward or conducting resistance of the diode is quite low. Thus, the charging rate of the capacitor will be such that the charge on the capacitor will closely follow the rise in input voltage. (The capacitor charges to very nearly the peak value of the input voltage.) When the input voltage is such as to cause the diode to be cut off, the TC of the circuit will be quite long since the resistance (R) is usually made quite large. Under these conditions, the capacitor discharges very slowly. Assuming that the period the diode is cut off is relatively short in comparison to the TC of the circuit, then the capacitor will lose very little

37

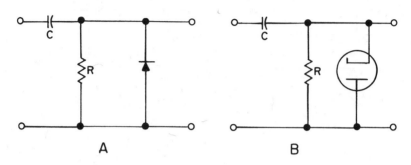

179.568
Figure 2-46.—Unbiased positive clamper.

of its charge between the end of one charge period and the beginning of the next. Thus, after a very few cycles the average charge on the capacitor will nearly equal the peak value of the input voltage and will remain at this level as long as the same input signal is applied. The output voltage now varies from zero to twice the peak value of the input voltage i.e., the voltage across R varies from near zero to that of the input in series with the voltage across C.

To better understand this action, assume that a 500-Hz square wave whose amplitude varies between the positive 50-volts level and the negative 50-volts level is applied as an input signal to one of the circuits in figure 2-46. It is further assumed that the forward resistance of the diode is 500 ohms and that the values assigned to the capacitor and resistor are 0.1 µfd and 1 megohm, respectively. Then, through the use of figure 2-47, the voltages which are present in the circuit at any given instant of time can be determined. At point A in figure 2-47, e_{in} is zero as are e_C and e_R. (Note—e_R equals e_{out}.) These are the conditions which exist in the circuit prior to the application of the signal.

At point B in figure 2-47, e_{in} equals 50 volts, e_R equals 50 volts, and e_C equals 0 volts. These are the conditions which exist in the circuit on initial application of a positive going voltage to the circuit. At this time the diode is cut off since its cathode is more positive than its anode or plate and the charge path for the capacitor will be through the resistance (R).

At point C in figure 2-47, e_{in} is still 50 volts, e_R has dropped to about 49.5 volts and e_C has charged to a negative 0.5 volts. These are the conditions which exist in the circuit just

prior to the time the input voltage switches to a negative 50 volts. Since the diode was cut off for the time period B to C, the TC for the circuit will be very long in comparison to this time period (100,000 usec as compared to 1000 usec shown for the period B to C), thus the capacitor is unable to assume much of a charge.

At point D in figure 2-47, e_{in} has switched from the plus 50 volts level to the negative 50 volts level, e_C equals negative 0.5 volts and e_R equals a negative 50.5 volts (e_{in} plus e_C). These are the conditions which exist in the circuit immediately after the switching action took place. At this time, the diode will be forward biased and start to conduct. When the diode conducts, the TC of the circuit drops to a very low value (50 usec). Under these conditions, the capacitor rapidly discharges to 0 volts and starts to charge in the direction opposite to which it was originally charged.

At point E in figure 2-47, e_{in} is still a negative 50 volts, e_C equals a positive 50 volts and e_R equals 0 volts. These are the conditions which exist in the circuit just prior to the beginning of the second cycle of input voltage. Since the TC is short in comparison to the period D to E, the capacitor has had plenty of time to charge to the peak value of the input voltage and little or no current now flows in the circuit.

At point F in figure 2-47, e_{in} has switched to the positive 50-volt level, e_C equals a positive 50 volts, and e_R equals 100 volts (e_{in} plus e_C). These are the conditions which exist in the circuit immediately after the switching action takes place.

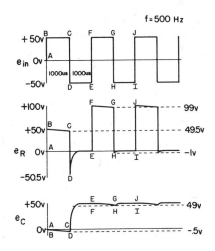

f = 500 Hz

179.569
Figure 2-47.—Unbiased positive diode
clamper waveforms.

of the input waveshape due to the reactive nature of the RC network. As we previously discussed, a square wave is assumed to be composed of a fundamental frequency plus an infinite number of odd harmonics of this frequency, and the RC network will not respond equally to all of these frequencies since the reactance of the capacitor and thus the total impedance of the circuit is dependent on the frequency of the applied voltage.

If the ratio of R to X_C is decreased in the charge path and/or increased in the discharge path, the shape of the output waveform will more nearly approach that of the input waveform. However, there is a practical limit on the ratios that can be obtained; thus, there will always be a certain amount of distortion in the output of this type of circuit and the amount of distortion which can be tolerated will be the primary factor to be considered when selecting circuit components.

POSITIVE DIODE CLAMPER WITH POSITIVE BIAS

The positive biased positive diode clampers (fig. 2-48) operate similarly to the unbiased positive clampers (fig. 2-46) with the exception that the most negative extremity of the output waveform is clamped to a reference which is the bias level. Observe that with positive bias, the negative terminal of the bias battery is grounded.

With zero input to the circuit (fig. 2-49, point A), the capacitor can be considered to be charged to the bias level. Hence, the output is +10 volts ($e_{out} = e_C + e_{in}$).

On the first positive alternation of the input cycle, the diode in nonconducting and the output level increases to +60 volts, resembling the appearance of the input waveform. Because of the long time constant circuit the capacitor discharges but slightly.

When the input becomes -50 volts, a change of 110 volts, the diode conducts and the capacitor rapidly charges to +60 volts ($e_{bias} + e_{in}$) before diode conduction ceases. This is the circuit condition at point B, figure 2-49. Observe that when e_C went to 60 volts, e_{out} returned to the reference level.

At point C on the succeeding positive alternation of the input cycle (fig. 2-49), the diode is nonconducting and e_{out} = +110 volts.

From the foregoing analysis, it can be observed that once e_C equals +60 volts, the output will vary between +10 volts and +110 volts as the input varies between -50 volts and +50 volts.

At point G in figure 2-47, e_{in} equals 50 volts, e_C equals a positive 49 volts, and e_R equals 99 volts. These are the conditions which exist in the circuit just prior to the time the input voltage switches to the negative 50-volt level for the second time. Since the diode is cut off for the period F to G, the circuit has a long TC and the capacitor loses little of its charge.

At point H in figure 2-47, e_{in} has switched to the negative 50-volt level, e_C equals positive 49 volts, and e_R equals a negative 1 volt (e_R and e_C added algebraically). The diode is again forward biased and the capacitor quickly charges back to a positive 50 volts.

At point I in figure 2-47, e_{in} is still a negative 50 volts, e_C is a positive 50 volts, and e_R equals 0 volts.

At point J in figure 2-47, e_{in} is again switched to the positive 50-volt level, e_C equals a positive 50 volts, and e_R = 100 volts.

As can be seen in figure 2-47, the circuit has now stabilized, e_C varies between a positive 49 and 50 volts, and e_R between a negative 1 volt and a positive 100 volts. Thus, for all practical purposes the negative peak of the input waveform has been clamped to 0 volts in the output. That is, once the circuit is stabilized, all major voltage excursions in the output will be positive with respect to ground or 0 volts. Note: The output waveshape will not be an exact replica

39

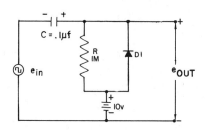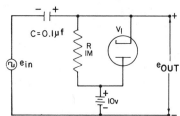

179.570

Figure 2-48.—Positive diode clamper with positive bias.

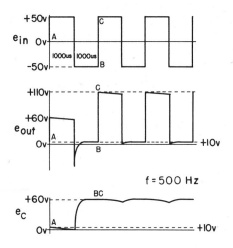

179.571

Figure 2-49.—Positive bias positive clamper waveforms.

POSITIVE CLAMPER
WITH NEGATIVE BIAS

The negative biased positive diode clamper (fig. 2-50) differs from the positive biased positive diode clamper (fig. 2-48) only in that the bias polarity is reversed, thereby relocating the reference. The reference on which the most negative extremity of the output waveform is positioned now becomes the value of negative bias used.

Referring to figure 2-51 point A, observe that the capacitor is charged to the bias level when e_{in} is zero. This is true because the generator (e_{in}) provides a charge path through its internal resistance. Therefore, the output voltage at this instant is -10 volts ($e_{out} = e_{in} + e_C$).

At point B (fig. 2-51) e_{in} rises to +50 volts, e_{out} increases from -10 volts to +40 volts or to the value of $e_{in} + e_C$. During this time the diode is nonconducting and the time constant is long, causing the capacitor to discharge only slightly.

At point C on figure 2-51, the input signal is -50 volts and the instantaneous output is -60 volts with respect to ground. At this time capacitor C charges through the conducting diode, due to the short time constant, and the 50 volt change across the capacitor occurs as shown in figure 2-51. At point D (fig. 2-51) the capacitor is charged to 40 volts, and the output is -10 volts.

When e_{in} again rises to +50 volts at point E (fig. 2-51), e_{out} rapidly increases from -10 volts to +90 volts. Due to the long time constant circuit during diode nonconduction, the capacitor will discharge but slightly. The circuit is now stabilized and as e_{in} varies between -50 and +50 volts, e_{out} varies between -10 volts and +90 volts.

NEGATIVE DIODE
CLAMPER (UNBIASED)

The principal difference between the positive and negative diode clamper circuits is the manner in which the diode is connected. In the positive diode clampers, the anode of the diode is connected to the reference. In the negative diode clampers, the cathode of the diode is connected to the reference as in figure 2-52. In this manner,

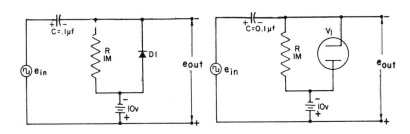

179.572

Figure 2-50.—Positive diode clamps with negative bias.

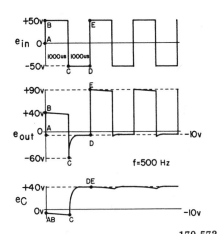

179.573

Figure 2-51.—Positive diode clamper with
negative bias waveforms.

the output waveform will have its most positive extremity clamped to the reference. Stated otherwise, the entire output waveform will lie in a more negative area than the reference. This is shown in figure 2-53. The diode in the negative clamper conducts upon application of the positive alternation of the input waveform, providing a path for rapid charging of the capacitor. This corresponds to a short time constant circuit. Upon application of the negative alternation of the input signal, the diode is nonconducting and the output voltage is the sum of the negative alternation (e_{in}) and the charge on the capacitor (e_C). A high value of resistance is utilized to prevent the capacitor from discharging appreciably during the period in which the

diode is nonconducting. This corresponds to a long time constant circuit.

The circuits in figure 2-52 and waveforms in figure 2-53 will help to clarify the action of the unbiased negative clamper. On the first positive alternation of the input signal, the diode conducts, and the capacitor charges rapidly to a potential of -50 volts. After the capacitor is charged, the output remains at zero volts ($e_{in} + e_C$). On the negative alternation of the input signal the diode is nonconducting. During this period, the capacitor does not discharge appreciably through the one-megohm resistor. Accordingly, the output voltage decreases to a negative 100 volts at times F, G, and H; and remains at this level during the entire duration of the negative input alternation. as in the basic unbiased positive diode clamper, the value of the resistor in the unbiased negative diode clamper must be large to avoid output waveform distortion.

NEGATIVE DIODE CLAMPER
WITH NEGATIVE BIAS

A negative biased negative clamper operates in the same manner as the negative biased positive clamper. A negative diode clamper with a -10 volts bias as shown in figure 2-54 will clamp the upper extremity of the waveform at a -10 volts rather than zero volts.

With zero volts input at point A of the input waveform shown in figure 2-55, the capacitor is initially charged to the value of bias voltage (-10 volts) through the source resistance. The output voltage (e_{out}) at that time is -10 volts ($e_{in} + e_C$). When the input waveform goes positive for the first time at point B, the capacitor rapidly charges through the low forward resistance of the conducting diode to -60 volts ($e_{in} + e_{bias}$). As indicated by figure 2-55,

41

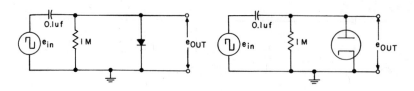

179.574

Figure 2-52.—Negative diode clampers (unbiased).

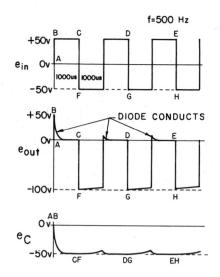

179.575

Figure 2-53.—Waveforms for negative unbiased clamper.

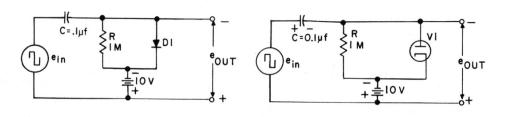

179.576

Figure 2-54.—Negative diode clampers with negative bias.

the capacitor is charged to a -60 volts long before point C (the end of the first positive alternation), and the output returns to -10 volts. On subsequent alternations when the input signal goes in a negative direction (fig. 2-55, points F, G, and H), the diode is nonconducting, and the charge on the capacitor remains relatively constant. The output voltage is approximately -110 volts during the negative input signal alternations. Whenever the input signal goes positive, the diode conducts, causing the capacitor to charge rapidly to negative 60 volts, and the output to return to -10 volts. As a result of this circuit action, the most positive extremity of the waveform is clamped to the negative 10-volt reference.

NEGATIVE DIODE CLAMPER WITH POSITIVE BIAS

Negative diode clampers with positive bias are shown in figure 2-56. The associated waveforms are shown in figure 2-57. With zero-volts input at point A, the capacitor charges rapidly to -40 volts, the sum of e_{in} and the bias voltage. The charge path is through the low forward resistance of the conducting diode. The output voltage after the capacitor has charged returns to a value of +10 volts. On all subsequent negative alternations of the input signal (fig. 2-57, points F, G, and H), the diode is nonconducting and e_{out} equals -90 volts (e_{in} + e_C). Whatever small voltage level the capacitor discharges during the negative alternations of the input is

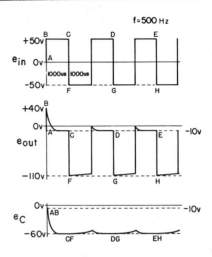

179.577
Figure 2-55.—Waveforms for biased
negative clamper.

quickly replenished when the diode conducts on the positive swing of the input voltage. Thus, with the input varying from -50 volts to +50 volts, the output varies from +10 volts to -90 volts as shown in figure 2-57. The most positive extremity of the waveform has been clamped to the +10-volts bias reference.

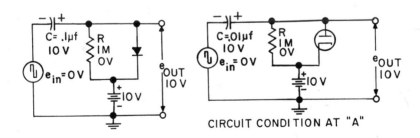

CIRCUIT CONDITION AT "A"

179.578
Figure 2-56.—Negative diode clamper with positive bias.

43

f = 500 Hz

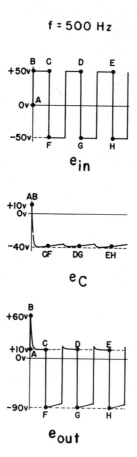

e_{in}

e_C

e_{out}

179.579
Figure 2-57.—Waveforms for positive biased
negative clamper.

CHAPTER 3

PULSE FORMING AND PULSE SHAPING CIRCUITS—PART II

This chapter is a continuation of the previous chapter. Where the previous chapter concerned itself with the basic pulse forming and shaping circuits, this chapter will deal with timing circuits. Timing circuits are used to start, stop, or synchronize various circuits in a system.

ECCLES-JORDAN TRIGGER CIRCUIT

The Eccles-Jordan trigger circuit, also referred to as a flip-flop, a bistable multivibrator, a binary scaler, or a scale-of-two circuit, is a bistable circuit—one which will remain in either of two stable states. It will change states very abruptly upon application of external excitation (a trigger). A complete cycle of operation requires two input pulses of sufficient amplitude. One is used to turn it on and another to turn it off again. The output pulse is commonly referred to as a GATE having fast rise and fall times and extreme flatness between transitions. Since triggers control the operation, the output frequency is directly related to the input trigger frequency. Two outputs, of opposite polarity, are available from the circuit.

MULTIPLE SOURCE VOLTAGE DIVIDERS

Before delving directly into an analysis of the Eccles-Jordan circuit, a brief review of multiple source voltage dividers is in order. Referring to figure 3-1, with switch S1 open, the following conditions exist in the circuit:

$$I_{total} = E_{total}/R_{total} = 350 \text{ volts}/350 \text{ kohms} = 1 \text{ milliampere (ma.)}$$

$$E_{R1} = R1 \times I_{total} = 50 \text{ kohms} \times 1 \text{ ma.} = 50 \text{ volts}$$

$$E_{R2} = R2 \times I_{total} = 200 \text{ kohms} \times 1 \text{ ma.} = 200 \text{ volts}$$

$$E_{R3} = R3 \times I_{total} = 100 \text{ kohms} \times 1 \text{ ma.} = 100 \text{ volts}$$

Therefore, the voltage at point P with respect to ground will be a positive 225 volts ($E_p = +V - E_{R1}$) and the voltage at point G with respect to ground will be a positive 25 volts ($E_G = -V + E_{R3}$).

Now focus on the voltage drop across R3. On one end the voltage with respect to ground is a negative 75 volts and on the other end it is a positive 25 volts. This means that at some point within the resistor there is a potential with respect to ground that is zero. This point is referred to as a FLOATING or PHANTOM GROUND.

When switch S1 is closed the conditions in the circuit will be:

$$E_{R1} = +V = 275 \text{ volts}$$
$$I_{R1} = +V/R1 = 275 \text{ volts}/50 \text{ kohms} = 5.5 \text{ ma.}$$
$$I_{R2} = -V/(R2+R3) = 75 \text{ volts}/300 \text{ kohms} = .25 \text{ ma.}$$
$$E_{R2} = I_{R2} \times R2 = .25 \text{ ma.} \times 200 \text{ kohms} = 50 \text{ volts}$$
$$E_{R3} = I_{R3} \times R3 = .25 \text{ ma.} \times 100 \text{ kohms} = 25 \text{ volts}$$

In this condition, point P is at ground potential because of the direct connection through the closed switch. The voltage at point G with respect to ground is a negative 50 volts. A comparison of the voltage levels when the switch is open and closed shows point P changing from a positive 225 volts to zero and point G changing from a positive 25 volts to a negative 50 volts respectively.

Figure 3-2A is a schematic composed of two of the voltage dividers just described. The switches are ganged together so that when one is open the other will be closed. Consequently, when the voltage at P1 with respect to ground is maximum, the voltage at P2 with respect to ground is minimum and vice versa. The voltages

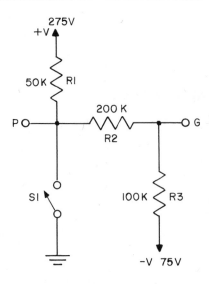

179.580
Figure 3-1.—Voltage divider.

referenced to ground at G1 and G2 behave in a similar manner. Using the potentials obtained from the single divider network, voltage level waveforms are shown in figure 3-2B at points P and G as a function of the switch settings.

BASIC FLIP-FLOP

The circuit diagram of figure 3-3A shows a basic flip-flop. The only difference between this circuit and the double voltage divider network is that the switches have been replaced by active devices performing essentially as switches. The active devices may be either transistors or tubes. NPN transistors or tubes can be used with the voltage polarities as shown. A PNP transistor may be used by reversing voltage polarities. The output of each amplifier is directly coupled to the input of the other amplifier. When power is applied, one of the devices, due to slight parameter differences, will be conducting, and the other will be cut off.

Although these devices are assumed to be identical, no two absolutely identical devices exist. This is due to small, many times immeasureable, differences which occur during manufacturing. Consequently, these devices never become absolutely identical as these and other differences will remain. A stable condition with

both devices conducting or both cut off cannot exist because of regenerative feedback between the stages. A detailed analysis of operation is in order, but to accommodate both transistor and tube circuits, specific voltage levels will not be used. In general +V and -V will be less than 20 volts for transistors and more than 100 volts for tubes.

Initially, assume that A1 is conducting and A2 is cut off. Due to this initial condition, collector or plate voltage, e_{A1-3}, will be low because of a large voltage drop across R1, as shown in figure 3-3B. This voltage will not be zero as previously described in the switch analogy, since the amplifying device will contain some minimum resistance. The base or grid voltage, e_{A1-2}, will not be as positive as in the switch analogy because the base or grid provides another conduction path. Conduction is minimal through this path because R4 is a large value. The increased voltage drop across R4 will result in e_{A1-2} going only slightly positive with respect to ground. Simultaneously A2 is cut off since a negative voltage (e_{A2-2}) is applied to the base or grid of A2. The amount of voltage is determined by the size of R1, R3 and R6 in relation to the level of +V and -V. The collector or plate voltage of A2 (e_{A2-3}) with respect to ground will be less than +V. This is due to the small base or grid current through A1, and the current through the voltage divider R5, R4 and R2.

When a negative trigger is applied, A2 will not be affected since this device is already cut off, however, the trigger will reduce the conduction level of A1. This reduced current causes the collector or plate voltage of A1, e_{A1-3}, to go more positive with respect to ground. This positive going voltage change will be coupled to the base or grid of A2, which brings A2 into conduction. A2's conduction causes its collector or plate voltage, e_{A2-3}, to decrease. This decrease is coupled to the base or grid of A1 and reduces the conduction of A1 even more. This process is accumulative until A1 is cut off and A2 is conducting heavily.

The state of A1 cut off and A2 conducting is maintained until another trigger is applied. Five successive triggers are shown in figure 3-3B with the resulting changes in circuit voltage levels. Due to the abrupt changes in conducting states the waveforms have virtually no rise or fall time.

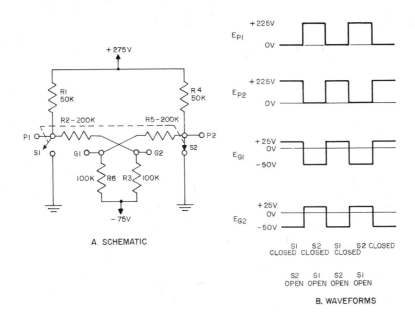

179.581
Figure 3-2.—Double voltage divider network.

Steering Diodes

The trigger pulses are usually obtained from differentiating a rectangular waveform. Therefore, both positive and negative trigger spikes are obtained. To allow only the desired trigger polarity to be applied to the amplifying devices, steering diodes are used. These diodes pass triggers of only one polarity. Figure 3-4 depicts an Eccles-Jordan trigger circuit employing steering diodes which permit only negative trigger signals to pass to the flip-flop.

Transition Time

It is desirable, and in most cases imperative, that the change from one state to the other be very rapid. The time interval which elapses when changing conduction state from one device to the other is called TRANSITION TIME. This time may be reduced by placing small capacitances in parallel with the resistors which couple collector or plate voltage changes to the base or grid. These capacitors, C1 and C2 (called SPEED UP CAPACITORS), are shown in figure 3-5.

To aid in understanding the parameters that affect transition time it must be remembered that both transistors and tubes have a finite quanity of input capacitance. To illustrate this, a more exact schematic of the basic divider network is shown in figure 3-6. Analysis of this circuit reveals that a change in voltage will not be felt across R3 until the input capacitance is changed. This input capacitance will charge at a rate determined by the ratio of input capacitance to the capacitance of the speed up capacitors. The voltage across R3 after transition is determined by the resistors. The effect of the input capacitance will be minimized to the greatest extent if R3 times the input capacitance equals R2 times the speed up capacitance. Perfect

47

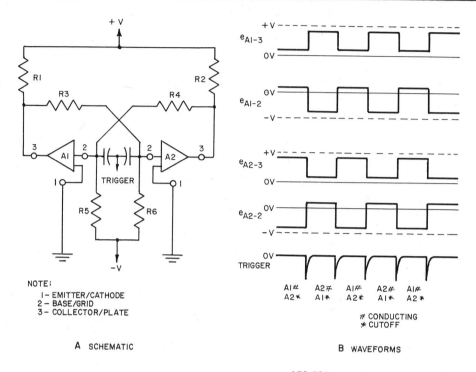

Figure 3-3.—Basic Eccles-Jordan circuit.

compensation does not occur because of the interaction effects between impedances of the driving source (active device) and the network impedance.

The optimum value of the speed up capacitors is a compromise between a short settling time and a short transition time. (Settling time is the time required for completing the recharging of the capacitors after transfer of conduction has taken place.) The reason for the compromise is that increasing the value of the speed up capacitors increases the settling time simultaneously with the desired decrease in transition time. Depending on circuit requirements the values of the speed up capacitors may vary from 10 pf. to 200 pf.

Resolving Time

Resolving time is the smallest allowable time between triggers—the reciprocal of the maximum frequency to which the flip-flop will respond. It is the sum of transition time and settling time.

Resolving time is improved in some circuits by triggering the active device out of conduction instead of out of cutoff. Binary circuits are more sensitive to a trigger whose polarity turns the active device off than a trigger whose polarity is such that it turns the active device on. For example, the NPN transistor or the tube is switched more rapidly by negative pulses than by positive pulses. This is due to these devices having greater gain near saturation (gain close to cutoff is considerably less).

Efficiency

Efficiency is improved in a flip-flop if it is not allowed to go into saturation. Electron tube circuits of this type do not draw grid current. The nonsaturation condition also reduces resolution time, especially in transistors, since the charge storage effect is greatly reduced. This method of operation can be accomplished by a critical selection of voltage divider components or by the use of reference diodes (Zener).

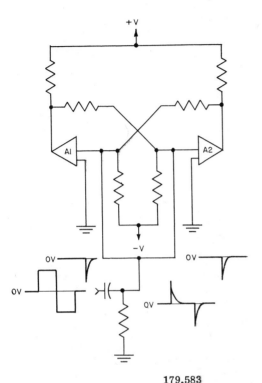

179.583
Figure 3-4.—Steering diodes.

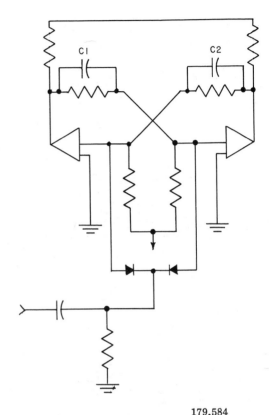

179.584
Figure 3-5.—Reducing transition time.

SCHMITT TRIGGER CIRCUIT

The Schmitt trigger is a two stage pulse shaping circuit. This circuit is amplitude sensitive and is designed to produce an output only when its input signal exceeds a prescribed reference level. The output of a Schmitt trigger circuit is a rectangular waveform of constant amplitude, whose pulse width is equal to that period of time during which the input signal exceeds a preset reference level. Examples of inputs to and corresponding outputs from a Schmitt trigger circuit are illustrated in figure 3-7.

Figure 3-8 illustrates a Schmitt trigger circuit. An examination of this circuit reveals that it is basically an emitter/cathode coupled bistable trigger circuit. Coupling from the second stage (A2) to the first stage (A1) is obtained by

the use of an unbypassed common emitter/cathode resistor rather than direct resistive coupling, as in the Eccles-Jordan trigger circuit. Also, there is no voltage divider to bias A1.

OPERATION (NO INPUT)

The polarities used in the following explanation are correct for NPN transistors or electron tubes. Operation with PNP transistors is the same except for the reversal of all polarities.

Resistors R1, R2, and R3 form a voltage divider network across the voltage supply. With ground as a reference, a positive potential will appear at the junction of R2 and R3. This positive potential is applied to the base/grid of A2, causing A2 to conduct. Since there is initially no

49

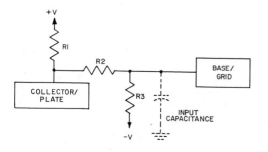

179.585

Figure 3-6.—Divider network with input
capacitance.

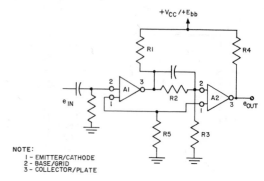

NOTE:
1 – EMITTER/CATHODE
2 – BASE/GRID
3 – COLLECTOR/PLATE

179.587

Figure 3-8.—Schmitt trigger circuit.

positive potential applied to the base/grid of A1,
A2 will always conduct first. The conduction of
A2 will develop a positive voltage with respect to
ground across the common emitter/cathode resistor R5. This voltage is sufficient to maintain
A1 in a cutoff condition. The collector/plate
voltage of A1 will be approximately $+V_{CC}/+E_{bb}$,
because A1 is cutoff, and the collector/plate
voltage of A2 will be low due to its heavy
conduction.

OPERATION (INPUT APPLIED)

All voltage polarities are with reference to
ground unless otherwise indicated. When an
input, e_{in}—figure 3-9, of the proper polarity
(positive for NPN transistors and electron tubes—
negative for PNP transistors) and of sufficient

amplitude to overcome the reference level (determined by the voltage drop across R5), is applied to the base/grid of A1, the device will conduct. As A1 conducts, its collector/plate voltage
(e_{A1-3}—fig. 3-9) decreases causing the base/grid
voltage of A2 to become less positive. A2, therefore, conducts less, which in turn results in a
decrease in the voltage across R5. The decreasing voltage across R5 allows A1 to conduct
appreciably more, causing its collector/plate
voltage and the base/grid voltage of A2 to become
much less positive. This cumulative and instantaneous action results in A1 conducting heavily
and A2 being cut off. A2's base/grid with respect
to emitter/cathode voltage has now become negative. At this time the collector/plate voltage of

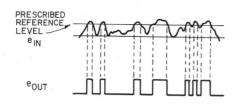

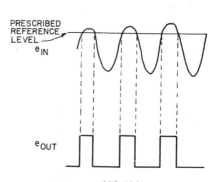

179.586

Figure 3-7.—Schmitt trigger input and output waveforms.

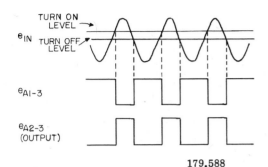

179.588
Figure 3-9.—Schmitt trigger waveforms.

A1 is low and the collector/plate voltage of A2
(e_{A2-3}—fig. 3-9 is approximately $+V_{CC}/+E_{bb}$.

When the input falls below a level which main-
tains the conduction of A1, A1's conduction decreases and its collector/plate voltage rises
toward $+V_{CC}/+E_{bb}$. This increasing collector/
plate voltage causes the base/grid of A2 to rise
positively with respect to the emitter/cathode,
which in turn brings A2 out of cutoff. The conduction of A2 increases the voltage drop across
R5 which in turn reduces the conduction of A1.
Again the action is cumulative and instantaneous and results in A1 being cut off and A2 conducting heavily. The circuit will remain in this
state until the input again rises above the reference level.

It would seem that the input level at which
A1 is driven into conduction (turn on level)
should be the same as the level below which A1
cuts off (turn off level). This, however, is not
the case. The circuit's turn on and turn off
levels are determined by the voltage drop across
R5. The voltage drop across R5 is lower during
the conduction of A1 than during the conduction
of A2. Therefore, the input level required to
cause A1 to conduct (i.e., turn on level) is greater
than the level below which the input must fall
in order to cause A1 to cutoff (i.e., turn off level).
This is depicted in figure 3-9.

MONOSTABLE MULTIVIBRATOR

The emitter/cathode coupled monostable multivibrator circuit (fig. 3-10) has only one stable
state. In this state, one active device conducts
while the other active device is normally cut off.
The circuit will function for only one complete

cycle of operation upon the application of one
trigger pulse. During this cycle, the circuit goes
to an unstable state, in which the active devices
reverse their condition of conduction or nonconduction. The time duration of the unstable state
is determined by the circuit constants. This circuit is commonly referred to as a ONE-SHOT
MULTIVIBRATOR.

A cycle of operation may be initiated in a
variety of ways. The normally conducting device
may be turned off, or the device cut off in the
stable state may be turned on. In many solid-
state circuits, turning off a conducting transistor
is accomplished more rapidly than turning on a
nonconducting transistor. Basically, a conducting
NPN transistor may be turned off by making the
base negative with respect to the emitter. (This
may be accomplished by applying a positive pulse
to the emitter or a negative pulse to the base.)
Of course, opposite polarities are required for
PNP transistors.

This simple process becomes more compli-
cated when it is necessary to achieve various
functions, such as faster switching time, purer
waveforms, shorter resolving times, etc. Therefore, in practical circuitry, several active devices
plus many passive ones may be used in conjunction with the basic multivibrator circuit to
achieve optimum performance in a given application.

OPERATION

To achieve the stable condition, the base/
grid of the normally conducting device (A2—fig.
3-10) is usually returned to its emitter/cathode
through a resistor (R2). The normally cutoff
device (A1) has its base/grid returned to ground
through resistor R1. In the case of transistor
circuits, the bases must also be biased from the
source (R_{BIAS}).

Because of the biasing, A will remain cutoff and A2 will conduct until a trigger is applied,
then A1 will conduct. When conducting, A1's
collector/plate voltage drop causes C2 to discharge. This discharge applies a high negative
voltage to the base/grid of A2 which effects cutoff. This unstable condition will remain until C2
discharges to a point where A2 conducts. The
conduction of A2 will produce a voltage drop
across the common emitter/cathode resistor R5
which will cut off A1. The circuit is now back
in a stable state, in which it will remain until the
next trigger pulse. The width of the pulse produced
during the unstable state is determined by the
discharge time constant of C2 and the cutoff
level of A2. However, the frequency of operation

51

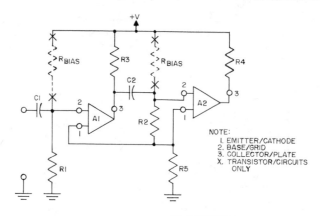

20.156(179)A

Figure 3-10.—Monostable multivibrator.

is determined by the trigger rate, since another pulse will not be produced until another trigger initiates the next unstable state, thus the term one-shot multivibrator is applied to this circuit.

Figure 3-11A shows the charge path for C2 to be from ground through R5, the conduction resistance of emitter/cathode to base/grid of A2, C2, R3 and back through the power source. The discharge path of C2 is illustrated in figure 3-11B. It is from C2, through R2, the conduction resistance from A1's emitter/cathode to collector/plate and back to C2. The resistance of R2 is much greater than the conduction resistance of A1, so that for rough analysis the discharge time constant may be calculated using only C2 and R2.

The unstable state may be adjusted so that it is half the duration of the time between triggers. This arrangement will produce a symmetrical output. Asymmetrical outputs may be obtained by longer or shorter time constants. The duration of the pulse may also be changed by varying the bias on A1.

Outputs may be obtained from either device. However, one advantage of this circuit is that a relatively pure rectangular output may be obtained from A2's collector. This is because the collector/plate of A2 is isolated from feedback (coupling) networks.

Waveform Analysis

Consider now the operation of this circuit (fig. 3-10) while referring to the waveforms shown in figure 3-12. When voltage is first applied, A2 conducts and A1 cuts off due to the biasing arrangement which allows A2 to conduct more heavily than A1. The collector/plate current flow of A2 through common-emitter/cathode resistor R5 makes the voltage at the top of the resistor positive with respect to ground, producing a bias voltage of sufficient amplitude to hold A1 cut off, and still permit A2 to conduct. Thus, at time t_0 on the waveform illustration, the base/grid voltage of A2 (e_{A2-2}) is positive, causing conduction through A2. As a result of A2's collector/plate current, there is a positive A1 emitter/cathode voltage (e_{A1-1}) developed across R5 which provides the bias to cut off A1. The positive going voltage at the collector/plate of A1 (e_{A2-3}) is coupled through C2 to the base/grid of A2, holding the base/grid positive. Also, at t_0, C2 charges through the low emitter/cathode to base/grid internal resistance of A2 and R3. At t_0, due to the effects occurring instantaneously and regeneratively, A2 conducts and A1 cuts off. The circuit will remain in this condition until a trigger pulse is applied.

Assume now that a positive trigger pulse of sufficient amplitude to cause conduction of A1 is applied through C1. The effect of this trigger pulse (applied at t_1) is to drive A1 into conduction, thus causing an increase in current through A1 and a decrease in the plate/collector voltage (e_{A1-3}) of A1. This negative going voltage is applied instantaneously through C2 to the base/grid of A2, driving A2 below cutoff. When A2's collector/plate current ceases, the voltage drop across R5 (e_{A1-1}) decreases to the level where it permits A1 to conduct more heavily. Thus,

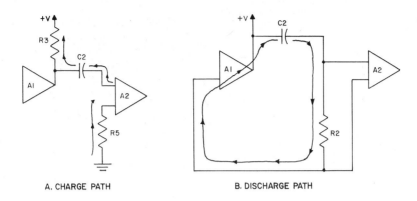

A. CHARGE PATH   B. DISCHARGE PATH

179.589

Figure 3-11.—Charge and discharge paths for capacitor C2
in monostable multivibrator.

at t_1 a switching action occurs and the multi-vibrator is in an unstable state, in which A1 conducts and A2 is cut off.

During the unstable state (t_1 to t_2), capacitor C2 begins to discharge, causing the base/grid voltage of A2 (e_{A2-2}) to become less negative. The discharge path of C2 is downward through R2 and then upward through the low emitter/cathode to collector/plate resistance of A1 (fig. 3-11B). At this time A1 alone is conducting; its collector/plate current is limited by its own emitter/cathode bias, which is not sufficient to cut off A1. As the voltage on the base/grid of A2 (e_{A2-2}) becomes more positive or less negative, because of the discharging of C2, it soon reaches the point, at t_2, where it is no longer of sufficient amplitude to hold A2 in cutoff. Consequently, A2 once again conducts.

Note that on waveform e_{A2-2} the zero reference is below the cutoff point in transistor circuits; i.e., the base must be positive with respect to ground in order for A2 to conduct. However, as shown in the illustration, the grid of an electron tube need not be positive with respect to ground to enable conduction of A2. The flow of A2's current through R5 increases the voltage drop (e_{A1-1}) across this resistor, again increasing the emitter/cathode voltage on A1 and reducing A1's current flow. As the conduction through A1 decreases, the collector/plate voltage (e_{A1-3}) of A1 rises toward the potential of the positive voltage supply, +V.

The positive going signal at the collector/plate of A1 is coupled through C2 to the base/grid of A2, driving this base/grid positive (waveform e_{A2-2} at t_2). Thus, C2 stops discharging

and again begins charging. The voltage waveform at the collector/plate of A1 (e_{A1-3}) is rounded off, and the voltage waveform at the base/grid of A2 (e_{A2-3}) has a small positive spike as a result of the charging of coupling capacitor C2 when A2 goes into conduction. The collector/plate voltage waveform of A2 (e_{A2-3}) has a small negative spike, and the voltage waveform across R5 (e_{A1-1}) has a small positive spike of the same time duration as the positive spike on the A2 base/grid waveform (e_{A2-2}). Hence, at t_2 the multivibrator reverts to its original stable condition in which A2 conducts and A1 is cut off; the circuit remains in this condition (t_2 to t_3) until another positive trigger pulse is applied at t_3.

Close examination of the waveforms reveals that the monostable multivibrator goes through one cycle of operation for each input trigger pulse. Also, the time of application of the trigger pulse determines when A1 is driven into conduction. The RC time constant of R2C2 and the bias on A2 determine when A2 is driven into conduction. Thus, the monostable multivibrator output frequency is determined by the input frequency, and the output gate width is determined by the discharging of C2 through R2 toward the cutoff potential of A2.

PHANTASTRON

The phantastron circuit is considered to be a relaxation oscillator similar to the multivibrator in operation. Whereas the multivibrator derives its timing waveform from an RC circuit, the phantastron uses a basic Miller sweep generator to generate a linear timing waveform, rather than

53

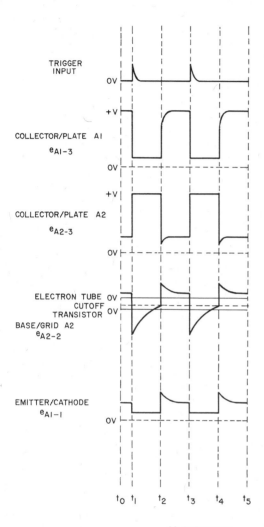

TRIGGER
INPUT

OV

+V

COLLECTOR/PLATE AI
e_{AI-3}

OV

+V

COLLECTOR/PLATE A2
e_{A2-3}

OV

ELECTRON TUBE OV
CUTOFF
TRANSISTOR OV
BASE/GRID A2
e_{A2-2}

EMITTER/CATHODE
e_{AI-1}

OV

t_0 t_1 t_2 t_3 t_4 t_5

20.156(179)B
Figure 3-12.—Monostable waveforms.

the exponential waveform developed by the RC circuit of the multivibrator. Thus, the output waveform is a linear function of the input (control) voltage, and the timing stability is improved.

The phantastron circuit is used to generate a rectangular waveform, or linear sweep, whose duration is almost directly proportional to a

control voltage. Because of its extreme linearity and accuracy, this waveform is used as a delayed timing pulse, usually in radar or display equipment. It is also used to produce time delayed trigger pulses for synchronizing purposes and movable marker signals for display. For example, it is used as a time modulated pulse, to indicate antenna position at any instant of rotation, or as a range strobe or delay marker.

OPERATION

The operation of the phantastron circuit is based on the use of a Miller linear sweep generator (fig. 3-13). The suppressor grid is normally biased (negative) to prevent plate current flow, while the screen conducts heavily. The control grid is returned to E_{bb} through a resistor so that it is effectively at zero potential, and the cathode is grounded. When a positive gate is applied to the suppressor grid (t_1), plate current flows and produces a voltage drop across the plate load resistor R1. This negative swinging plate voltage (e_p) is fed back through a small capacitor (C) to the control grid, and quickly drives the grid negative; thus maintaining the plate current at a small value, and greatly reducing the screen current. Reduction of the heavy screen current produces a large positive swing on the screen (e_{sc}) and the tube essentially remains in this condition, producing a positive screen gate. Meanwhile the plate current flows under control of the feedback voltage applied to the control grid until no further feedback is produced. During this time the plate current increase is linear, and the plate voltage continues to drop. (The normal discharge of C through R_g would cause the current through the tube to increase in an exponential manner, thereby causing the plate voltage to drop exponentially. However, any exponential change is fed back to the grid 180 degrees out of phase with the normal discharge of C, thereby causing a linear increase in plate current.) At a point a few volts above ground (t_2) however, no further plate swing is possible, and the screen again conducts heavily, returning almost to the initial operating point (t_3). When the suppressor gate ends, the plate current is cut off, the screen returns to its initial operating point, and the cycle is ready to be resumed under control of the next gate.

Screen Coupled
Phantastron

A typical monostable screen coupled pentode circuit is shown in figure 3-14. This circuit is started by a positive trigger applied to the

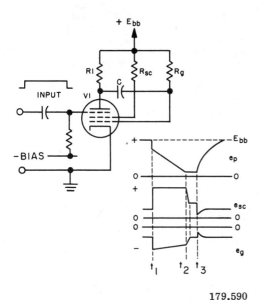

179.590

Figure 3-13.—Miller linear sweep generator.

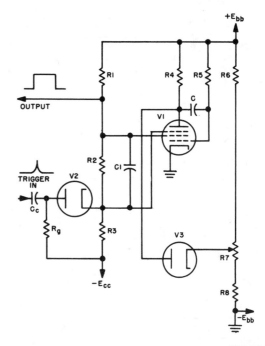

179.591

Figure 3-14.—Basic screen coupled phantastron.

suppressor grid, and at the end of operation it returns to the initial starting condition, ready to repeat the cycle of operation when the next trigger arrives. The output taken from the screen is a rectangular positive gate whose duration, or length is controlled by R7. In the illustration, tube V1 is the basic phantastron, and diode V2 acts as a trigger injector and also as a disconnecting diode to effectively isolate the trigger circuit after the action is started. Diode V3 sets the maximum level of plate voltage as controlled by the position of R7, and since the turn off level is fixed, it effectively controls the time during which the circuit produces the linear gate or sweep. Operation occurs at the rate fixed by the discharge of C through R5. Feedback capacitor C provides regenerative feedback from the plate to the control grid, to allow quick response to any changes in the plate circuit. Capacitor C1 couples the positive gate from the screen to the suppressor grid thereby holding the tube in a condition where plate current can flow.

Circuit operation can best be understood by referring to the waveforms shown in figure 3-15. Three steps are involved in the circuit action— turn on, linear sweep development, and turn off. Before initiation of action, the circuit is resting with the plate current cut off, because a negative

voltage is applied to the suppressor element through R3. Resistors R1, R2, and R3 form a combined suppressor and screen grid voltage divider connected between E_{bb} and E_{cc}. The values are such that the screen is positive and the suppressor is sufficiently negative to cut off plate current. Since they are directly connected, both elements are d.c. coupled. They are also a.c. coupled through capacitor C1. Therefore, both d.c. and a.c. voltages appearing on one element also affect the other. Since the cathode is grounded and the control grid is connected through R5 to E_{bb}, the grid remains near zero bias. Thus, although the plate current is cut off, the screen current is heavy. When a positive trigger is applied to the plate of disconnecting diode V2 through coupling capacitor C_c, the diode conducts, and the positive trigger appears across R3 and is applied to the suppressor grid of V1 (t_1—fig. 3-15). The trigger is large enough to overcome the fixed negative bias and drive the suppressor positive. Therefore, the plate current

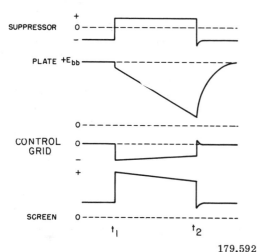

179.592

Figure 3-15.—Waveforms of screen coupled phantastron.

flows through R4. Since R4 is a relatively large value resistor, the plate current quickly goes from zero to a low value, and simultaneously the negative swing produced across R4 is applied through C to the control grid, driving it from zero to a negative value of only a few volts, but sufficiently negative to reduce the total cathode current. The control grid is now in full control, and the reduction of screen current produces a large positive increase in screen voltage. Through C1 the positive going screen voltage is fed back to the suppressor so that the action is regenerative; as a result, the tube is quickly triggered from the static condition to the operating condition, which produces a screen waveform with a sharp leading edge. The linear sweep development or timing cycle now begins, with the plate current of V1 increasing steadily. Since the control grid is returned to E_{bb}, the control grid voltage attempts to reach the zero bias level; however, it can change only slowly because the plate side of feedback capacitor C is steadily decreasing, so that any positive control grid swing is immediately counteracted by a negative plate swing fed back to the control grid. Therefore, capacitor C starts to discharge and electrons flow from the plate of V1, discharging C through R5. Thus C discharges at a rate determined by the time constant of C and R5.

In discharging, the control grid side of capacitor C gradually becomes more positive, causing an increased current flow through R4 and producing a constant decrease in plate voltage. The positive increment on the control grid is always slightly greater than the negative plate swing it produces; therefore, the control grid potential gradually rises, and the plate potential gradually drops. When the plate reaches the point where a voltage change on the control grid will produce no further plate voltage change, the turn off point is reached. Up to this time a positive gate has been produced in the screen circuit and coupled to the suppressor grid. A negative gate has been developed on the control grid which is smaller in amplitude from the leading edge to the trailing edge by about a volt. (The amount is dependent upon the gain of the tube.)

Since the tube plate voltage is only a volt or so above zero and the plate current can no longer increase, but the control grid voltage is still rising toward zero, causing an increase in current flow, the screen current increases. The moment the screen current increases, the screen voltage drops and feeds back a negative swing through C1 to the suppressor grid. The suppressor grid then resumes its original negative condition, stops any flow of plate current, and assumes control again (t_2). Since this action is regenerative, a sharp trailing edge is produced. Simultaneous with plate current cutoff, the plate voltage swings positive and feeds back to the control grid a positive voltage, which helps the control grid to return to normal zero bias condition. Since the charge on C cannot change instantly, the plate swing tapers off exponentially and the tube is not ready for another trigger cycle until it has completely recovered.

Diode V3, a PLATE-CATCHING DIODE (because of the way it catches or arrests, the positive excursion of the plate voltage), operates to catch the plate at a specific voltage, so that with a fixed bottoming point the length of the output pulse and its time duration depend on the plate voltage fixed by V3. The cathode of V3 is biased positive by the voltage divider consisting of R6, R7, and R8, as controlled by potentiometer R7. When the plate voltage of V1 is greater than the positive voltage applied to the cathode of V3, the diode conducts and quickly brings the plate voltage of V1 down to the level of the cathode voltage. Thus, the linear sweep action always starts at the voltage set by R7, and the duration of the phantastron gate (or the length of the sweep) is thereby determined. Because the amplitude of the plate sweep depends on the level at which it starts, R7 directly controls the amplitude. The amplitude of the screen gate is determined by the voltage applied and the screen current, and is only slightly affected by R7.

Cathode Coupled
Phantastron

A typical cathode coupled pentode is shown in figure 3-16. This circuit is also started by a positive trigger applied to the suppressor grid and turns itself off automatically like the screen coupled circuit. The output can be taken from either the screen or cathode, or both. The screen output is a positive gate, and the cathode output is a negative gate.

In this circuit, tube V1 is the basic phantastron, with diode V3 operating to control the plate voltage level and determine the duration of the output gate. To minimize overshoot on the control grid, cathode, and screen grid (positive on the control grid and cathode, negative on the screen grid), diode V2 is connected between R1 and R2 on a voltage divider network consisting of R1, R2, and R3. (The voltage on the cathode of V2 is normally 1 or 2 volts less than the cathode voltage of V1.) Note that no negative supply is needed to bias the suppressor grid. Cathode bias is used, and the suppressor is held at a lower positive potential than the cathode; thus, it is effectively biased negative with respect to the cathode, cutting off plate current. The screen grid is drawing current, which produces a positive voltage on the cathode.

When a positive input trigger appears across R3, it is applied to the suppressor grid. This trigger is prevented from affecting the control grid by automatically reverse biasing diode V2. The control grid is normally biased near zero, being held by diode V2 at a potential determined by the voltage divider (R1, R2, and R3) connected between E_{bb} and ground, and the cathode is positive with respect to the control grid by approximately 1 or 2 volts.

When the trigger is applied, it overcomes the bias between the suppressor grid and cathode, and plate current flows. The decrease in plate voltage, due to the flow of plate current, is fed back to the control grid through capacitor C, causing the tube current to decrease and the cathode voltage to drop. This drop in cathode voltage further decreases the bias between the suppressor grid and cathode, and plate current increases further. Since the total tube current is decreasing and the plate current is increasing, the screen grid current must decrease. This

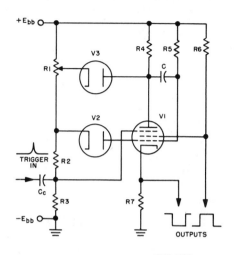

179.593
Figure 3-16.—Basic cathode coupled
phantastron.

action is regenerative, and plate current will jump from zero to a value determined by the tube characteristics. (Note that the bias between the cathode and suppressor grid is decreasing, which is regenerative, causing the plate current to increase. The bias between the cathode and control grid is increasing. This action is degenerative, which decreases the total tube current. Therefore, there must be a point where these two effects are equal and the current will stabilize for an instant.)

At this instant there is no further change in plate voltage, and the control grid voltage increases in the positive direction at a rate determined by C and R5, since it is returned to E_{bb} through resistor R5. This causes the plate current to increase. As the plate current increases, the plate voltage decreases, and this negative change is coupled through C to the control grid. It can be seen that this signal is degenerative, and prevents the plate current from increasing rapidly. This action continues, providing a linear sweep until the plate voltage drops to a level at which it can no longer cause an increase in plate current. At this time, degenerative feedback to the control grid stops, and the grid will go in a positive direction more quickly. This causes an increase in the total tube current, and thus an increase in cathode voltage, an increase in the cathode to suppressor grid bias, and a decrease in plate current. With the total tube current increasing and the plate current decreasing, the screen grid current must be increasing. It can be seen that the action taking place is regenerative, as the plate will go positive, causing the control grid to go positive, and the plate current will go rapidly to cutoff, leaving the tube in its pretriggered condition. The positive swing in the grid is limited by diode V2. Before the circuit is ready for the next cycle of operation, capacitor C must recharge through R3, R2, V2, and R4.

As stated previously, when the phantastron is triggered there is a large drop in the screen grid current. This produces a positive waveform on the screen grid with a steep leading edge. As the tube current gradually increases, producing the linear sweep in the plate circuit, the screen grid current increases in the same manner, but by a smaller amount, in proportion to the plate current. The screen grid waveform will therefore decrease linearly by a small amount until plate current cutoff (described previously) is reached. At plate current cutoff, the screen grid current increases abruptly, causing a steep trailing edge. Negative overshoots at the trailing edge of the waveform will be limited by the action of diode V2.

The resultant waveform across the cathode resistor can be visualized from the previous description of tube operation, by taking into account the changes in the total tube current. This waveform will be a negative gate with steep leading and trailing edges and with the flat portion falling off in amplitude at a linear rate. Any positive overshoot at the trailing edge will be limited by diode V2. R1 is variable, and is connected to the plate of V1 through diode V3, thereby setting the level of plate voltage at which the phantastron begins its action (when triggered). It can be seen that this will determine the amplitude and thus the duration of the plate waveform. R1 is usually an external control to vary the width or delay.

In contrast to the screen coupled phantastron, the cathode coupled phantastron has a smaller range of operation. The maximum plate amplitude swing of V1 is limited by the value of the cathode resistor, in that bottoming of the plate voltage occurs at a more positive potential than in the case of the screen coupled phantastron.

ASTABLE MULTIVIBRATORS

Astable multivibrators have two unstable states between which the circuit makes successive transitions without the necessity of external triggering. This is possible as the active devices conduct and are cut off for intervals of time determined by circuit constants. These circuits are commonly called FREE RUNNING MULTIVIBRATORS since there is no requirement for trigger signals. Astable circuits may be used to generate square or rectangular waves. Although there are various types of free running multivibrators, this treatment of the subject covers emitter/cathode coupled and collector/plate coupled types.

EMITTER/CATHODE COUPLED ASTABLE MULTIVIBRATOR

The emitter/cathode coupled astable multivibrator has two unstable states so that it continues to switch states even with the absence of trigger pulses.

The active devices in figure 3-17 may be either transistors or tubes. NPN transistors or tubes may be used with the voltage polarities shown. PNP transistors can be used by reversing voltage polarities.

The amplifying devices (A1 and A2) are of the same type. Capacitor C1 provides the coupling from the collector/plate of A1 to the base/grid of A2. R1 and R4 are the base/grid resistors of A1 and A2, respectively; R2 and R5 are the collector/plate load resistors. R3 is the common emitter/cathode resistor providing bias for both devices, as well as direct coupling from A2 to A1. The resistors labeled R_{BIAS} provide the required forward bias in the solid-state version. Unless it is otherwise specifically designated, all voltages referenced in these descriptions are measured with respect to ground.

Charge and
Discharge Paths

C1 and R4 form an RC network which establishes the time constant in the base/grid circuit of A2. The initial charge path for C1 is (as shown in fig. 3-18A) from ground, through R3, through the emitter/cathode to base/grid conduction resistance of A2, through C1, and then through R2 to the +V supply. A small portion of the charge current flows through R4 but it is almost negligible compared to the current flow through R3 and the input resistance of A2. Actually R2 is the resistor which, primarily, determines charge time. Figure 3-18B depicts the discharge path for C1 to be through R4, R3, the low conduction resistance of A1's emitter/cathode to collector/plate, and to the other side of C1. The charge time constant and the cutoff level of A1 determine the length of time

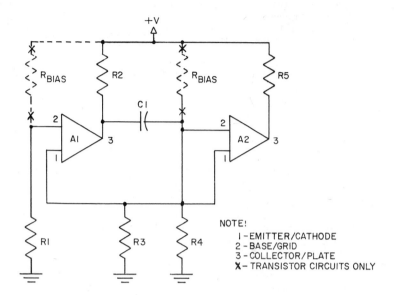

179.594

Figure 3-17.—Emitter/cathode coupled astable multivibrator.

NOTE!

1 – EMITTER/CATHODE
2 – BASE/GRID
3 – COLLECTOR/PLATE
X – TRANSISTOR CIRCUITS ONLY

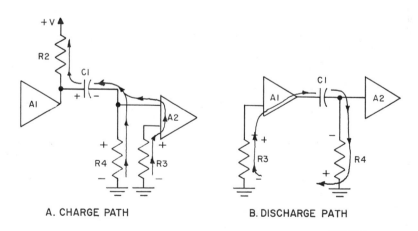

A. CHARGE PATH

B. DISCHARGE PATH

179.595

Figure 3-18.—Charge and discharge paths for C1 in emitter/cathode
coupled astable multivibrator.

that A1 is cut off. The discharge time constant and the cutoff level of A2 control A2's cutoff time. If the time constants for charge and discharge are equal, a symmetrical square-wave output is produced by the circuit. Differing time constants result in an asymmetrical, or unsymmetrical, rectangular output wave. Although the conduction resistances of A1 and R3 are in the discharge path, their resistance is small compared with that of R4, and may be neglected in discharge time constant calculations in this treatment.

Operation

The circuit operation of figure 3-17 will now be analyzed with the aid of the waveforms illustrated in figure 3-19. When voltage is first applied, the base/grid potentials of both devices are such that collector/plate current begins to flow through R2 and R5. Voltage applied to the collector/plate of A1 (e_{A1-3}) causes C1 to begin to charge along the path previously outlined in figure 3-18A. As C1 charges, the base/grid voltage of A2 (e_{A2-2}) becomes more positive. Since there is no coupling capacitor from the collector/plate of A2 to the base/grid of A1, A2's collector/plate voltage (e_{A2-3}) has no effect on the conduction of A1. The collector/plate current of A2 flowing through R3 makes the voltage at the top of R3 (e_{A1-1}/e_{A2-1}) positive with respect to ground. This voltage is of sufficient amplitude to cut off A1 and still permit A2 to conduct, since at this time the base/grid of A1 is at a lower positive potential than the base/grid of A2 (e_{A2-2}). Thus, the initial conditions of circuit operation at time t_0 are established; that is, A2 is conducting and A1 is cut off.

Note that the base/grid of A2 (e_{A2-2}) is driven positive enough to cause heavy conduction through A2. At this instant, the collector/plate voltage of A2 (e_{A2-3}) decreases as a result of collector/plate current through R5. Also, the same current flowing through common-emitter/cathode resistor R3, produces a positive voltage (e_{A1-1}) which provides a bias sufficient to cut off A1. In this cutoff condition, A1's collector/plate voltage (e_{A1-3}) rises toward +V. The positive going voltage at the collector/plate of A1 is instantaneously coupled through capacitor C1 to the base/grid of A2, driving this base/grid still further positive (e_{A2-2}). All of the action described is instantaneous and cumulative, so that the high positive potential on the base/grid of A2 causes this device to conduct heavily while A1 is cut off.

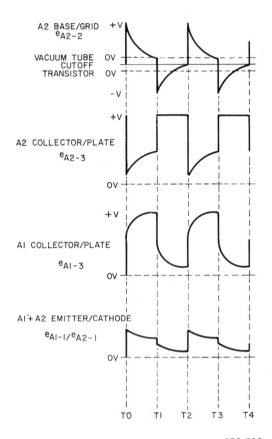

179.596

Figure 3-19.— Waveforms for emitter/cathode coupled astable multivibrator.

It is possible, especially in electron tube circuits due to uneven cathode heating, that A1 would conduct first (at t_0) and A2 would be cut off. When power is applied, C1 would begin charging as described previously. However, should A1 conduct appreciably more than A2, A1's collector/plate voltage will drop toward zero and C1 will discharge, cutting off A2. This alternation will be shorter because C1 does not charge to the total +V source level before A1's collector/plate voltage drops. Now that the possibility of opposite initial circuit conditions has been explored, the rest of this discussion assumes A1 proceeding toward cutoff and A2 conducting at t_0.

Since A2 is conducting at t_0, its collector/plate voltage (e_{A2-3}) drops. With A1 cutoff, its collector/plate is near +V, and C1 charges toward this value. The waveform at the collector/plate of A1 (e_{A1-3}) is rounded off, and the waveform at the base/grid of A2 (e_{A2-2}) has a positive spike of the same duration. Both result from the charging of C1.

As C1 charges, the potential at the base/grid of V2 (e_{A2-2}) decreases. This is illustrated during time interval t_0 to t_1. The decreasing base/grid voltage causes a reduction in the collector/plate current of A2, which results in a decreasing voltage drop across R3. This action continues until the voltage drop across R3 (e_{A1-1}) decreases to the level where A1 is no longer held below cutoff. At this time, t_1, A1 conducts and rapidly cuts A2 off because the large negative going signal at its collector/plate is coupled through C1 to the base/grid of A2. Thus, the first switching action occurs; that is, A2 is cut off and A1 is conducting.

When A1 is conducting, C1 discharges through R4, R3, and the collector/plate resistance of A1 as shown in figure 3-18B. The base/grid voltage of A2 approaches cutoff from below as C1 discharges; this is illustrated by the base/grid voltage waveform of A2 during time of interval t_1 to t_2. At t_2 the base/grid voltage of A2 just reaches the cutoff level, permitting A2 to conduct. An important concept to remember at this point is that (NPN) transistors require a positive base voltage, with respect to the emitter, to emerge from cutoff. However an electron tube may begin conducting while the grid is still negative with respect to the cathode. In the circuit shown the emitter/cathode is always positive with respect to ground. Therefore, waveform e_{A2-2} of figure 3-19 shows cutoff for the transistor to be above the zero volt base potential with respect to ground, while for the electron tube, cutoff is shown below the zero volt level on the grid with respect to ground.

When the collector/plate current of A2 increases, the voltage across R3 (e_{A1-1}) also increases; this takes A1 toward cutoff and thereby reduces the conduction of A1. The decreasing collector/plate current of A1 produces a positive going signal across its load resistor, R2, which, in turn is coupled through C1 to the base/grid of A2, causing this base/grid to become highly positive. Thus, the second switching action occurs and the cycle is complete as the initial conditions once again are reached; that is, A2 is conducting and A1 is cut off.

The multivibrator just discussed is considered to be a symmetrical type; the periods for conduction and cutoff of the devices are identical. An asymmetrical, or unbalanced, output may be obtained by adjusting the size of R3 (emitter/cathode resistor). This effect is achieved since it is the voltage across R3 that determines when A1 begins conduction. A variation of this emitter/cathode coupled circuit uses separate emitter/cathode resistors, with A2 coupled to A1 by capacitor coupling rather than direct coupling.

FREE-RUNNING COLLECTOR/PLATE COUPLED MULTIVIBRATOR

The basic free-running collector/plate coupled multivibrator circuit, shown in figure 3-20, is a simple two stage resistance coupled amplifier, with the output of the second state coupled back into the input of the first stage. Since each stage inverts the signal, a voltage change occurring at the base/grid of either device will be amplified, inverted, and coupled to the base/grid of the other, where it will again be amplified, reinverted, and coupled back in phase to the base/grid from which it originated.

Operation

When collector/plate potential is applied, both active devices begin to conduct, and capacitors C1 and C2 begin to charge with the polarities shown. Initially the collector/plate currents of the active devices are nearly equal, however, there is always a slight difference in the circuits which results in an unbalanced condition. This unbalanced condition causes one device to begin conducting more neavily than the other. This unequal condition brings about a cumulative or regenerative switching action, which, in this example, is assumed to end with A1 conducting and A2 cut off. Although described as if it occurred slowly, this switching occurs with extreme rapidity (in much less than a microsecond in a well designed multivibrator). This action is followed by a relatively long period in which the stages are quiescent. During this interval one capacitor charges and the other capacitor discharges.

Assume that initially the collector/plate current of A1 begins to rise more rapidly than the collector/plate current of A2. The collector/plate voltage of A1 will begin to decrease (because of the increased drop across R3). As A1's collector/plate voltage decreases, C2 will begin to discharge. The discharge path, as shown in figure 3-21, will be from the negative terminal of C2,

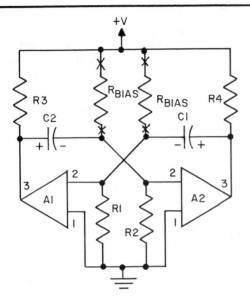

NOTE:
1. EMITTER/CATHODE
2. BASE/GRID
3. COLLECTOR/PLATE
X. TRANSISTOR CIRCUITS ONLY

179.597
Figure 3-20.—Free-running collector/plate
coupled multivibrator schematic.

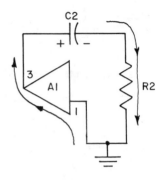

179.598
Figure 3-21.—C2 discharge path in free-running
collector/plate coupled multivibrator.

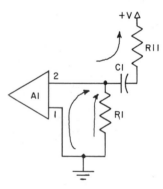

179.599
Figure 3-22.—C1 charge path in free-running
collector/plate coupled multivibrator.

through R2, A1, and back to the positive terminal
of C2. This discharge will cause a negative potential to be applied to the base/grid of A2 which will,
in turn, cause a decrease in the current of A2
and an increase in its collector/plate voltage.
An increase in the positive potential at the col-
lector/plate of A2 will cause C1 to increase its
charge. The charge path for C1, as shown in figure 3-22, will include R4, the +V supply, and the
base/grid to emitter/cathode resistance of A1 in
parallel with R1; and will thereby increase the
current of A1. This action is cumulative and will
result in cutting off A2.

The circuit is now in one of its two stable
states of operation, and will remain so as long as
A2 is in a cutoff condition. The negative potential
at the base/grid of A2 will decrease towards the
cutoff value at an exponential rate as C2 discharges. At the instant the cutoff value of A2 is
reached and it begins to conduct, the actions just
discussed will be reversed, and switching will

result in the circuit changing to its second stable
state of operation. During this state of operation,
A1 will be cut off and A2 will be conducting. C1
will be discharging through R1 and A2; and C2 will
be charging, its path being the parallel combination of R2 and the base/grid to emitter/cathode
resistance of A2, R3, and the +V supply.

Waveform Analysis

For the purpose of discussion, it is assumed
that the circuit has been operating and at time t_1
in figure 3-23, the switch from one operating

62

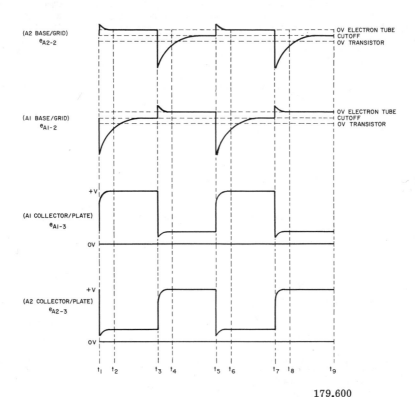

179.600
Figure 3-23.—Collector/plate coupled multivibrator waveforms.

state to the other has just occurred. At t_1, A2 is conducting. The charging current of C2 through the parallel combination of R2 and the base/grid to emitter/cathode resistance of A2 causes the base/grid voltage (e_{A2-2}) to go above cutoff.

Since a capacitor (in a circuit containing resistance) cannot instantaneously change its voltage, the entire change on A2's collector/plate is coupled through C1 to the base/grid of A1 causing a large negative level (e_{A1-2}). Although this negative on the base/grid of A1 causes the active device to cut off, its collector/plate voltage (e_{A1-3}) does not rise instantly to +V because the charging current of C2 causes a voltage drop across R3.

Due to the relatively low value of resistance in its charge path, the charging time constant of C2 will be fairly short. Time t_1 to time t_2 in figure 3-23 represents C2 charge time. As C2 assumes its charge, A2's base/grid voltage

(e_{A2-2}) decreases, as shown between t_1 and t_2. During this time, the exponential decrease of C2's charge current causes less voltage drop across R3 and thereby, raises A1's collector/plate voltage until (at some time between t_1, and t_2) it equals +V. Notice that, during the period t_1 to t_2, A2's collector/plate voltage also increases. This is caused by the fact that, as the base/grid voltage, e_{A2-2}, decreases, the collector/plate current decreases. From the instant switching occurred the base/grid voltage of A1 (e_{A1-2}) begins increasing towards the cutoff value. The relatively large resistance in the discharge path causes the discharge time constant to be fairly long.

From t_2 until t_3 the only change taking place in the circuit is that of the grid voltage e_{A1-2}, with the other voltages remaining in a quiescent state. When C1 has been discharging for a time, A1 will

63

come out of cutoff. At the instant the base/grid voltage e_{A1-2} equals the cutoff threshold (t_3), A1 will begin to conduct and the switching action of the circuit will be initiated. It is seen by the nearly vertical lines on the waveforms at t_3 that the switching action is practically instantaneous. However, in the following discussion this instant will be expanded to ensure a more thorough understanding of the action. In order to accomplish this expansion, the action of the circuit will be halted immediately after cutoff is attained.

Just previous to t_3 the circuit conditions are as follows:

The base/grid voltage of A1 (e_{A1-2}) is at cutoff.
The collector/plate voltage of A2 (e_{A2-3}) is very low.
The base/grid voltage of A2 (e_{A2-2}) is way below cutoff.
The collector/plate voltage of A1 (e_{A1-3}) is equal to +V.

The action of the circuit will now be examined when the base/grid of A1 is slighly above cutoff. This base/grid level will cause a very small collector/plate current to flow. This small current will cause a small drop across R3, which in turn lowers A1's collector/plate voltage by the same amount. This change in A1's collector/plate voltage is coupled through C2 to R2, causing a negative voltage at the base/grid of A2. If an amplification of 10 is assumed for the device in this circuit, the change at the base/grid will cause 10 times this voltage change in A2's collector/plate voltage. The increase in A2's collector/plate voltage will cause C1 to draw a small charging current up through R1.

The change in A2's collector/plate voltage also appears on A1's base/grid. This means that A1's base/grid goes instantly from cutoff to a positive voltage. Due to amplification, this base/grid voltage change will appear at A1's collector/plate as a larger change. This is a regenerative feedback action and will continue with the collector/plate voltage change being coupled to A2's base/grid, driving it into cutoff.

Due to this regenerative action, it can be seen that even the slightest change in the state (nonconduction to conducting) of the cutoff device will result in an almost instantaneous switching of the circuit. The circuit is now in its second condition of operation, with A1 in conduction and A2 cut off.

The action of the circuit from t_3 to t_4 is the same as that described for t_1 to t_2, except the role of the components is reversed (where C1 was

discharging it is now charging, etc.). The action of the circuit between t_4 and t_5 is the same as that described between t_2 and t_3. At t_5 the voltage at the base/grid of A2 reaches cutoff, and switching is again initiated. Thus, time one to time five represents one complete cycle of operation. The output from this circuit may be taken from either collector/plate, although the base/grid waveform is sometimes useful.

Changing Parameters

The effects on circuit operation of increasing the value of various components are shown by the waveforms in figure 3-24. Part A of the illustration shows the normal waveforms of the circuit when all comparable components are equal in value. Part B shows the waveforms of the circuit when the value of R3, collector/plate load resistor of A1, is increased. The time of one complete cycle is longer, since the duration of the first alternation of all the waveforms in part B is longer than the first alternation of the normal waveform. Due to the increased resistance of R3 the collector/plate voltage of A1 will experience a larger change. Thus, the amplitude of e_{A1-3} will be increased. The duration of the first alternation is longer than the second alternation, an indication that A1 is conducting for a longer period of time than it is cut off. Notice the increase in the rounding of the leading edge, caused by the increased charging time of C2. The long duration of the first alternation of e_{A2-3} indicates that the base/grid of A2 (e_{A2-2}) is driven further into cutoff than the base/grid of A1 (e_{A1-2}) due to the larger change at the collector/plate of A1. Therefore, C2 will discharge for a longer period of time until the base/grid voltage reaches the cutoff value of the device, thereby holding A2 cut off longer than A1. Since the sum of the two alternations is now longer than for normal operation, the frequency of the circuit will be decreased.

Part C of figure 3-24 illustrates the waveforms observed when R1, the base/grid resistor of A1, is increased. All other components have the same value as the normal circuit. The first alternation of the waveform has the same duration as the normal waveform. However, increasing R1 has lengthened the duration of the second alternation considerably. The collector/plate voltage waveform, e_{A1-3}, indicates that A1 is cutoff for a much longer period of time than it is in conduction. Waveform e_{A2-3} indicates that A2 is conducting for a longer period of time than it is cut off. Notice the increased rounding of the leading edge due to the increased charging time of C1.

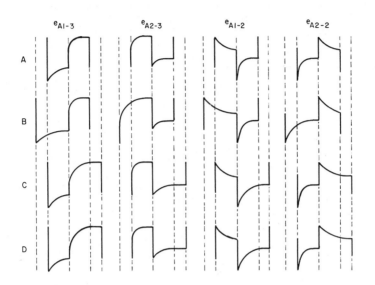

179.601

Figure 3-24.—Waveforms showing effect of increasing component values.

Increasing R1 increases the discharge time constant of C1, thereby applying a cutoff potential to the base/grid of A1 for a longer period of time (as shown by the e_{A1-2} waveform). Since one alternation is longer than normal, the complete cycle will encompass a longer period of time. Thus, the frequency of operation will be lower than normal.

Part D of figure 3-24 shows the waveforms observed when the value of coupling capacitor C1 is increased. The effect of a larger capacitance is to increase the RC discharge time constant, with the same results as were obtained when the resistance was increased. Thus, the waveforms in part D will be the same as the waveforms in part C, and the same explanation applies.

Frequency Stability

The frequency stability of a free-running multivibrator is rather poor. In order to achieve a better understanding of the manner in which supply voltage and component value variations affect the frequency stability, figure 3-25 will be used with the following text.

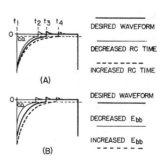

179.602

Figure 3-25.—Effect of voltage and component value variation on frequency stability.

Part A of the figure shows the effects of a decrease in the value of either the base/grid resistor or the coupling capacitor on the frequency of operation. The heavy line shows the desired base/grid voltage waveform. Notice that the base/grid voltage approaches the cutoff value of the device at a very shallow angle. At time t_3 it reaches the cutoff value and switching occurs. Due to the shallow angle of approach, slight variations in the RC discharge time can cause the cutoff value to be reached at some time other than the desired time. For instance, if the value of the base/grid resistor or coupling capacitor decreases for some reason, the RC discharge time will be decreased. The light line in figure 3-25A shows the waveform with the RC discharge time slightly decreased. Notice that even the slight decrease in the RC time will cause the cutoff value to be reached at time t_2 rather than time t_3. The dotted line shows the waveform for a slight increase in the RC discharge time. This will cause the cutoff value to be reached at time t_4 rather than time t_3. Thus, a slight variation in the RC time will cause a relatively large variation in the duration of the alternation, resulting in poor frequency stability.

Part B of figure 3-25 shows the effect of a varying supply voltage on the frequency of a free-running multivibrator. The heavy line represents the desired waveform and t_1 to t_3 is the desired duration of the alternation. A decrease in the supply voltage will cause a lower change on the coupling capacitor and a smaller change in collector/plate voltage coupled to the opposite base/grid. This results in the base/grid being driven less negative during cutoff (shown by the light line in figure 3-25B). The coupling capacitor will then discharge for a shorter period of time before reaching the cutoff value (time t_1 to t_2). The cutoff value will also decrease a slight amount, but this will not change the operation a significant amount.

Increasing the supply voltage will cause a larger charge of the coupling capacitor and a larger change in collector/plate voltage during switching. This will result in the base/grid being driven further negative and a longer period of discharge before the base/grid voltage reaches cutoff. This condition is shown by the dotted line in figure 3-25B. The cutoff value will be increased slightly in this case, but again will not significantly affect the operation. Increasing the supply voltage will increase the duration of the cutoff alternation, and thereby decrease the frequency of operation. The shallow angle or approach to cutoff will result in a slight change in supply voltage causing a relatively large change in frequency.

Synchronization

The various multivibrator circuits find extensive used in radar and test equipment. However,

due to the poor frequency stability of free-running multivibrators, emitter/cathode coupled or collector/plate coupled types are seldom used in this mode of operation. They are usually synchronized with another more stable frequency which forces the period of multivibrator oscillation to be some function of the synchronizing frequency.

Sharp pulses are generally used for synchronizing purposes, although waveforms of almost any shape could be used. After several time constants, the discharge slope of a capacitor is not very steep. Therefore slight circuit differences from cycle to cycle significantly affect the time duration of a waveform and its reciprocal, frequency. Figure 3-26A shows the waveform on the emitter/cathode of an emitter/cathode coupled multivibrator. It should be recalled that at points A and B, A1 of figure 3-20 switches into conduction and A2 cuts off. Note the shallow angle of approach to the switching points and visualize how a significant time period occurs for only a very small change in voltage. Figure 3-26B shows A1's base/grid waveform of the collector/plate coupled multivibrator. Again the small rate of change characteristic is prevalent near the switching point. Now, if a synchronizing pulse were applied before transition occurs, so that the pulse caused the circuit to change states, frequency of operation would be determined by the synchronizing signal. The synchronization frequency is always higher than the free-running frequency. The free-running base/grid waveform shown in figure 3-27A is driven by the pulses shown at B. This results in the transition and duration of the first alternation being determined by the pulse rate. Figure 3-27C shows waveform A after it is synchronized.

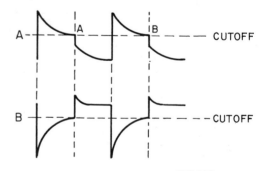

179.603

Figure 3-26.— Unsynchronized waveform.

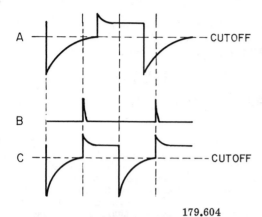

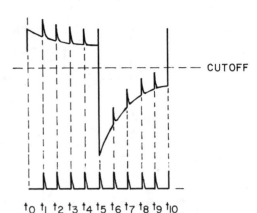

179.604

Figure 3-27.—Synchronized waveform.

t_0 t_1 t_2 t_3 t_4 t_5 t_6 t_7 t_8 t_9 t_{10}

179.605

Figure 3-28.—Amplitude and polarity
synchronization.

The synchronizing frequency may be at a harmonic of the desired stabilization frequency. Figure 3-28 shows how the first nine pulses are of insufficient amplitude or polarity to bring about transition. The first five pulses are of the opposite polarity required to switch the device in the condition it is in at this time, so no effect will occur. Pulses occurring at t_6 through t_9 are of insufficient amplitude to switch the device out of cutoff. At t_{10}, the polarity is correct for the condition of the device at this time and it is of sufficient amplitude to bring the device out of cutoff (i.e., the capacitor in the circuit has discharged to a level close to the ·cutoff potential and the additional trigger amplitude is enough to reach above cutoff).

Synchronizing pulses may be applied to any element of the switching device. However, pulses applied to collectors/plates will be of higher amplitude, in general, than those applied to other elements. A common sync application of the emitter/cathode coupled multivibrator is to the free or unused base/grid. Since there are no other transient voltages present at this element, this multivibrator circuit lends itself especially well to synchronization. For this reason many applications which formerly used other types of astable multivibrators, and consequently high values of sync voltages, now use the emitter/cathode coupled circuit. The polarity of a trigger is determined by the element on which it is to be applied and whether it is used to switch on or to switch off the device to which it is applied.

CHAPTER 4
PULSE FORMING AND PULSE SHAPING CIRCUITS—PART III

Continuing the discussion of pulse forming and pulse shaping circuits this chapter covers sawtooth generators, blocking oscillators, counters, and coincidence circuits.

SAWTOOTH GENERATORS

Ideally, the sawtooth waveform consists of a voltage which increases linearly with time until it reaches a predetermined final value, instantaneously returns to zero, and immediately increases again as the cycle repeats. One application of this waveform is to produce a linear time base for use in oscilloscopes. Since, in this application, the sawtooth waveform causes an electron beam to sweep across the oscilloscope screen it is also called a SWEEP VOLTAGE.

All linear voltage sweep circuits operate in a similar manner. A capacitor is allowed to charge through a high resistance until a predetermined voltage is reached. At that point a low resistance discharge path is provided and the capacitor is rapidly discharged. At some point in the capacitor discharge time, the discharge path is opened and the cycle is repeated. Variations in sweep circuits are produced by varying the values of the elements controlling the time of charge and discharge.

A basic sweep or sawtooth generating circuit is illustrated in figure 4-1. When the switch is placed in position 1 the capacitor charges through R1 toward E_{bb} (fig. 4-1A). When the voltage across the capacitor reaches some predetermind value, E_{max}, the switch is thrown to position 2, discharging the capacitor through R2. Upon reaching a predetermined minimum value, E_{min} (fig. 4-1B), the switch is returned to position 1 and the cycle repeats. Although, in the basic circuit of figure 4-1 the capacitor could have been discharged to an E_{min} of zero volts, it will be found that in a practical circuit this is not done.

Before considering more practical sweep circuits a closer examination of the circuit and waveforms of figure 4-1 is in order. Note that what is

being generated is a portion of an exponential charging curve. To achieve an essentially linear sweep, operation must be restricted to no more than the first ten percent of the total exponential charge curve. This portion of the curve, as shown in figure 4-1B, is nearly linear. The smaller the percentage of the total charging curve used, the better the linearity. Thus, E_{bb} must be made much greater than the desired output voltage swing (E_{max} $-E_{min}$) to ensure good linearity. Also, note that the retrace time of the sawtooth (t_2) is a function of the resistance of R2. Reducing R2 to a small enough value will cause the retrace time to be insignificant.

If the mechanical switch of figure 4-1A is replaced by a device which automatically performs the switching function a practical sawtooth generator results. One device which is used to fulfill the function of an automatic switch is a Shockley diode.

The Shockley diode is a four layer semiconductor device which blocks forward current flow until a specific value of potential is applied between its anode and cathode. This potential, positive on the anode with respect to the cathode, is called the BREAKOVER VOLTAGE. Once the breakover voltage is reached, the diode will continue to conduct until its anode voltage is decreased below a specific minimum voltage, called the DIODE TURN OFF VOLTAGE. Once the anode voltage has been decreased below the diode turn off voltage, the diode will not conduct until the anode voltage is again raised to the breakover voltage.

A pictorial diagram, schematic symbol, and equivalent transistor circuit of a Shockley diode are illustrated in figure 4-2. P1, N1, and P2 (fig. 4-2A) form the equivalent of one transistor, while N1, P2, and N2 form a second equivalent transistor. The mechanism of operation of the Shockley diode is reverse breakdown of junction 2 (J2, fig. 4-2A).

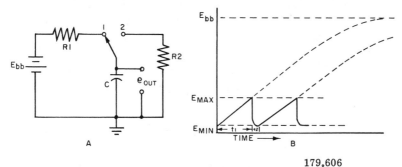

179.606

Figure 4-1.—Simplified sawtooth generator and waveforms.

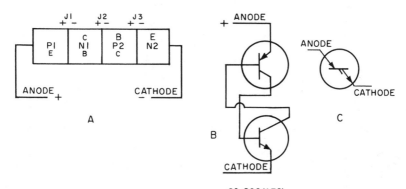

20.399(179)

Figure 4-2.—Shockley diode diagrams.

SHOCKLEY DIODE SAWTOOTH GENERATOR

Figure 4-3 illustrates a Shockley diode saw-tooth generator. At the instant S1 (fig. 4-3A) is closed, the voltage across the parallel combina-tion of D1 and C1 is zero. Hence, D1 is cut off and acting as an open switch. C1 now begins to charge through R1 toward E_{bb} (solid arrows fig. 4-3A). When the potential across C1 reaches the breakover voltage (V_{BO}) of D1 (fig. 4-3B), it causes D1 to conduct, thereby providing a dis-charge path for C1 (dotted arrows fig. 4-3A). D1 will continue to conduct until the potential across C1 drops below the turn off voltage (V_{TO}) of D1. When D1 cuts off, C1 again charges through R1 toward E_{bb}, and the cycle repeats.

The output frequency of a Shockley diode sweep generator is primarily dependent upon the breakover and turn off voltages of the diode used, upon the value of the RC combination (i.e. RC time constant), and the value of source voltage. An increase in V_{BO} causes a decrease in output frequency. This is depicted in figure 4-4. Note also, that this causes an increase in output amplitude and a decrease in waveform linearity. The decrease in linearity occurs because the out-put represents a greater percentage of the total charging curve. A decrease in V_{BO} has the op-posite effects. V_{BO} may be changed only by inserting a different diode into the circuit.

If either the resistance or capacitance is increased the charge time of the capacitor is increased (so, to a lesser degree, is the discharge

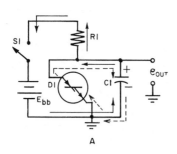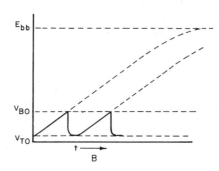

179.607

Figure 4-3.—Shockley diode sawtooth generator and waveforms.

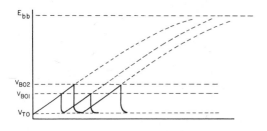

Figure 4-4.—Effect of increasing V_{BO}.

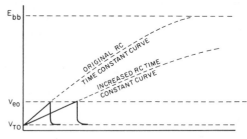

179.609

Figure 4-5.—Effect of increasing RC time constant.

time). This means that it will take the capacitor a longer time to charge to V_{BO}, resulting in a lower operating frequency. This is depicted in figure 4-5. Note that a change in RC time does not affect linearity or amplitude. The same percentage of the charging curve is used in both instances. A decrease in RC time has the opposite effect on frequency.

An increase in source voltage will cause an increase in output frequency. This is due to the capacitor charging more rapidly in its attempt to reach a higher source voltage in the same amount of time (RC time constant is unchanged). Figure 4-6 depicts this action. Although output amplitude is unchanged, linearity is improved since the output represents a smaller percentage of the total. A decrease in source voltage has the opposite effects.

In a practical circuit, source voltage and the type of diode are not variable. Only the resistance and capacitance may be varied to accomplish a

change in frequency. Figure 4-7 illustrates such a circuit. Coarse frequency control is achieved by switching in the desired value of capacitance. Fine frequency control is obtained by varying R2. R1 sets a minimum value of resistance in the capacitor charge path.

A cold cathode gas filled voltage regulator (VR) tube may be used in place of the Shockley diode. Circuit operation and factors affecting the operation of both circuits are identical. Only the amplitude of potentials and the mechanism of operation of the VR tube differ from the Shockley diode circuit. The mechanism of operation of a cold cathode gas filled VR tube has been discussed previously and will not be covered here. Figure 4-8 illustrates a cold cathode gas filled VR tube sawtooth generator.

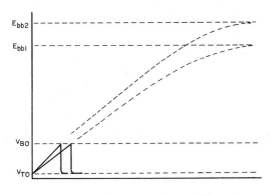

179.610
Figure 4-6.—Effect of increasing source
voltage.

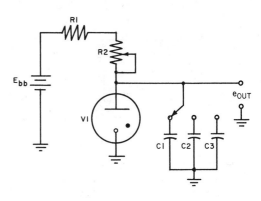

179.612
Figure 4-8.—VR tube sawtooth generator.

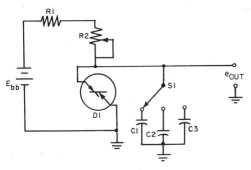

179.611
Figure 4-7.—Shockley diode sawtooth generator
with variable output.

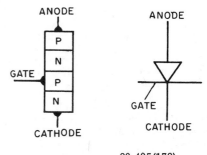

20.405(179)
Figure 4-9.—Silicon controlled
rectifier (SCR).

SCR SAWTOOTH GENERATOR

One disadvantage of the Shockley diode sawtooth generator is that the output amplitude is fixed for any particular circuit. This disadvantage may be overcome with the use of silicon controlled rectifier (SCR).

The SCR is similar in construction and operation to the Shockley diode. However, the SCR has a gate terminal connected as shown in figure 4-9. The potential on the gate with respect to the cathode determines the breakover (breakdown)

potential of the SCR (V_{BO}). If the potential on the gate is made positive with respect to the cathode, the required breakover potential is decreased. The more positive the gate to cathode voltage the smaller the required V_{BO}. Of course, if the gate potential is made too large the device may be destroyed. Once V_{BO} is reached and while the SCR conducts heavily, the gate has no further control of the SCR conduction. The gate regains control when the anode voltage is decreased to V_{TO}.

Figure 4-10 illustrates an SCR sawtooth generator. The operation and factors affecting the operation of this circuit and the Shockley diode sawtooth generator, are identical, with one exception. In the SCR circuit the amplitude of output

71

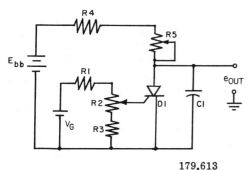

179.613
Figure 4-10.—SCR sawtooth generator.

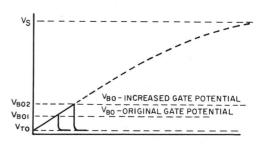

179.614
Figure 4-11.—Effects of an increase in gate potential.

may be controlled by the gate to cathode potential.

Figure 4-11 depicts the effects of a change in gate to cathode potential on the output. V_{BO1} results from a particular setting of the wiper arm or R2 in figure 4-10. If the wiper arm is moved downward, making the gate less positive with respect to the cathode, the breakover voltage is raised to V_{BO2}. Note that this results in a decreased frequency and linearity, as well as an increase in amplitude. Making the gate more positive with respect to the cathode would have an opposite effect.

The SCR sawtooth generator may be synchronized by positive pulses applied between gate and cathode (fig. 4-12A). To accomplish this, the free-running frequency of the generator is set slightly lower than the desired output frequency The synchronizing pulses will occur at the desired frequency or at an integral multiple of the desired frequency and cause the circuit to operate at the desired frequency.

The positive sync pulses cause an instantaneous reduction of V_{BO} and allow the SCR to conduct slightly sooner than it would if it were free-running. The dotted lines (fig. 4-12B) indicate the free-running waveform. The solid lines indicate the synchronized output. Note that in the case of the waveform synchronized by the third harmonic of the desired frequency, only every third pulse finds the anode voltage of sufficient amplitude to cause breakover. The capability of the SCR sweep generator to be synchronized is another advantage of this circuit over the Shockley diode sweep generator.

THYRATRON SAWTOOTH GENERATOR

The electron tube equivalent of the SCR sawtooth generator is the thyratron sawtooth generator. The thyratron is a gas filled triode whose operation is similar to that of the gas filled diode, with two exceptions. The thyratron utilizes a heated cathode. Also, the ionizing potential of the thyratron is determined by the difference of potential between its control grid and cathode. The more negative the control grid with respect to the cathode the higher the required plate to cathode ionizing potential. As in the SCR, the control grid has no further control over tube conduction once the tube ionizes. It regains control when the tube deionizes.

Figure 4-13 illustrates a thyratron sawtooth generator circuit. Making the grid of the thyratron more negative with respect to its cathode corresponds to making the gate of an SCR less positive with respect to its cathode. This is true because both actions increase the potential required on the anode/plate with respect to the cathode to cause the device to conduct heavily. The effects attending the changes in gate or control grid potential are identical in both circuits. Naturally, making the control grid of the thyratron less negative with respect to its cathode corresponds to making the gate of the SCR more positive with respect to its cathode.

Sychronization of the thyratron sawtooth generator is identical to that of the SCR sawtooth generator. That is, it is accomplished by applying positive pulses between the control grid and cathode.

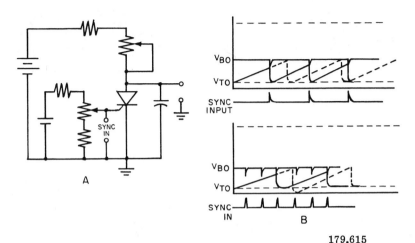

179.615

Figure 4-12.—Synchronization of an SCR sawtooth generator.

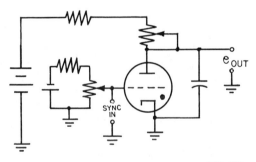

179.616

Figure 4-13.—Thyratron sawtooth generator.

TRANSISTOR/ELECTRON TUBE SAWTOOTH GENERATOR

Figure 4-14A depicts a transistor/electron tube sawtooth generator circuit. The operation of this circuit differs from that of the circuits previously discussed in that an input gate, or pulse, is required in order to produce an output. It is, therefore, more properly termed a WAVE SHAPING CIRCUIT than an oscillator or generator.

The operation of the circuit of figure 4-14A may be understood by referring to the waveforms depicted in figure 4-14B. The polarities used are

correct for NPN transistors and electron tubes. The polarities for a PNP transistor circuit would be opposite to those depicted.

With no input signal applied, A1 is conducting heavily, and its collector/plate voltage is therefore very low. C2 (fig. 4-14A) is charged to this low potential. When a negative gate pulse of sufficient amplitude is applied between base/grid and emitter/cathode of A1, A1 cuts off. This causes C2 to charge through R2, toward the source voltage. The RC charge time of C2 is made much longer than the period of the input gate. This means that C2 will charge for only a small portion of its total charge curve, providing good linearity. When the gate voltage returns to zero, A1 again conducts, and C2 discharges through the low conduction resistance of A1 (from emitter/cathode to collector/plate).

The output frequency of this circuit is determined by the frequency of the input. Note that there is a time delay between the end of retrace of one sweep and the beginning of the trace of the next sweep. This delayed sweep is often used in oscilloscopes and radar indicators. The length of delay depends upon the time interval between negative gates. This is depicted in figure 4-15. Note that the width of the input gate has not been changed. Only the time between gates has been changed.

The amplitude of the output waveform is determined by three factors: the value of supply voltage, the RC time constant, and the width of the input

73

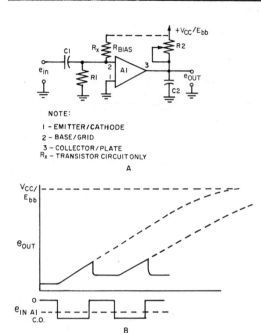

179.617
Figure 4-14.—Transistor/electron tube
sawtooth generator and waveforms.

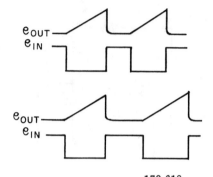

179.618
Figure 4-15.—Delayed sweep.

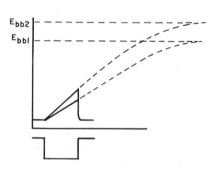

179.619
Figure 4-16.—Effects of a change in
source voltage.

gate. If the value of the supply voltage is decreased, the output amplitude will increase (all other variables held constant). This is depicted in figure 4-16. Note that although the amplitude of output is increased, the percentage of the total charge curve used remains the same. This is due to the fact that the percentage of source voltage increase and the percentage of amplitude increase are equal. Linearity, therefore, remains the same. The increase in amplitude is caused by the increased rate of charge of the capacitor. This increased rate of charge is due to the fact that the capacitor is attempting to charge to a greater supply voltage in the same amount of time. Decreasing the supply voltage would have the opposite effect.

An increase in the charge time constant of C2 (with all other variables held constant) will result in a decrease in output amplitude and an increase in waveform linearity. This is illustrated in figure 4-17. Decreasing the RC charge time constant would have the opposite effects.

If the width of the input gate is increased, the amplitude of the output will increase. Figure 4-18 depicts this action. Note that this also decreases waveform linearity. Decreasing gate width would have the opposite effects.

Usually it is undesirable to have the amplitude change with changes in input gate width. To compensate for such changes, R2 (fig. 4-14A) is usually varied automatically with changes in gate width.

BLOCKING OSCILLATORS

A blocking oscillator is an oscillator that conducts for a short period of time and is then cut off (blocked) for a much longer period. This process produces an intermittent sawtooth waveform. Blocking oscillators are usually of the SELF-PULSING or SINGLE-SWING type.

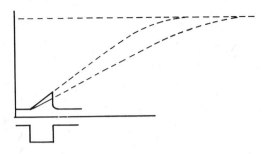

179.620

Figure 4-17.—Effects of increasing RC
charge time.

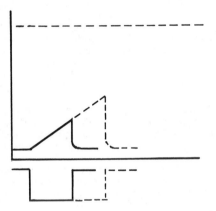

179.621

Figure 4-18.—Effects of changes in gate width.

SELF-PULSING
BLOCKING OSCILLATOR

The self-pulsing blocking oscillator is a modified sinusoidal LC oscillator which generates a predetermined number of cycles of output and then cuts itself off for a predetermined rest time.

Any properly modified sinusoidal LC oscillator may function as a self-pulsing blocking oscillator. A shunt fed Hartley oscillator is used in the following explanation. The voltage polarities, referred to are correct for NPN transistors and electron tubes. For operation with PNP transistors, all polarities would be opposite.

Figure 4-19 depicts a Hartley self-pulsing blocking oscillator. The only difference between the self-pulsing and conventional Hartley oscillator is the size of the base/grid leak bias components, $C_{b/g}$ and $R_{b/g}$. (Resistor R_x in dotted lines is the base bias resistor necessary when a transistor is used as the amplifying device.) In the self-pulsing blocking oscillator, these components are of much greater electrical size than their counterparts in a conventional oscillator. (For a review of grid-leak bias and LC oscillators refer to the chapter on LC oscillators.)

The build up of oscillations in the self-pulsing blocking oscillator is identical to that action in a conventional oscillator. However, due to the large value of $C_{b/g}$ the amplitude of oscillations increases much more rapidly than the bias voltage, since $C_{b/g}$ cannot be charged quickly. Therefore, as shown in figure 4-20, the oscillations reach maximum amplitude (t_2) before the base/grid bias voltage has been appreciably built up.

After t_2, base/grid current continues to flow during the positive peaks of the tank voltage, charging $C_{b/g}$. Therefore, the base/grid waveform is a series of oscillations about an axis which is becoming increasingly more negative. The collector/plate waveform is a series of pulses. The duration and amplitude of these pulses decrease as the bias voltages go further below cutoff.

As the collector/plate current pulses decrease, a time (t_3) is reached at which the pulses are not of sufficient amplitude to supply the power taken from the oscillator tank by the load. This causes the amplitude of oscillation to decrease. Since the base/grid bias can change only slowly (due to the large values of $C_{b/g}$ and $R_{b/g}$), the decrease of oscillation amplitude causes the conduction period of the amplifying device to become still shorter. Therefore less power is supplied by the amplifying device, and the oscillations damp out. The Q of the tank is such that the oscillations damp out very rapidly once the amplifying device is cut off.

The circuit will remain cut off (blocked) until the charge on $C_{b/g}$, discharging through $R_{b/g}$, has diminished sufficiently to allow the amplifying device to conduct. At that point the cycle repeats itself as shown in figure 4-21.

The frequency of the oscillations contained within the output pulse is known as OSCILLATION FREQUENCY. This frequency is determined by the value of inductance and capacitance of the tank. The number of times per second the circuit produces periodic bursts of these RF oscillations is called the PULSE REPETITION FREQUENCY (PRF).

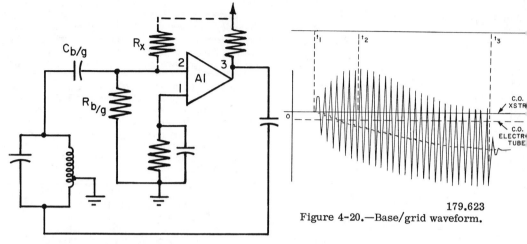

179.623
Figure 4-20.—Base/grid waveform.

NOTE: I – EMITTER/CATHODE
2 – BASE/GRID
3 – COLLECTOR/PLATE
Rx – TRANSISTOR CIRCUIT
ONLY

179.622
Figure 4-19.—Hartley self-pulsing blocking
oscillator.

The PRF is mainly determined by the value of $C_{b/g}$ and $R_{b/g}$. Increasing the value of $C_{b/g}$ will increase both the pulse width and rest time since both the charge and discharge time of $C_{b/g}$ will be increased. This, in turn will result in a decrease in the PRF. Increasing the value of $R_{b/g}$ will increase $C_{b/g}$'s discharge time, resulting in a longer rest time. This also increases the PRF. However, $R_{b/g}$ has no effect on the charge time of $C_{b/g}$. Therefore, variations of $R_{b/g}$ do not affect the pulse width of the output.

SINGLE-SWING
BLOCKING OSCILLATOR

The single-swing blocking oscillator generates a very narrow pulse and is often used as a master oscillator in radar systems. Basically, the single-swing blocking oscillator is a sinusoidal oscillator whose free-running period of operation is interrupted by the build-up of a disabling voltage across two of its controlling elements.

The duration of the pulse or spike is in the order of .05 to .25 microseconds. The schematic diagram and output waveform of the single-swing

blocking oscillator are illustrated in figure 4-22.

The circuit of the single-swing blocking oscillator consists of R_b/R_g—base/grid resistor, C_b/C_g—base/grid coupling capacitor (also called base/grid-leak capacitor), T1—pulse transformer, and A1—amplifying device (transistor/tube). Resistor R_x, the base bias resistor (in dotted lines), is necessary when a transistor is the amplifier device.

Due to the very narrow pulse width involved, special transformers are used, so as not to distort the waveform. These transformers are called PULSE TRANSFORMERS. Pulse transformers have ferromagnetic cores. Their windings are closely coupled and have relatively few turns. T1 (fig. 4-22) is a pulse transformer which consists of L1 (primary winding), L2 (secondary winding), and L3 (tertiary or third winding—output winding). The dots shown in T1 are called POLARITY or PHASING DOTS. They indicate which leads of the transformer T1 have the same instantaneous polarity. For example, when the primary lead number 5 becomes positive then leads 2 and 4 will also be positive with respect to the opposite end of the winding concerned.

Circuit Operation

Assume that C_b/C_g (fig. 4-22) is discharged and A1, the switching device, is an NPN transistor or an electron tube. At the instant (time t_0) power supply voltage is applied, A1 begins to conduct.

In the transistor circuit, the transistor is forward biased through R_b-R_x and in the electron tube circuit grid bias is zero.

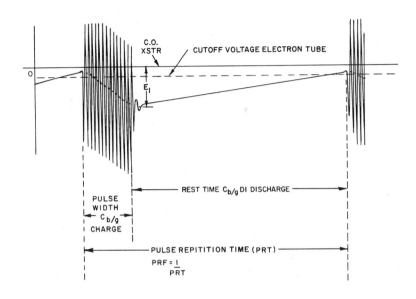

179.624
Figure 4-21.—Output waveform.

The path for collector/plate current is from the negative supply (ground) through A1 and the primary (L1) of the pulse transformer, then back to source (positive).

As collector/plate current flows, it causes the magnetic field about L1 to expand and induces a voltage across windings L2 and L3 of the polarities shown in figure 4-22.

The voltage of L2 is instantaneously felt on the base/grid of A1 due to the coupling action of C_b/C_g. At this time, three other important actions are occurring: (1) A1's conduction is increasing (I_c/I_p is increasing); (2) Output voltage is increasing. Note: Polarity of the output voltage may be either positive or negative depending on which of the tertiary (L3) leads is grounded; and (3) C_b/C_g (base/grid-leak capacitor) is charging.

The charge path of C_b/C_g is from ground, through A1 (emitter to base or cathode to grid), through L2 (acting source) and back to ground. The actions described are cumulative and will continue until the collector/plate current ceases to increase due to the transistor/tube approaching saturation.

At time t_1 (fig. 4-23A), collector/plate current ceases to change, collector/plate voltage is minimum, base/grid voltage is maximum, and the magnetic field of the primary no longer induces a voltage into L2 and L3. The coupling capacitor starts to discharge (fig. 4-23B), driving A1 into cutoff (t_1 to t_2). The magnetic field of the primary winding (L1) collapses since there is no longer a current flow through A1. This collapsing field induces a voltage of opposite polarity across L2 and L3. The voltage developed across L2 is now aiding the voltage of the coupling capacitor and drives A1 far below cutoff (t_2 to t_3).

The period t_2 to t_3 is the decay time of the magnetic field, and its effect on the base to ground or grid to ground voltage in the base/grid circuit (v_b/e_g waveform) is shown in figure 4-23A. After the magnetic field has collapsed, the coupling capacitor will continue to discharge through R_b/R_g holding the circuit in cutoff as indicated by the waveform between times t_3 and t_4. At time t_4, A1 comes out of cutoff and the cycle is repeated. The combination of R_b/R_g and C_b/C_g determines the rest duration. Increasing R_b/R_g or C_b/C_g increases the rest

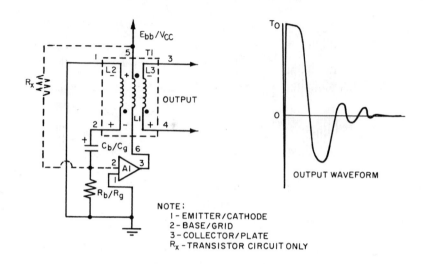

179.625

Figure 4-22.—Single-swing blocking oscillator and output waveform.

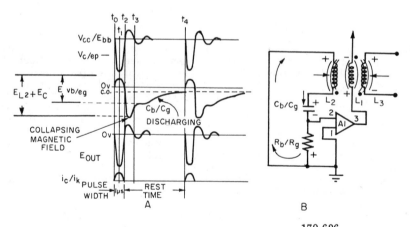

179.626

Figure 4-23.—Waveforms and discharge path of C_b/C_g .

time, and vice versa. Although both of the components can be variable, R_b/R_g has a greater effect on the circuit operation, and, therefore, is usually made variable while C_b/C_g is a fixed value. Pulse width of the circuit is mainly determined by the inductance of the pulse transformer as shown in figure 4-24. The circuit will remain

in cutoff until the capacitor has discharged to the cutoff value of A1. A1 will then start conducting and the cycle will be repeated.

The desired output from the single-swing blocking oscillator is a very narrow, high amplitude pulse. Due to interelement capacitances of the transformer, however, damped oscillations

1. ORIGINAL
   REST TIME

2. NEW REST TIME

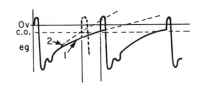

R_b/R_g decreased – Rest time decreased

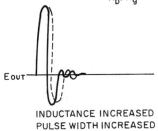

E_{OUT}

INDUCTANCE INCREASED
PULSE WIDTH INCREASED

$$\text{PULSE WIDTH} = T = \frac{L}{R}$$

WHERE L = INDUCTANCE OF PULSE
TRANSFORMER

R = RESISTANCE OF A1

179.627

Figure 4-24.—Effects of varying R_b/R_g and inductor.

occur and are felt in the output, as shown in figure 4-25. A resistor placed across the output winding will reduce some of the unwanted oscillations. A better method of reducing these undesired oscillations is to connect a damping diode across the output winding.

When the output pulse, as shown in figure 4-26, goes positive it simply reverse biases the diode, so that it has no effect on the output waveform. As the output voltage passes through zero toward the negative direction it forward biases the diode and the diode conducts, effectively short circuiting the output.

DRIVEN BLOCKING
OSCILLATOR

Stability of the output waveform occurring at a specific time is the prime concern in the output waveform of the driven blocking oscillator. The driven blocking oscillator is simply the basic single-swing blocking oscillator with the addition of a bias network and an input coupling capacitor, as shown in figure 4-27A.

Function of the bias network (R1 and R2) is to keep A1 cut off until triggered. The trigger coupling capacitor passes the trigger signal from the previous stage to the base/grid of A1 and starts the cycle of operation.

For explanation purposes, A1 is considered to be an electron tube to which all proper operating voltages have been supplied. The cutoff voltage for this particular tube is -18 volts (grid with respect to cathode) as shown in figure 4-27B. The voltages across the voltage divider bias networks (R1 and R2) are 30 and 220 volts respectively.

R1 is the cathode bias resistor and its purpose is to develop self bias. In this circuit, however, R2 is in series with R1 and both are connected across the plate supply. Thus, R1 will develop a voltage proportional to its resistance, in this case 30 volts with respect to ground. Bias (measured from grid to cathode) now is also -30 volts.

Since the cutoff voltage of the tube is -18 volts, then the tube will be in a state of nonconduction and the plate voltage at this time will be the same as source voltage. The circuit will remain in this nonconducting state until triggered.

When a trigger of sufficient amplitude to overcome the fixed bias is coupled to the grid, the tube begins to conduct and the operation of the driven blocking oscillator is the same as for

79

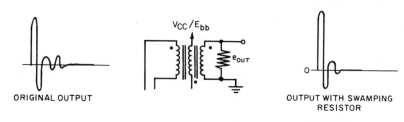

ORIGINAL OUTPUT

V_{CC}/E_{bb}

e_{OUT}

OUTPUT WITH SWAMPING RESISTOR

179.628

Figure 4-25.—Effects of swamping resistor across output.

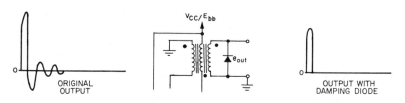

ORIGINAL OUTPUT

V_{CC}/E_{bb}

e_{out}

OUTPUT WITH DAMPING DIODE

179.629

Figure 4-26.—Effects of the damping diode across output.

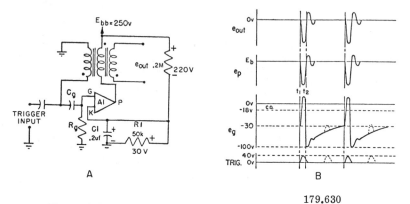

A

B

179.630

Figure 4-27.—Driven blocking oscillator and waveforms.

the single-swing blocking oscillator. That is, tube current increases rapidly; the action of the expanding magnetic field charges the coupling capacitor; plate current through the tube ceases to increase; the coupling capacitor begins to discharge and cuts the tube off; the magnetic field about T1 collapses and the cycle of operation

has been completed. The circuit will remain in this nonconducting state until triggered again.

RINGING OSCILLATOR

A circuit which also produces periodic oscillations, is the ringing oscillator. This circuit (fig. 4-28) produces a short series of oscillations each time an input gate is applied. The oscillations are normally used as distance marks in radar indicators.

The ringing oscillator utilizes a parallel-resonant LC tank circuit to produce an output. An active device is used as a switch to gate the oscillations. The frequency of oscillations in the tank is determined by the values of inductance and capacitance in the tank circuit. The pulse repetition frequency is determined by the rate at which the switch is opened and closed. The pulse width is determined by the time lapse between opening and closing the switch.

In the quiescent condition the device is biased by R1 (fig. 4-28) causing A1 to conduct heavily (near saturation). Current flow is from ground through L1 and A1 to source voltage, building up a magnetic field around the inductor. The circuit will remain in the quiescent state, with no output, until the negative gate is applied.

The negative input gate, applied through C1 to the base/grid of A1, instantaneously drives the device into cutoff. A1 will remain cut off for the duration of the input gate. Current flow through A1 ceases, causing the magnetic field around L1 to collapse. The collapsing field induces a voltage in the inductor of a polarity that keeps current flowing in the same direction through the coil. Since A1 is cut off this continuing current charges C2 negative with respect to ground. When the magnetic field has completely collapsed, no further voltage is induced in the inductor, and current flow ceases. Consequently, there is no longer an induced voltage to maintain the charge

on C2 and it discharges through L1. The discharge current builds up a magnetic field around L1 which is of opposite polarity to the original field. When C2 is completely discharged current flow attempts to cease, causing the field around L1 to collapse, charging C2 in the opposite direction with respect to ground. This flywheel action continues for the duration of the negative gate.

The tank circuit is designed to have a very high Q (low loss), so that damping of the waveform will be negligible. If the negative gate was sufficiently long, damping of the waveform would occur due to the d.c. losses inherent in the tank. The maximum possible pulse width obtainable from a given oscillator would be determined by the Q of the tank. During normal operation the tank will only oscillate for the duration of the negative gate. The frequency of the RF oscillations in the tank is determined by the values of L and C.

When the negative input gate ceases, A1 will again conduct heavily causing a steady current through L1, preventing tank oscillations. The only output at this time will be a slight positive voltage with respect to ground due to the d.c. resistance of the inductor.

COUNTERS

The positive diode counter circuit is used in timing or counting circuits which depend upon a proportional relationship between the output voltage and the number of input pulses. It may indicate frequency, it may count the r.p.m. of a shaft or other device, or it may register a number of

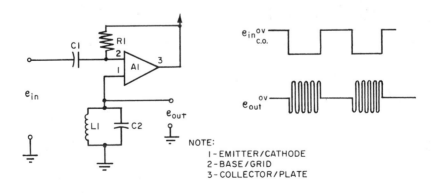

NOTE:
1—EMITTER/CATHODE
2—BASE/GRID
3—COLLECTOR/PLATE

179,631
Figure 4-28.—Ringing oscillator and waveforms.

operations. The diode counter establishes a direct relationship between the input frequency and the average d.c. voltage. As the input frequency increases the output voltage also increases; conversely, as the input frequency decreases the output decreases. In effect, the positive diode counter counts the number of positive input pulses by producing an average d.c. output voltage proportional to the repetition frequency of the input signal. For accurate counting, the pulse repetition frequency must be the only variable parameter in the input signal. Therefore, careful shaping and limiting of the input signal is essential to insure that the pulses are of uniform width, or time duration, and that the amplitude is constant. When properly filtered and smoothed, the d.c. output voltage of the counter may be used to operate a direct reading indicator.

The basic solid-state counter circuit is shown in figure 4-29A. Solid-state and electron tube counters operate sufficiently similar to warrant simultaneous treatment. (To change figure 4-29A to an electron tube circuit, diode tube symbols would be used in place of CR1 and CR2 and would be labeled V1 and V2.)

Capacitor C1 is the input coupling capacitor. Resistor R1 is the load resistor, across which the output voltage is developed. For the purpose of circuit discussion, it is assumed that the input pulses are of constant amplitude and time duration, and that only the pulse repetition frequency changes.

As shown in figure 4-29B, at time t_0 the positive going input pulse is applied to C1 and causes the anode of CR2 to go positive. As a result, CR2 conducts and current i_c flows through R1 and CR2 to charge C1. Current i_c develops an output voltage (e_O) across R1 as shown.

The initial heavy flow of current produces a large voltage across R1, which tapers off exponentially as C1 charges. The charge on C1 is determined by the time constant of load resistor R1 and the conducting resistance of the diode, in series, times the capacitance of C1. For ease of explanation, it is assumed that C1 is charged to the peak value before t_1.

At time t_1 the input signal reverses polarity and becomes negative going. Although the charge on the capacitor C1 cannot change instantly, the applied negative voltage is equal to or greater than the charge on C1 so that the anode of CR2 is made negative, and conduction ceases. When CR2 stops conducting, output pulse e_O is at zero, and C1 quickly discharges through CR1 since its cathode is now negative with respect to ground (anode is grounded). Between times t_1 and t_2

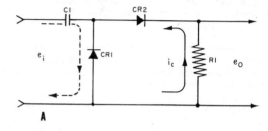

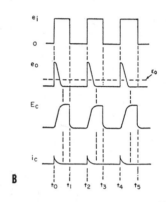

179.632

Figure 4-29.—Positive diode counter and waveforms.

the input pulse is again at zero level, and CR2 remains in a nonconducting state. Since the very short time constant offered by the conduction resistance of CR1 and C1 is much less than the long time constant offered by CR2 and R1 during the conduction period, C1 is always completely discharged between pulses. Thus, for each input pulse there is a precise level of charge deposited on C1. For each charge of C1 an identical output pulse is produced by the flow of i_c through R1. Since this current flow always occurs in the direction indicated by the solid arrow (fig. 4-29A), the d.c. output voltage is positive with ground.

At time t_2 the input signal again goes positive, and the cycle repeats. The time duration between pulses is the interval represented by the period between t_1 and t_2 or between t_3 and t_4. If the input pulse frequency is reduced, these time periods become longer. On the other hand, if the frequency is increased, these time intervals become shorter. With shorter periods, more pulses occur in a given time and a higher average (d.c.) output voltage is produced; with longer periods, fewer pulses occur and a lower average output voltage is produced. Thus, the d.c. output is directly proportional to the repetition frequency of the input pulses. If the current and voltage are sufficiently large, a direct reading meter can be used to indicate the count; if they are not large enough to actuate a meter directly, a d.c. amplifier may be added. In the latter case, a pi-type filter network is inserted at the output of R1 to absorb the instantaneous pulse variations and produce a smooth direct current for amplification.

Consider now some of the limits imposed on solid-state circuit operation. Since the semiconductor diode has a finite reverse resistance, there is a flow of reverse current during the periods when the diode is supposedly in a nonconducting condition. Although this reverse flow is small at normal temperatures (on the order of microamperes), it increases as the temperature rises. Therefore, at high temperatures the average output voltage will tend to decrease because of the effects of diode CR2.

From the preceding discussion it is evident that the voltage across the output varies in direct proportion to the input pulse repetition rate. Hence, if the repetition rate (frequency) of the incoming pulses increases, the voltage across R1 also increases. In order for the circuit to function as a frequency counter, some method must be employed to utilize this frequency to voltage relationship to operate an indicator. The block diagram in figure 4-30A represents one simple circuit which may be used to perform this function. In this circuit, the basic counter is fed into a low pass filter, which controls an amplifier with a meter calibrated in units of frequency.

A typical schematic diagram is shown in figure 4-30B. The positive pulses from the counter are filtered by C2, R2 and C3. The positive d.c. voltage from the filter is applied to the input of A1. This voltage increases with frequency, and as a consequence the current through the device increases. Since emitter or cathode current flows through M1, an increase in amplifier current causes an increase in meter deflection. The meter may be calibrated in units of time, frequency, revolutions per minute, or any function based upon the relationship of output voltage to input frequency.

The step-by step counter is used as a voltage multiplier when it is necessary to provide a stepped voltage to any device which requires such an input. The step-by-step (or step) counter provides an output which increases in one-step increments for each cycle of input. At some predetermined level, the output voltage reaches a point which causes a circuit, such as a blocking oscillator, to be triggered.

A schematic diagram of a positive step-by-step counter is shown in figure 4-31A. With no signal applied to the input, there is no output. As the input signal is applied, and increases in a positive direction, the anode of CR2 or V2 becomes more positive than its cathode, and the diode

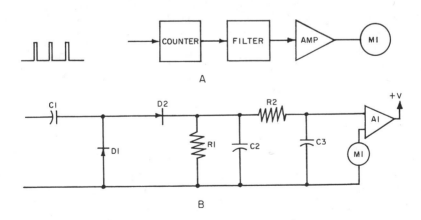

A

B

179.633
Figure 4-30.—Basic frequency counter.

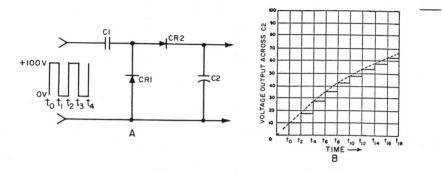

179.634
Figure 4-31.—Basic step counter circuit and waveforms.

conducts. When CR2 or V2 conducts, capacitors C1 and C2 begin charging.

The action of the counter can best be understood by referring to figure 4-31B. Assuming C2 to be ten times as large as C1 and the peak to peak input voltage to be 100 volts, C1 assumes nine-tenths of the positive input voltage swing at t_0, while C2 assumes only one-tenth or, in this example, 10 volts. At time t_1, the input drops in a negative direction, and CR2 or V2 is driven into cutoff. The cathode of CR1 becomes more negative than its anode, and the diode conducts, discharging C1. The charge on C2 remains, however, because it has no discharge path. Thus, there is a d.c. voltage at the output which is equal to one-tenth of the input.

At time t_2, the input again increases positively, but this time CR2 cannot conduct until the input becomes greater than 10 volts, the charge on C2. At this level, CR2 conducts and C2 again charges to one-tenth of the total available voltage. The total voltage available at this time, however, is no longer 100 volts, but 100 volts minus the 10-volt charge on C2.

Thus, the first cycle of input produced a 10-volt charge on C2, but the second cycle added only an additional 9 volts, which is one-tenth the quantity of 100 volts minus the 10-volt charge on C2. Each additional cycle provides an exponential increase in the same manner. It is for this reason that the accuracy decreases as the ratio increases. Thus, as the ratio becomes too great, the higher steps become almost indiscernible. When the counter is used to trigger a blocking oscillator, such as the one shown in figure 4-32, the oscillator bias is adjusted by the potentiometer. This changes the amount of emitter or

cathode voltage which determines the step that will cycle the blocking oscillator. When the oscillator draws current, it discharges C2 and the cycle repeats. The step counter, therefore, becomes a frequency divider supplying one output trigger for a number of input triggers.

Explicit explanation of this circuit varies with driving source reference levels and type of output load circuit. For example, the input waveform may only vary above and below a zero reference in contrast to the waveform in figure 4-31 which only varied in a positive direction from the zero reference. Figure 4-33 illustrates this input and the resulting output.

Note that step 2 is higher in amplitude than step 1 in contrast to the successive decrement in amplitude illustrated in the description of figure 4-31. This is because at t_0, 50 volts is applied and C2 charges to one-tenth this value. At t_1, C1 not only discharges its 45-volt charge, but it charges to 50 volts in the opposite direction, through D1. At t_2, the input again goes to the positive 50-volt level and will be additive to the 50 volts across C1. C2 will charge an additional 9.5 volts, since this is 10% of the 100 volts applied minus the 5-volt charge already on C2. Thus, at t_2, the charge on C2 reaches 14.5 volts.

Succeeding charges will be 10% of the 100-volt charge minus the existing charge on C2 at that time. Therefore, the output is affected because the initial input charge was only half the total peak to peak charge. However, if C2 is discharged by a circuit such as the blocking oscillator just mentioned, the second staircase waveform will be like that in the initial description (fig. 4-32). This is because C1 now has a

84

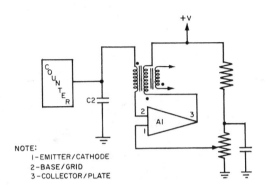

179.635

Figure 4-32.—Voltage control discharge
circuit.

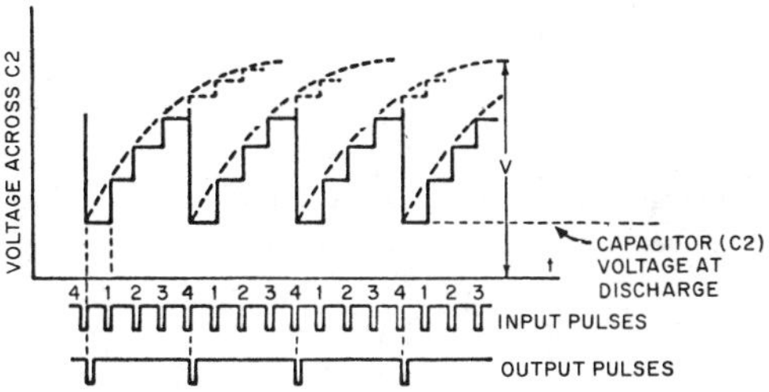

NOTE:
 I - EMITTER / CATHODE
 2 - BASE / GRID
 3 - COLLECTOR / PLATE

179.637

Figure 4-34.—Voltage across C2 in a 4:1
counter.

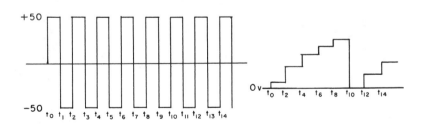

179.636

Figure 4-33.—Input waveform varying about zero and
resulting output.

50-volt negative charge from preceding alternations so that on the next positive going charge after C2 is discharged, the full 100-volt swing is felt across the circuit.

Another condition may exist when the input swings both positive and negative around a zero reference. This condition occurs if the initial alternation swings in the negative direction, charging C2 through D1, thus providing the entire 100-volt potential on the first positive swing or

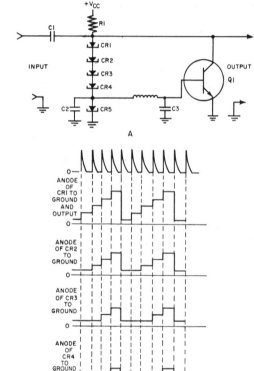

A

B

179.638

Figure 4-35.—Positive tunnel diode step-by-step counter and waveforms.

second alternation. Other first cycle conditions will exist if the input only goes negative with respect to a zero reference. These examples illustrate the variety of first cycle conditions that may be encountered.

As previously mentioned, counting stability is dependent upon the exponential charging rate of C2. When it is desired to count by a large number, for example—24, a 6:1 counter and a 4:1 counter connected in cascade may be used. A more stable method of counting 24 would be to use a 2:1, a 3:1 and 4:1 counter in cascade. Most step counters operate on ratios of 5:1 or less. The waveforms of a 4:1 frequency divider are shown in figure 4-34.

A solid-state step-by-step counter which uses tunnel diodes is shown in figure 4-35A. Resistor R1 sets the bias for CR1 through CR5 in their low voltage state. Positive pulses coupled through C1 sequentially switch the diodes to their high voltage state. The first pulse turns on CR1, the second CR2, the fourth CR3, the fourth CR4, and the fifth CR5. When CR5 goes to its high voltage state, Q1 is biased into conduction. Consequently, the voltage from the collector of Q1 to ground is reduced to a value which will cause the diodes to revert to their low voltage state, allowing the cycle to repeat.

Figure 4-35B shows the various voltage waveforms in the circuit. C2, C3 and L1 provide a slight delay which allows CR5 to complete switching to its high voltage state before Q1 conducts and recycles the operation. A five-step positive going stairstep voltage waveform is developed at the output.

COINCIDENCE CIRCUITS

A coincidence circuit produces an output only when each of its selected inputs is of sufficient amplitude, proper polarity, and concurrent in time with its desired mate or mates. Coincidence circuits are also known as AND GATES when used in digital logic applications.

Figure 4-36 depicts a three-input coincidence circuit and its associated waveforms. The polarities shown are correct for NPN transistors and electron tubes. For PNP transistors all polarities would be reversed. All polarities are referenced to ground unless otherwise noted.

All the devices are connected in series, and are biased into cutoff. There can be no current

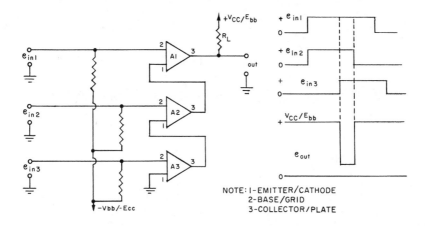

179.639

Figure 4-36.—Basic coincidence circuit and waveforms.

flow until all the devices are caused to conduct. During the period of time all inputs are sufficiently positive, the devices conduct heavily. This conduction produces a negative going pulse at the output. If any one input is absent, no amplifier current flow is possible, and the output will be simply some constant value of positive voltage.

Figure 4-37A illustrates a pentode electron tube coincidence circuit. The control grid and suppressor grid are biased sufficiently negative with respect to the cathode to allow either grid to hold the tube cut off. As shown in figure 4-27B, positive voltages must exist simultaneously, on both the control grid and suppressor grid, in order for the tube to conduct.

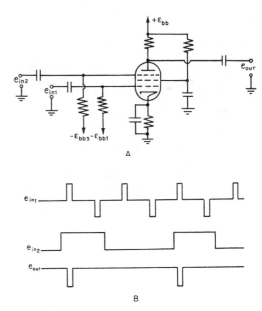

179.640

Figure 4-37.—Pentode electron tube coincidence circuit and waveforms.

87

CHAPTER 5

PRINCIPLES OF UHF COMMUNICATIONS

The frequencies from 300 MHz to 3000 MHz constitute the ULTRAHIGH FREQUENCY (UHF) BAND. The short wavelength of the signals within the UHF band permits construction of antennas that can concentrate the signal energy into a narrow beam. For example, this may be accomplished by the use of a parabolic antenna which operates on the same principle as the reflector in a flash light. Parabolic antennas for lower frequencies (longer wavelengths) would be too large to be practical. The concentration of UHF signal energy into a narrow beam allows the use of lower transmitter power since the signal is not weakened by spreading out over a large area.

The type of propagation which is used in UHF systems is the direct wave. In this type of propagation the radiated energy travels directly from the transmitting antenna to the receiving antenna in the atmosphere adjacent to the earth's surface. This type of propagation lends itself to the concentration of the UHF signal energy that was discussed earlier. Since the UHF energy, concentrated in a narrow beam, travels in approximately a straight line, the theoretical range of transmission is known as line of sight communication. The actual range of communication is determined mainly by antenna height, since this would determine the distance to the theoretical horizon.

Again, due to the short wavelength (high frequency) of UHF signals, much smaller values of circuit inductance and capacitance are required. This, coupled with the lower power requirements, for UHF equipment, permits a reduction in the physical size of the individual components and, therefore, the overall physical size of the equipment. This compactness makes UHF equipment highly suitable for air to air, ship to ship, air to ship, or ship to air communication.

UHF TRANSCEIVER

Often, to achieve a greater compactness and efficiency, UHF equipment may be in the form of a transceiver. A transceiver consists of a transmitter and receiver arranged and connected to function as a single unit. Some of the circuits in a transceiver serve a dual purpose, that is, perform one function while in transmit and another while in receive.

DOUBLE CONVERSION

Figure 5-1 is a simplified block diagram of the receiver section of a basic UHF transceiver. This receiver uses DOUBLE CONVERSION. Double conversion describes the process of converting the incoming signal to the final intermediate frequency in two steps. The output of the RF strip, along with the frequency multiplied signal from the 1st local oscillator, is fed to the 1st mixer. The 1st local oscillator and frequency multiplier are made variable in order to provide tracking between the oscillator and the input signal.

The output of the 1st mixer is a modulated IF, which is amplified in the 1st IF strip and used as the RF input to the second mixer. Heterodyning of this signal with that of the second local oscillator will produce a lower IF signal which in this case is the final IF. If a lower final IF were desired, a third conversion system would be added merely by incorporating a third mixer, local oscillator, and IF strip. This would be called triple conversion. Notice that only the first local oscillator is variable, since the output of the 1st mixer is a constant frequency.

The purpose of multiple conversion is maximum image rejection, increased selectivity, and improved gain. In a double conversion receiver, the first intermediate frequency supplies the wide separation between the desired signal and its image. The second intermediate frequency supplies the increased gain and selectivity. The higher gain and better selectivity are more easily acquired in the second IF due to the lower frequency of these circuits.

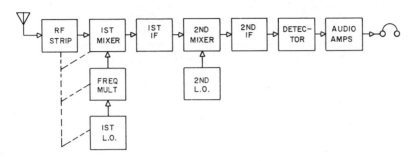

179.641

Figure 5-1.—Block diagram of basic UHF receiver.

FREQUENCY MULTIPLICATION

Figure 5-2 illustrates a basic UHF transceiver transmitter section. A crystal oscillator is used to achieve frequency stability. However, cutting a crystal for UHF would be impractical, since the crystal would be extremely thin and fragile. For this reason, stages of frequency multiplication are necessary. The final power amplifier (FPA) in a UHF transmitter may sometimes be used for frequency multiplication. This is possible since high power is not a major concern, and the lower efficiency of an FPA frequency multiplier is tolerable.

By employing a frequency conversion system using several crystal oscillators, a maximum number of different frequency channels may be obtained with minimum crystal switching. In the functional UHF transmitter block diagram shown in figure 5-3 the use of 38 crystals yields a capability of 1750 channels spaced .1-MHz apart. To achieve this capability with only one master oscillator and using frequency multiplication only, many more crystals and stages of frequency multiplication would be required.

Figure 5-4 is the functional block diagram of a UHF transceiver. The receiver is a triple conversion type and the transmitter utilizes frequency synthesis. Many of the circuits are used on both transmit and receive functions.

PUSH-PUSH FREQUENCY MULTIPLIER

As has been explained, most UHF transmitters utilize frequency multiplication. There are three basic configurations used for frequency multiplication: SINGLE ENDED, PUSH-PULL, and PUSH-PUSH. The single ended frequency multiplier can be used to multiply to any integral multiple of the input (2 times, 3 times, etc.), but is relatively inefficient. Where increased efficiency is required a push-pull configuration may be used, but it is limited to odd order harmonic multiplication.

When a high power frequency doubled output is required, a push-push configuration is used. Such a circuit will produce this type of output without appreciable multiplication losses. While push-push circuits also operate as quadruplers (since only even order harmonics can be produced with a significant output), the same output may be more easily obtained by employing frequency doublers as they require slightly less than half the driving power of a quadrupler.

Figure 5-5 shows a typical push-push doubler stage. A push-pull input is provided by RF transformer T1. L1 is the primary, and the secondary, L2, is tuned to the fundamental frequency by split-stator capacitor C1. The output tank (C3 and L3) is tuned to twice the input (fundamental) frequency.

All voltages are referenced to ground unless otherwise noted, and all polarities are correct for NPN transistors and electron tubes. The polarities would be opposite for PNP transistors.

The devices are biased class C by virtue of a fixed bias supply. When the input signal drives the top of L2 positive and the bottom of L2 negative, A1 is caused to conduct while A2 remains cutoff. On the next alternation of the input (top of L2 negative and the bottom of L2 positive) A1 will be cutoff and A2 will conduct when the bottom of L2 becomes sufficiently positive.

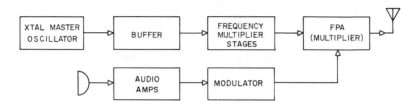

179.642

Figure 5-2.—Block diagram of basic UHF transmitter.

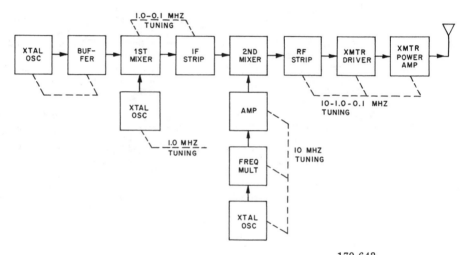

179.643

Figure 5-3.—Functional block diagram of UHF transmitter.

The relationships between base/grid voltage, collector/plate current, and collector/plate voltage are depicted in figure 5-6. Since the collectors/plates of A1 and A2 connect to the same end of the output tank a pulse of current will flow in the same direction through the output tank twice each input cycle. This is an ideal condition for sustaining oscillation of the output tank at the second harmonic of the input frequency (or any even multiple of the input frequency). The tank circuit flywheel action is reinforced, by a pulse of current flow, at the negative peak of each cycle. (If the output tank was tuned to the fourth harmonic of the input the flywheel action would be reinforced every second cycle.)

Due to the relationship between the pulses of collector/plate current and the phase of the collector/plate voltage, the push-push multiplier cannot be used to produce odd order harmonics or to operate at the fundamental frequency.

NOISE

One of the limitations on UHF amplifiers is the generation of noise. Noise may be generated in either the amplifier tube or in the conductors

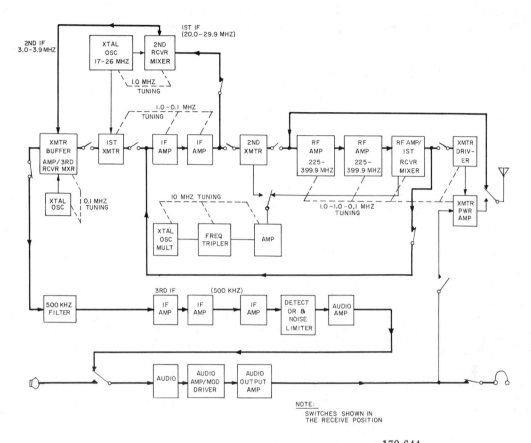

179.644

Figure 5-4.—Functional block diagram of UHF transceiver.

which form the circuit. The latter noise is due to thermal agitation of the electrons within the conductor, and is called THERMAL NOISE.

Noise generated within the tube, is a result of irregularities in electron flow through the tube. There are several classifications of tube noise and each warrants an individual discussion.

SHOT EFFECT is a type of noise due to variations in emission of electrons from the cathode. It can be minimized by operating the tube well below emission saturation.

PARTITION NOISE is present in multigrid tubes. It is caused by some of the electrons that leave the cathode, and move through the control grid and toward the plate, reaching the plate while others strike one of the additional grids and do not arrive at the plate at all. Therefore, a random distribution of electrons between plate and other positive potential elements will occur and produce random variations in plate current.

INDUCED GRID NOISE is produced by the passage of electrons through the grid of a tube in their journey from cathode to plate. As the electrons approach the grid, they induce (in the grid) a movement of charges in the opposite direction. Since the flow of electrons from cathode to plate

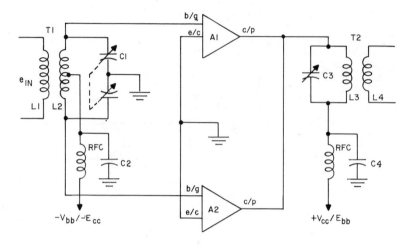

179.645
Figure 5-5.—Push-push frequency multiplier.

is random, and these electrons must pass the grid on their way to the plate, random variations in grid current occur.

GAS NOISE is generated by the random rate of production of gas ions due to electron collision. This type of noise is not a problem in high vacuum tubes.

SECONDARY EMISSION NOISE is caused by the bombardment of the plate by high velocity electrons. Electrons liberated from the plate by this bombardment are called secondary emission electrons. When these electrons return to the plate they cause secondary emission noise.

FLICKER EFFECT is a type of noise caused by a low frequency variation in emission that occurs with oxide-coated emitters.

Shot effect noise, partition noise, and induced grid noise are the most troublesome types of noise. It should be noted that partition noise and induced grid noise are caused by grid structures, and, therefore, to minimize this noise, tubes containing many grids are not used where noise is critical.

GROUNDED-GRID AMPLIFIER

UHF requires amplifiers with good noise characteristics and good gain. Multigrid tubes such as pentodes cannot normally be used for the reasons previously discussed. A circuit which meets these requirements is a grounded-grid amplifier. Due to the grounding of the grid, grid noise is practically eliminated as is the need for neutralization.

The schematic of figure 5-7 illustrates a typical grounded-grid circuit. For convenience, the tank circuit is shown as a conventional LC parallel-tuned circuit; in actual practice, however, coaxial lines or cavities are used at the high frequencies where this circuit is often used. Input coupling capacitor C_{C1} functions as both a coupling capacitor and a d.c. blocking capacitor to isolate the input circuit from the antenna or previous stage. Thus, the cathode bias is not affected by the input circuit. Radio frequency choke RFC keeps the cathode above ground, since the grid is grounded to the chassis. Resistor R1 is a conventional but unbypassed cathode-bias resistor which supplies class A bias for V1. The plate of V1 is series fed through voltage dropping and decoupling resistor R2 and tank coil L1. Bypass capacitor C2 keeps the lower end of tank coil L1 at RF ground potential, and acts as an RF bypass for R2. C1 is the tank tuning capacitor. The rotor of C1 is at a.c. ground to eliminate body capacitance effects when tuning. The output is capacitively coupled through C_{C2} to the next stage.

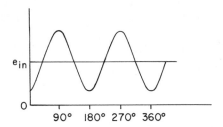

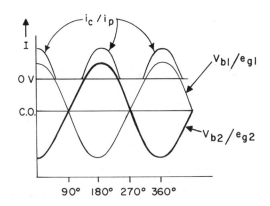

179.646

Figure 5-6.—Collector/plate and base/grid waveforms and their relationships.

When an RF signal appears at the input, the low reactance of C_{C1} allows it to appear on the cathode of V1 without any appreciable attenuation. The input signal may be from an antenna or a preceding RF stage, and in some cases it may be the output of a tuned tank circuit. With the grid at ground potential, V1 is biased by the total cathode current flow through R1. With a positively biased cathode, the grid is effectively biased negative, and only quiescent class A plate current flows. With no input signal there is no change in plate current and, consequently, no output.

Assume that an unmodulated RF signal of constant amplitude appears at the cathode of V1. Since this signal appears between the cathode and ground, it can be considered as being supplied by a generator connected in series between the V1 cathode and ground. The RF choke presents a high impedance to ground, and prevents shunting of the input signal to ground through bias resistor R1. On the positive RF half-cycle the cathode is momentarily more positive resulting in the grid becoming more negative, so that a reduction of plate current occurs. In the plate circuit, the tuned parallel tank circuit, L1 C1, appears as a high impedance to the RF component of the plate current. With less plate current flowing through the tank impedance, less voltage drop is developed across it and the plate voltage rises toward the source voltage (becomes positive swinging). Thus, a positive output signal is developed and fed through C_{C2} to the next stage. It is evident that the grounded-grid circuit

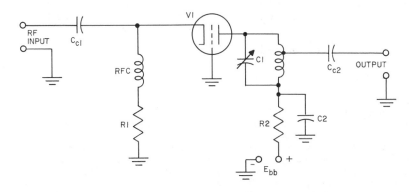

179.647

Figure 5-7.—Typical grounded-grid RF amplifier circuit.

produces an output signal which is in phase (of the same polarity) with the input signal producing it.

During the negative half-cycle of operation, the cathode becomes less positive (is driven in a negative direction). A negative cathode swing causes the plate current to increase, and produces a large voltage drop across the output load impedance (tank circuit). Since the voltage drop across the tank causes the effective plate voltage to be less, a negative output swing is developed. Again the output signal is in phase with the input signal. (This action is opposite the conventional 180-degree phase shift produced in the grounded-cathode circuit, and corresponds with the action of the common (grounded)-base circuit in semiconductors.) With the grounded-grid amplifier operated class A, and with equal positive and negative swings, the average value of plate current does not change. The current flow from the cathode remains steady and occurs during the entire cycle; thus, cathode bias can be used, since the plate current is never interrupted. At the same time, the instantaneous signal changes cause larger (amplified) instantaneous RF pulses of plate current, producing the voltage drop across the tank impedance and an amplified output voltage. Note that the tank circuit must be tuned to the RF signal to produce a high impedance and develop an output. Thus, signals with a frequency outside the tuned circuit passband are not amplified, or are greatly discriminated against.

From the above description of circuit functioning, it can be seen that the basic functioning of the grounded-grid circuit is similar to that of other types of RF or audio amplifier circuits. Further consideration is necessary to understand the actions that are peculiar to this circuit alone.

In the simplified equivalent of the grounded-grid circuit (fig. 5-8) the input signal is shown as an a.c. generator connected in series with input resistance R_k. Actually, the cathode input impedance is inherently very low, and electron flow is from ground to the cathode, through the grid to the plate, and back into the supply, producing the polarities shown in the simplified circuit. Since the grounded grid is placed between the cathode and the plate, it acts as a shield which divides the circuit into two parts—an input circuit and an output circuit, both at above-ground potentials. Hence, any coupling is effectively minimized by the grounded grid. The grounded-grid amplifier requires more drive than the conventional grounded-cathode amplifier.

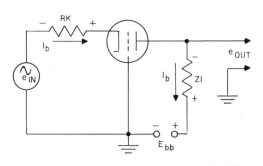

179.648
Figure 5-8.—Simplified equivalent circuit.

Since feedback resulting in oscillation normally occurs from capacitive coupling between the output and input circuits, the good shielding of the grounded grid reduces this effect to a minimum. In addition, the interelectrode capacitances are reduced. The output capacitance is the grid-to-plate capacitance, which is usually the lowest in an electron tube; thus, capacitive shunting effects on the output are reduced at the higher radio frequencies to provide better performance. In addition, the plate-to-cathode capacitance is reduced, since it is the series capacitance produced by the plate-to-grid and grid-to-cathode interelectrode capacitances. Actually, in practical tubes it is reduced to a value on the order of 0.2 picofarad, which is negligible, so that neutralizing is not normally required.

Since signal voltage e_{in} is connected between the cathode and ground, it is effectively in series with the tube plate circuit; thus, in tuned RF voltage amplifiers the output voltage is produced as though the circuit were driven in the normal manner (grounded cathode), but had an increased amplification factor of $\mu + 1$. Hence, high voltage gain is obtained.

It is important to remember, however, that the matter of gain is relative. A low amplification factor tube will not give as much amplification as a high amplification factor tube. Nor will a triode give as much gain as a pentode at the lower frequencies. Thus, even though one speaks of the grounded-grid circuit as providing high gain, it does not mean that the gain is as great as that provided by the grounded-cathode circuit using the same tube and voltages. At the ultrahigh frequencies where this circuit is most useful, the performance and gain are better because of the poor performance of the pentode. At the lower radio frequencies, it usually requires two stages of grounded-grid amplification to obtain results equivalent to those obtained with a single grounded-cathode pentode stage.

In power amplifier applications, low power gain is obtained because of the increased drive requirement and the low input impedance. The low input impedance, however, does not absorb all of the input (driving) power and cause a complete loss. Instead, the driving power is fed into the plate circuit (it is connected in series with the plate and cathode circuit), and adds to the total plate power (less the amount needed to drive the tube). The additional plate power supplied by the driver is distributed between the tubes internal plate resistance and the tank circuit, so that only a portion is lost or dissipated in the tube plate. The total output power in watts is equal to $I_p(e_{in} + E_p)$, where e_{in} is equivalent to the r.m.s. value of grid voltage (E_g).

In RF power amplifiers with directly heated filaments, since the filament is also the cathode, it is necessary to use RF chokes in the filament leads, or provide some other arrangement to keep the filament above ground. Otherwise, the filament and grid would be short circuited and the circuit would not operate.

When employed as a modulated power amplifier, the small portion of drive power which is inserted into the plate circuit remains unmodulated, making it practically impossible to obtain 100-percent modulation when plate modulation alone is used.

PARALLEL TYPE
TUNED LINES

Circuits composed of lumped inductance and capacitance elements (conventional tank circuits) can be made to resonate from very low frequencies to hundreds of megahertz. At UHF, however, the inductive and capacitive elements required become too small to be practical. In addition, skin effect within the coils introduces reactance which reduce the Q of the circuit. To avoid these conditions, the tuned circuit for UHF

oscillators usually consists of a quarter-wavelength transmission line called a tuned line. A segment of transmission line one-quarter wavelength long and shorted at the remote end, has the characteristics of a parallel resonant circuit at the near end. This action can be analyzed by referring to the quarter-wave shorted transmission line segment shown in figure 5-9A. At the shorted end, the current will be maximum and the voltage will be minimum. The maximum current condition is characteristic of a series resonant circuit.

Part B of figure 5-9 shows the equivalent circuit conditions of the quarter-wavelength line at various points between the shorted end and the source. At the shorted end part B shows the line to be equal to a series resonant circuit. One-eighth wavelength back toward the source, at the point where the current and voltage waveforms cross, the circuit will appear inductive. (This is shown in the equivalent circuit by the inductance connected across the line.) The circuit will appear inductive in varying degrees between the source and the shorted end. At the source, the current will be minimum and the voltage will be maximum—a condition characteristic of a parallel resonant circuit. Since the parallel resonant circuit is the first element that the source sees, it may be said that the shorted quarter-wave line will act like a parallel resonant circuit as far as the source is concerned. The tuned

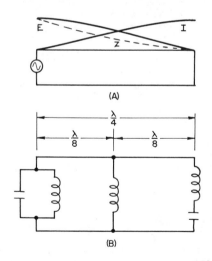

179.649
Figure 5-9.—Quarter-wave shorted tuned line.

lines (mentioned previously as a replacement for a UHF tank circuit) are adjustable to facilitate a change in their operating frequency. This adjustment is made by means of a shorting bar as shown in figure 5-10. The function of the shorting bar is to change the physical length of the line. This will, of course, change the wavelength of the line and thereby the frequency at which the line displays parallel resonant characteristics to the source. Therefore, assuming the tube to be the source, connecting tuned lines between plate and cathode and grid and cathode and adjusting the shorting bars for the proper frequencies will effectively cause the effect of resonant tanks in the tubes plate and grid circuits.

Coupling from the plate circuit tuned lined may be accomplished by various methods. One of the most popular methods is called LOOP COUPLING. Loop coupling is shown in figure 5-11. The loop is merely a conductor placed in position with the tuned line so it will act as a transformer, and energy is coupled to the loop by the transformer action.

UHF CIRCUIT LIMITATIONS

As the frequency to be amplified increases, the gain which can be achieved through the use of conventional electron tubes decreases until a frequency is reached where such gain is unity. The reduction of gain is caused by dielectric losses, finite values of lead inductance, interelectrode capacitance, and transit times.

Dielectric losses have been decreased by tube designs which confine the dielectric material (bases, insulators, envelopes) to portions of the tube where dielectric stresses are minimized, and by the use of dielectric materials with the lowest possible losses.

As the operating frequency is increased, the inductances and capacitances inherent in the tube structure become an increasing portion of the tuned circuit of an amplifier stage. This continues until, for all practical purposes, the external tuned circuit disappears or is obliterated by the tube, and no further tuning is possible. Thus, a limiting frequency is reached. In addition to this, as the operating frequency is increased, even the reactance of relatively short leads in the tube becomes great enough to decrease the magnitude of the driving signal appearing across the tube element.

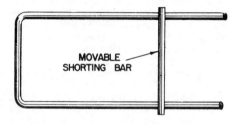

179.650
Figure 5-10.—Tuned line with shorting bar.

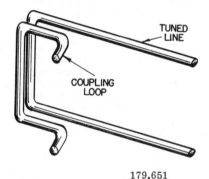

179.651
Figure 5-11.—Loop coupling.

Transit time effects are lessened, in tube design, by decreasing the interelectrode spacing in the tube, and increasing the plate voltage. However, decreasing the interelectrode spacing increases the interelectrode capacitances; therefore, electrode dimensions must be decreased in order to maintain low capacitance. Reducing the size of the electrodes reduces the heat dissipation capabilities of the tube. However, the use of heat sinks will allow the operation of relatively small-electrode triode tubes at dissipation levels which allow considerable useful output.

LIGHTHOUSE (DISK SEAL) TUBES

The effects of lead inductance (and also skin effect) must minimized in tubes designed for UHF operation. This is accomplished by the use of large diameter leads, multiple leads, and planar element construction (the arrangement of cathode, grid, and plate in parallel planes), which allows connections to external circuitry to be made around the periphery of contact disks or over the entire surface of cylinders. Figure 5-12 illustrates a tube utilizing planar construction. Figure 5-13 shows the same tube with construction details.

Lighthouse tubes were designed in the octal-tube era and have this style of base. However, only heater leads and d.c. cathode connections are brought to this base. All RF connections are made to the sleeve, disk, and cap which connect to or support the cathode, grid, and plate respectively. This allows low-inductance leads which when combined with the close spacing and low interelectrode capacitances make the tubes useful at UHF and VHF. The plate is relatively small, and connection to a heat sink is required to take advantage of the rated plate dissipation.

179.652
Figure 5-12.—Lighthouse tube.

MICROWAVE LINK

As may be recalled from chapter 22 of Volume I, the number of signals that can be used to modulate a carrier is dependent upon the carrier frequency. In the microwave frequencies it is possible to modulate the carrier with as many as 600 channels.

Line of sight systems are made up of one or more links having a clear path between the antennas at the ends of the link. They usually use frequencies above 300 MHz. The properties of such links are usually quite stable. Wideband transmission suitable for 24 to 60 voice channels has been obtained by proper system planning.

The length of a single link is determined by the terrain. Most systems are composed of links of 30 miles or less, except where especially favorable sites can be found. Repeater stations may be used to connect one link to another to form long chains thereby setting up long paths for many voice channels where needed. For example, chains of more than 40 links cross the United States carrying voice and television signals. With proper engineering, excellent quality may be preserved through many sequential links.

Microwave link is a reliable and economical method of communication. The installation, operation and maintenance costs per channel mile are relatively low. Also, the security of microwave link is more easily maintained due to the fact that all equipment is concentrated to a smaller area.

Microwave links are often used for carrying signals from one site to another at large HF stations. The design of links for this purpose is the same as for long haul systems, but, since the number of links in a chain is usually only one or

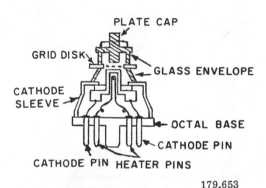

179.653
Figure 5-13.—Lighthouse tube construction.

two, some easing of specifications may be allowed, compared to links in long haul systems.

MICROWAVE LINK REPEATERS

Microwave link repeaters are classified according to how the signal is detected and separated into individual channels in the repeating process. Three types of repeaters will be discussed here.

Heterodyne Repeater

In a heterodyne repeater the incoming signals are heterodyned (entire received signal shifted in frequency as a block) to another band, amplified, and retransmitted. In an RF heterodyne repeater, this shifting is done directly to the desired transmitting channel in one step. Amplification will normally be used both before and after the frequency changing process. In an IF heterodyne repeater, the incoming signals will be shifted to a relatively low intermediate frequency at which most of the amplification is done. Then they will be shifted again to the desired channel for transmission.

Baseband Repeater

The baseband is the band of frequencies in which the signal is provided to the transmitter for modulation. This signal may be a television signal, or a group of voice channels that have been combined by any of the standard types of voice multiplex equipment. This is the signal that is fed to the transmitter to modulate the outgoing carrier, and also the signal delivered from the receiver output terminals after being demodulated.

In a baseband repeater, the incoming signal is converted to an intermediate frequency and then demodulated to produce the baseband. This baseband is fed to the input terminals of the associated transmitter to modulate the outgoing signal. This arrangement is often referred to as a transmitter and receiver back to back.

Audio Repeater

In the two repeaters mentioned previously, the signal is handled as a single entity. Whether it consists of one voice channel or many is not a determining factor. All the signals go through the same path. Sometimes it is desired to split certain channels from others at the repeater point. (There may be branching of the traffic flow; some channels may end at this point while others go further down the chain of repeaters.) When this is necessary, not only must the basic signal be demodulated to provide a baseband, but the baseband must also be filtered to select particular groups of individual channels. These groups of channel signals are then connected individually to the proper outgoing circuits.

This form of repeater provides individual channel access for flexibility, but adds distortion. This distortion shows up as interference between channels. In practice, the number of audio repeaters must be kept as small as possible in any given long distance chain. Thus, audio repeaters usually are employed only at major switch centers and on relatively short circuits.

MICROWAVE LINK
PROPAGATION FADING

Even where a clear line of sight exists between the two antennas of a link, the signals are disturbed in their passage through atmospheric layers. The following effects are all causes of fading.

ATTENUATION is caused by oxygen particles that absorb and scatter energy. Water vapor clouds, droplets of rain, and other forms of precipitation absorb some of the signal. The loss in signal strength from these causes varies from time to time, causing fading of the signal at the receiving end of the link.

REFRACTION is the change in direction a traveling wave experiences as it crosses layers of atmosphere having different properties. This effect fluctuates widely, and adds to the fading caused by absorption. It should be noted that some cases occur where refraction causes the waves to bend in a manner that improves the link performance. This is one reason that paths slightly beyond true line of sight are sometimes usable. The refraction in the atmosphere may sometimes fluctuate very rapidly causing the received signal to flutter. This rapid random fading is called SCINTILLATION.

REFLECTION occurs when the microwave signal meets a large obstruction. This may be a large cloud or a cluster of ionized air molecules (such as a meteor trail). When reflection occurs, signals arrive at the receiving end by more than one path. When these paths differ in length, the signals arrive with different phases. Depending on the phase relationship, the signals are either added to or subtracted from one another. If the length or attenuation of either path varies, the cancellation effect produces fading. More than two paths will be found in many cases.

POLARIZATION is the property of a wave that relates to the direction of the electric waves and magnetic components in space. Polarization is determined at the transmitting antenna by the direction of the antenna feed horn or feed dipole. The receiving antenna feed should match the polarization of the incoming signals. At lower frequencies, the polarization of the wave shifts as the wave passes through ionized regions. This makes it difficult to decide on the proper polarization for the receiving antenna. At microwave frequencies, however, this shift is negligible. Therefore, it may be assumed that the wave is received with the same polarization as was transmitted. Thus, the effect of polarization on fading is minimal.

To prevent fading effects from impairing the operation of the link, diversity reception (discussed later) is used. Fading caused by the above effects does not occur at the same time for different paths. Even for paths that are almost the same, great differences are found. By providing two receiving antennas spaced 50- to 100-wavelengths apart, each receiving from the same transmitter, we can get an improvement. Signals that differ by a small amount in carrier frequency will show different patterns of fading, even when sent through the same antennas.

The signals received by one path do not fade at the same time as those from the other path. To make the best use of these signals they must be combined so that one fills the gaps left in the other. Two or four signals may be combined in the receiver in any one of a number of ways. Two basic types of combining methods may be used: PREDETECTION (or IF) COMBINING, and POST DETECTION COMBINING. Only the former type can be used with RF heterodyne repeaters.

MICROWAVE LINK TROPOSPHERIC SCATTER TRANSMISSION

Tropospheric scatter transmission is used for point to point, beyond the horizon communication. Frequencies of about 350 to 8000 MHz are commonly used for this purpose. Reliable multichannel communication can be obtained over paths up to 400 miles. In the frequency range under consideration, the signals are not refracted by the ionosphere. They do not return to earth beyond the horizon by bending as lower frequency waves do. Some other means of propagation must be considered.

The tropospheric region of the atmosphere extends up to an altitude slightly over 6 miles. Microwave signals transmitted through this region are scattered in the forward direction. Some of this radiated energy reaches the receiving antenna. If directive transmitting and receiving antennas are used and these antennas are both aimed at the same point in the troposphere, reliable communication is possible even though the received signal fluctuates widely. The amplitude of the received signal mainly depends on transmitter power, antenna gain, distance between stations, altitude of the scattering volume, transmitter frequency, ank the scatter angle between the transmitted and received beams. Figure 5-14 illustrates the scatter angle, scattering volume altitude, and distance factors.

At the receiver, the energy arrives over a number of paths. The signal from each path has a different attenuation, phase, and polarization. This multipath effect leads to fast or slow fading of the received signal. Fast fading may take place in fractions of a second. It results from signals arriving randomly at the receiving antenna. Slow fading occurs over time intervals from minutes

to several hours. Signal variations of the order of 15 db are to be expected. In addition, long term strength variations occur due to daily and seasonal changes in propagation.

Diversity Reception

To obtain a steady signal, energy may be combined from each of a number of fluctuating signals. This is called DIVERSITY RECEPTION. All tropospheric scatter systems used diversity reception. To obtain signals over different paths that fade and vary independently, some or all of the following methods may be used.

SPACE DIVERSITY, which is comprised of receiving antennas separated by 50 wavelengths or more at the signal frequency (usually 10 to 200 feet is sufficient).

FREQUENCY DIVERSITY, or transmission on different frequencies. Different frequencies fade independently even when transmitted and received through the same antennas.

ANGLE DIVERSITY, which is two feedhorns producing two beams from the same reflector at slightly different angles. This results in two paths based on illuminating different scatter volumes in the troposphere.

Signals obtained over two or four independent paths by the above methods are combined in the receiver in such a way as to make use of the best signal at all times.

The path loss of a tropospheric system is high. To overcome this, high transmitter power and high gain antennas are used. Highly sensitive receivers with low internal noise are also employed. Antennas range in size from 8 feet in diameter for mobile units to 120 feet for fixed installations. The size of the antenna depends on the operating frequency, required gain, environmental conditions, and required mobility.

In line of sight systems, the bandwidth that can be handled through a given link is dependent only on the equipment provided. Links carrying 1800 voice channels through a single transmitter and receiver have been built. In tropospheric scatter systems this is not so. The multipath effects vary over the spectrum so that very wide channels do not work well. This is because distortion occurs in the path between the antennas. Combining several signals reduces this effect, but does not eliminate it. The number of channels that can be transmitted over a given link depends on the degree of distortion the user can accept. For links that are part of long haul telephone systems, the distortion must be kept low. Typical tropospheric scatter link capacities are illustrated in table

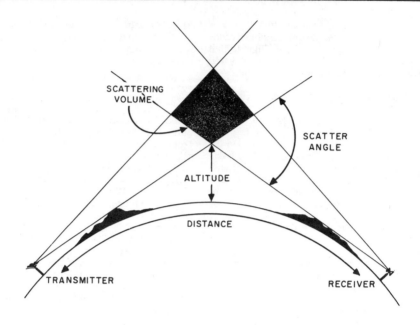

93.21
Figure 5-14.—Geometry of tropospheric scattering.

5-1. Television signals have substantial bandwidth, but can tolerate more distortion. Conventional television signals in a bandwidth up to 6-MHz have been successfully transmitted up to 200 miles by tropospheric systems.

The major factors that affect the received signal level are free space loss (loss occurring between the transmit antenna and the receive antenna), scatter loss, transmitter power, receiver gain, and the angle between the transmit and receive beams.

Both free space and scatter loss can be predicted. Scatter loss, however, depends on more variables. As the distance between the transmitter and receiver is increased, scatter volume height increases. These changes reduce the amount of received signal. An increase in the vertical angle of the receiving or transmitting antenna also reduces the amount of received signal.

Table 5-1.—Tropospheric Scatter Link Capacities.

Distance	No. Voice Channels
0-100 miles	up to 252
100-200 miles	up to 132
200-300 miles	up to 72
above 300 miles	12-24 (quality usually limited)

CHAPTER 6

MICROWAVE APPLICATIONS

Microwave frequencies are defined by the Federal Communications Commission as those frequencies of 890 MHz and above. The Navy, however, generally considers microwave frequencies to be 1000 MHz and above. Microwave communications are discussed briefly in chapter 5. This chapter discusses the application of these basic principles to a radar system.

The word RADAR is an acronym formed from the words RAdio-Detection-And-Ranging. Radar is a means of employing radio waves to detect and locate material objects such as aircraft, ships, and land masses. Location of an object is accomplished by determining the distance and direction from the radar equipment to the object. The process of locating objects requires, in general, the measurement of three coordinates; range, angle of azimuth, and angle of elevation.

A radar set consists fundamentally of a transmitter and a receiver. When the transmitted signal strikes an object (target) some of the energy is sent back as a reflected signal. The receiving antenna collects a portion of the returning energy (called the ECHO SIGNAL) and sends it to the receiver. The receiver detects and amplifies the echo signal. The information is then sent to the indicator.

RADAR WAVEFORM

A representative radar pulse (waveform) is shown in figure 6-1. The number of these pulses transmitted per second is called the PULSE REPETITION FREQUENCY (PRF) or PULSE REPETITION RATE (PRR). The time from the beginning of one pulse to the beginning of the next pulse is called the PULSE REPETITION TIME (PRT). The PRT is the reciprocal of the PRF (PRT = 1/PRF). The duration of the pulse

(the time the transmitter is radiating energy) is called the PULSE WIDTH (PW). The time between pulses is called REST TIME or RECEIVER TIME. The pulse width plus the rest time equals the PRT (PW + Rest Time = PRT).

RANGE DETERMINATION

The distance to the target (range) is determined by the time required for the pulse to travel to the target and return. The velocity of electromagnetic energy is 186,000 statute miles per second, or 162,000 nautical miles per second. (A nautical mile is the accepted unit of distance and is equal to 6076 feet.) However, in many instances, measurement accuracy is secondary to convenience, and as a result a unit known as the RADAR MILE is commonly used. A radar mile is equal to 2000 yards or 6000 feet. The small difference between a radar mile and a nautical mile introduces an error of about one per cent in range determination.

For purposes of calculating range, the two-way travel of the signal must be taken into account. It can be found, that it takes approximately 6.18 microseconds for electromagnetic energy to travel one radar mile. Therefore, the time required for a pulse of energy to travel to a target and return is 12.36 microseconds per radar mile. This takes into account the two-way travel time. The range, in miles, to a target may be calculated by the formula— Range = $\Delta t/12.36$, where Δt is the time between transmission and reception of the signal in microseconds. However, for shorter ranges and greater accuracy, range is measured in yards. Electromagnetic energy travels 328 yards per microsecond. Therefore, range in yards may be calculated by the formula— Range = $\frac{328 \Delta t}{2}$ = $164 \Delta t$, where Δt is in microseconds.

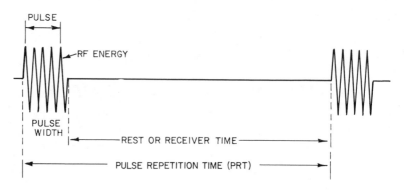

179.655
Figure 6-1.—Radar pulses.

To illustrate one basic method of measuring range by radar, refer to figure 6-2. Assume that a target ship is 20-miles away from the radar. Using the formula for range, it is calculated that it takes a total of 244 microseconds to reach the target and return to the receiver.

Figure 6-2 shows how the range to the target is determined: ① the transmitted pulse is just leaving the antenna. A part of the generated energy is fed to the vertical deflection plates of a cathode-ray tube (scope) at the instant the pulse is transmitted. The potential on the vertical deflection plates cause a vertical line (pip) to appear at the zero-mile mark on the scope; ② 61.8-microseconds later, the transmitted pulse has traveled 10 miles toward the target. Note however, that the trace has only reached the 5-mile mark on the scope, that is, one-half the distance the transmitted pulse has traveled (the sweep frequency is timed to indicate one-half distance); ③ 123.6 microseconds after the initial pulse left the transmitter, the transmitted pulse has reached the target, 20 miles away, and the echo has started back. The scope reading is now 10 miles; ④ 185.4-microseconds after the start of the initial pulse, the echo has returned half the distance from the target, and the scope reading is 15 miles; ⑤ 247-microseconds after the initial pulse, the echo has returned to the receiving antenna. This relatively small amount of energy, called an ECHO, is amplified in the radar receiver, and applied to the vertical deflection plates of the indicator unit which contains the display scope. The echo pip (of smaller amplitude than the initial energy burst) is finally displayed on the scope at the 20-mile mark.

If two or more targets are in the path of the transmitted pulse, each will return a portion of the incident energy as echoes. The targets farthest away (assuming they are similar in size and type of material) will return the weakest echo.

RADAR SYSTEM PARAMETERS

Once the pulse of electromagnetic energy is emitted by the radar, a sufficient length of time must elapse to allow any echo signals to return and be detected before the next pulse is transmitted. Therefore, the PRT of the radar is determined by the longest range at which targets are expected. If the PRT were too short (PRF too high) signals from some targets might arrive after the transmission of the next pulse. This could result in ambiguities in measuring range. Echoes that arrive after the transmission of the next pulse are called SECOND RETURN ECHOES (also SECOND TIME AROUND or MULTIPLE TIME AROUND ECHOES). Such an echo would appear to be at a much shorter target range than actually exists and could be misleading if not identified as a second return echo. The range beyond which targets appear as second return echoes is called the MAXIMUM UNAMBIGUOUS RANGE. Maximum unambiguous range may be calculated by the formula: maximum unambiguous range = PRT/12.2; where range is in miles and the PRT is in microseconds. Figure 6-3 illustrates the principles of the second return echo.

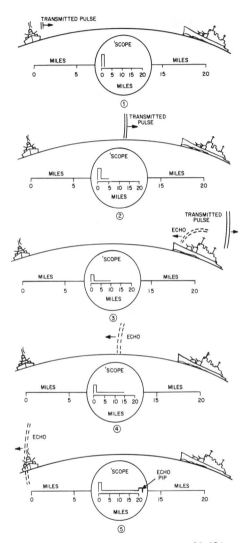

20.282

Figure 6-2.— Radar range determination.

takes 793 microseconds to return. However, this is 183-microseconds after the next pulse was transmitted; therefore, target number 2 will appear to be a weak target 15-miles away. Thus, the maximum unambiguous range is the MAXIMUM USABLE RANGE and shall be referred to from now on as simply MAXIMUM RANGE. (It is assumed here, that the radar has sufficient power and sensitivity to achieve this range.)

If a target is so close to the transmitter that its echo is returned to the receiver before the transmitter is turned off, the reception of the echo will be masked by the transmitted pulse. In addition, almost all radars utilize an electronic device to block the receiver for the duration of the transmitted pulse. However, DOUBLE RANGE ECHOES are frequently detected when there is a large target close by. Such echoes are produced when the reflected beam is strong enough to make a second trip, as shown in figure 6-4. Double range echoes are weaker than the main echo, and usually appear at twice the range.

Minimum range is usually measured in yards and may be calculated by the formula: minimum range = 164 PW where range is in yards and pulse width is in microseconds. Typical pulse widths range from fractions of a microsecond for short range radars to several microseconds for high power long range radars.

A radar transmitter generates RF energy in the form of extremely short pulses with comparatively long intervals of rest time. The useful power of the transmitter is that contained in the radiated pulses and is termed the PEAK POWER of the system. This power is normally measured as an average value over a relatively long period of time. Because the radar transmitter is resting for a time that is long with respect to the pulse time, the average power delivered during one cycle of operation is relatively low compared with the peak power available during the pulse time.

A definite relationship exists between the average power over an extended period of time and the peak power developed during the pulse time. The overall time of one cycle of operation is the reciprocal of the pulse repetition frequency. Other factors remaining constant, the greater the pulse width the higher will be the average power; and the longer the pulse repetition time, the lower will be the average power. Thus,

$$\frac{\text{average power}}{\text{peak power}} = \frac{\text{pulse width}}{\text{pulse repetition time}}$$

Figure 6-3 shows a signal with a PRT of 610 microseconds, which results in a maximum unambiguous range of 50 miles. Target number 1 is at a range of 20 miles. Its echo signal takes 244 microseconds to return. Target number 2 is actually 65-miles away, and its echo signal

103

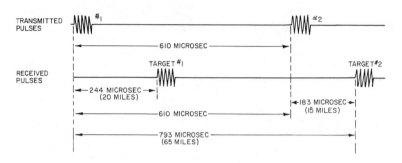

179.656
Figure 6-3.—Second return echo.

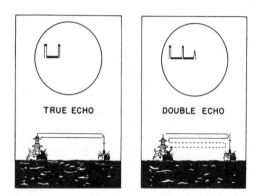

59.36
Figure 6-4.—Double range echo.

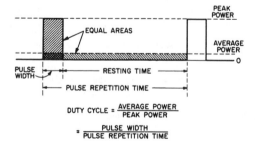

20.288
Figure 6-5.—Relationship of peak and average power.

These general relationships are shown in figure 6-5.

The OPERATING CYCLE of the radar transmitter can be described in terms of the fraction of the total time that RF energy is radiated. This time relationship is called the DUTY CYCLE and may be represented as:

$$\text{duty cycle} = \frac{\text{pulse width}}{\text{pulse repetition time}}$$

(Remember that PRT equals the reciprocal of PRF.) For example, the duty cycle of a radar having a pulse width of 2 microseconds and a pulse repetition frequency of 500 hertz is $\frac{2 \times 10^{-6}}{2 \times 10^{-3}}$ or .001.

Likewise, the ratio between the average power and peak power may be expressed in terms of the duty cycle. In the following example it is assumed that the peak power is 200 kilowatts. Therefore, for a period of 2 microseconds a peak power of 200 kilowatts is supplied to the antenna, while for the remaining 1998 microseconds the transmitter output is zero. Because average power equals peak power times duty cycle, the average power equals $(2 \times 10^5) \cdot (1 \times 10^{-3})$ or 200 watts.

High peak power is desirable in order to produce a strong echo over the maximum range of the equipment. Conversely, low average power enables the transmitter tubes and circuit components to be made smaller and more compact. Thus, it is advantageous to have a low duty cycle. The peak power that can be developed is dependent upon the interrelation between the peak and average power, and the pulse width and pulse repetition time, or, in other words, the duty cycle.

RANGE AND BEARING ACCURACY

If a scope were to have a presentation like the one illustrated in figure 6-6, it might be thought that the object reflecting the pips between the numbers 7 and 9 range scale marks is much larger than the object reflected between the 1 and 2 marks. This would be true if the transmitted pulse duration was extremely slight. However, the range equivalent of the pulse duration (being greater than zero) increases the apparent extent of each target. No matter how small an object is, its return pulse cannot be of shorter length than the transmitted pulse.

Because of this, if a transmitted pulse is 1 microsecond long, a target which was 1000 feet in range would cause an echo to extend 1500 feet on the range scale. (For example, the pip extending between 2 and 3.5 in fig. 6-6 would actually only be 1000-feet long.) Also, targets less than 500-feet apart cause one pip on the scope as the echo pulses would run together. (Thus, the echo between the range marks of 3.5 and 4.5 in fig. 6-6 might actually be two targets.) RANGE RESOLUTION is the term applied to the ability to separate close objects and is determined by the duration of the transmitted pulse, steepness of the transmitted pulse leading edge,

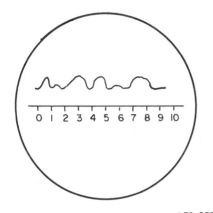

179.657
Figure 6-6.— Radar scope showing range scale.

the receiver bandwidth, and the indicator characteristics. Of these, the pulse width is the primary factor; the narrower the pulse width, the better the range resolution.

Measurement of azimuth and/or elevation angles of an object is accomplished by the use of a highly directional antenna. (Most radars use the same antenna for transmitting and receiving. This is accomplished with the use of a device called a duplexer and will be discussed in a succeeding chapter.) The orientation of the antenna when it is receiving maximum echo signal will indicate the azimuth or elevation of the target. The narrowness of the beam in the horizontal and vertical planes determines the bearing and elevation resolution respectively. The narrower the beam widths the better is the resolution. As will be illustrated in a later chapter, antenna beam widths are a function of the type of antenna used.

BASIC RADAR BLOCK DIAGRAM

A block diagram of a basic radar system is shown in figure 6-7. The pulse repetition frequency is controlled by the TIMER (also called TRIGGER GENERATOR or SYNCHRONIZER) in the modulator block. The pulse forming circuits in the modulator are triggered by the timer, and generate high voltage (HV) pulses of rectangular shape and short duration. These pulses are used as the anode supply voltage for the transmitter, and, in effect, turn it on and off. The modulator, therefore, determines the pulse width of the system. The transmitter generates the high frequency, high power RF carrier and determines the carrier frequency. The duplexer is basically an electronic switch which allows the use of a common antenna for both transmitting and receiving. The receiver section is basically a conventional superheterodyne receiver. In older radars no RF amplifier is found, due to noise problems with the RF amplifiers of that era.

RADAR MODULATOR

The modulator is considered to be the heart of the radar system. It produces accurately timed high voltage pulses of the proper width, amplitude, and polarity which actuate both transmitter and indicators.

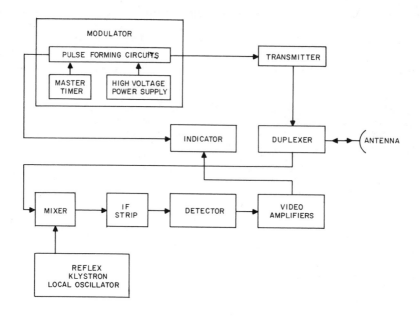

179.658
Figure 6-7.—Basic radar block diagram.

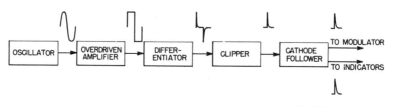

179.659
Figure 6-8.—Modulator timing and shaping circuits.

Modulator Timing And
Shaping Circuits

Figure 6-8 illustrates the timing and shaping circuits used in the modulator. The timer, usually a sine wave oscillator such as a phase-shift or Wein bridge, determines the system PRF (PRT). Timing circuitry must generate stable outputs in both amplitude and frequency. Operational frequency is usually less than 1500 Hz, and is determined by the purpose and type of radar concerned.

The timer output is applied to shaping circuits which are usually found in the modulator. Here the sinusoidal voltage is formed into a very narrow pulse which corresponds to the 0° point of each cycle produced by the timer. The shaping circuits and associated waveforms are shown in figure 6-8.

Pulse Forming Section

The pulse forming circuitry of the modulator is discharged by the timing and shaping section.

Thus a high voltage pulse is developed, which is used to supply anode voltage to the power oscillator/power amplifier in the transmitter section. In addition, this pulse is used to synchronize the indicator/repeater circuits. The pulse width is critical since it determines the length of time the transmitter will be producing an output. The length of the pulse width is determined by the frequency and application of a particular radar system.

Modulator Types

The block diagrams of two types of modulator systems are illustrated in figure 6-9. The LINE PULSING MODULATOR (fig. 6-9A) consists of the following components:

1. SOURCE — provides high voltage to the storage element in the modulator.
2. Z_{ch} — The total impedance in the storage network charge path.
3. SWITCH — allows the storage network to charge when open; allows the storage network to discharge when closed.
4. PULSE FORMING NETWORK (PFN) — Stores the high voltage provided by the source. Also determines the shape and width of the high voltage pulse applied to the load.
5. LOAD — Primary winding of the pulse transformer.

The DRIVER HARD TUBE MODULATOR (fig. 6-9B) differs from the line pulsing type in many respects. (A hard tube is an electron tube which has been evacuated to approach a perfect vacuum as opposed to a tube which contains an appreciable amount of gas (soft tube).) Although the source load (Z_{ch}), the source, and the load are similar to the line pulsing type, the storage network in the hard tube system does not shape the output pulse. The hard tube modulator consists of the following components:

1. Z_{ch}, SOURCE LOAD — same as in the line pulsing type.
2. STORAGE NETWORK — Usually a large capacitor. (Does not determine pulse width.)
3. HARD TUBE SWITCH — Controls the charge and discharge of the storage unit. When the switch tube is cut off, the storage unit charges; when the switch tube conducts, the storage unit discharges.
4. DRIVER — Input is a trigger from the master timer; output is a HV pulse used to drive the switch tube into conduction. Since the width of this pulse determines how long the switch tube will conduct, the driver pulse width will determine the width of the HV pulse applied to the load. This is discussed in more detail at the end of this chapter.

ARTIFICIAL TRANSMISSION LINES

Some of the characteristics of a transmission line, such as time delay, pulse shaping, and energy storage, can be used to advantage in a radar set. The physical length required of a real transmission line is too excessive for practical use. In this case an ARTIFICIAL TRANSMISSION LINE may be constructed by first determining the values of L and C in the real line, and lumping these component values into an artificial line. The artificial line will then have the same electrical characteristics as the real line. The construction of an artificial line is illustrated in figure 6-10. Note that the lumped values of L and C are used to replace the distributed L and C values which occur in a real line.

The artificial line will also produce the same delay characteristics as a real line. A voltage applied to the artificial line terminals at point A will appear at the terminals at point B delayed by a time interval which is determined

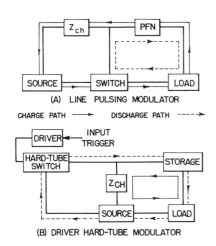

(A) LINE PULSING MODULATOR

CHARGE PATH ———▶   DISCHARGE PATH ----▶

(B) DRIVER HARD-TUBE MODULATOR

179.660
Figure 6-9. — Two types of modulators.

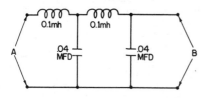

179.661

Figure 6-10. — Artificial transmission line.

by the values of L and C. The time delay may be calculated by using the formula:

$$t_d = N\sqrt{LC}$$

where: t_d is the delay time in seconds, N is the number of artificial line sections; L the inductance in henrys per section, and C the capacity in farads per section.

The line (artificial line may be referred to as line in the remainder of this text) illustrated in figure 6-10 will have a delay time of:

$$t_d = N\sqrt{LC} = 2\sqrt{1 \times 10^{-4} \times 4 \times 10^{-8}}$$

$$= 2\sqrt{4 \times 10^{-12}}$$

$$= 2 \times 2 \times 10^{-6}$$

$$= 4 \times 10^{-6} \text{ seconds or 4 micro-seconds}$$

An EMF applied to the input terminals of this line will NOT appear at the output terminals until 4 microseconds have elapsed.

Since the line presents the same series-parallel LCR circuit characteristics as a real line; the method of determining the characteristic impedance described in chapter 28 of volume I is used. The formula for determining characteristic impedance is:

$$Z_0 = \sqrt{\frac{L}{C}}$$

where Z_0 is the characteristic impedance in ohms; L is the inductance per section in henrys and C is the capacity per section in farads. The Z_0 of the circuit illustrated in figure 6-10 is:

$$Z_0 = \sqrt{\frac{L}{C}} = \sqrt{\frac{1 \times 10^{-4}}{4 \times 10^{-8}}}$$

$$= \sqrt{0.25 \times 10^4}$$

$$= \sqrt{25 \times 10^2}$$

$$= 5 \times 10 \text{ or 50 ohms}$$

Note that the addition of more line sections will not change the Z_0 since the L to C ratio remains constant.

The line is similar to the two wire transmission line, and does have a high-frequency limitation. Since the line is composed of inductance and capacitance; filtering action will take place. However, at a very high frequency, the square wave pulse will become distorted.

CHARGING OPEN ENDED LINES

The d.c. charging circuit of an open ended artificial transmission line is illustrated in figure 6-11A. R_{ch} is the internal resistance of the source. The Z_0 of the line can be found by the formula:

$$Z_0 = \sqrt{\frac{L}{C}} = \sqrt{\frac{1 \times 10^{-4}}{1 \times 10^{-8}}} = \sqrt{1 \times 10^4} = 100 \text{ ohms}$$

The time delay for one way travel is:

$$t_d = N\sqrt{LC} = 2\sqrt{1 \times 10^{-4} \times 1 \times 10^{-8}}$$

$$= 2 \times 10^{-6} \text{ or 2 microseconds}$$

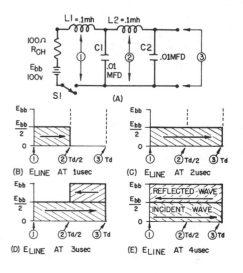

179.662

Figure 6-11. — D.c. charging circuit for an open-ended artificial transmission line.

The d.c. charging sequence starts when S1 is closed. Since $Z_o = R_{ch}$ one-half of E_{bb} will be applied to the line at point ①. The first section of the line, composed of L1 and C1, charges to $E_{bb}/2$ (50 volts) at $t_d/2$ or 1 microsecond (fig. 6-11B). At this time 50 volts is measured at points ① and ②, but the voltage at point ③ is zero. Two microseconds after E_{bb} is applied to the circuit, the second section of the line, L2 and C2, charges to $E_{bb}/2$ or 50 volts. At this instant 50 volts can be measured at points ①, ②, and ③ (fig. 6-11C). The voltage wave which travels from the input terminals to the output terminals of the line is called the INCIDENT WAVE.

While the line is charging to $E_{bb}/2$, charge current flows through L1 and L2. The instant C2 charges to 50 volts, the magnetic field around L2 collapses, increasing the charge on C2 to E_{bb} or 100 volts. At this instant the voltage measured at point ③ is 100 volts; this is the start of the REFLECTED WAVE. The open end of the line now acts as a 100 volt source, and the reflected wave travels back to the input terminals, increasing the charge on the line from 50 volts to 100 volts. After 3 microseconds, 50 volts is measured at point ① while 100 volts is measured at points ② and ③ (fig. 6-11D). After 4 microseconds, 100 volts is measured at points ①, ②, and ③, and the line is fully charged to E_{bb} with zero current (fig. 6-11E).

It is necessary that Z_o and R_{ch} be equal if the line is to be fully charged after one incident and reflected wave. If R_{ch} is less than Z_o, the voltage applied to the line is greater than $E_{bb}/2$, and the line will oscillate above and below the value of E_{bb} several times before charging to E_{bb}. If R_{ch} is greater than Z_o it will take more than one incident and one reflected wave to charge the line to E_{bb}. The line will charge in a stair step fashion and the number of steps will be determined by the degree of mismatch.

A shorted end line will charge in the same manner as the line in figure 6-11. In the shorted end line, the incident and reflected waves will be current waves rather than voltage waves. When the shorted end line is fully charged, current will be maximum and voltage will be zero.

D.C. RESONANCE CHARGING

Another method of charging an artificial transmission line is illustrated in figure 6-12A.

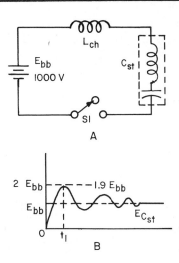

179.663

Figure 6-12.—D.c. resonance charging circuit.

In this circuit, the line will be used for charge storing and designated C_{st}. Charge current flows through series inductor L_{ch}. The inductance of L_{ch} is much greater than the inductance present in the C_{st}. When S1 is closed, C_{st} starts to charge to E_{bb}. The charge current of C_{st} flowing through L_{ch} produces a magnetic field.

When C_{st} is charged to E_{bb}, the field around L_{ch} collapses, increasing the charge on C_{st} to approximately 1.9 E_{bb} (fig. 6-12B). At this point, the field surrounding L_{ch} has completely collapsed, and C_{st} will discharge through the source. Since C_{st} and L_{ch} form a series resonant circuit, the voltage appearing across C_{st} will take the form of a damped sine wave with E_{bb} as the reference level. The frequency of oscillation will be dependent on the values of L_{ch} and C_{st}; the damping time will be dependent on the circuit I^2R losses.

The advantage of this charging method lies in the magnitude of C_{st} charge occuring at t_1 in figure 6-12B. If, at this instant, C_{st} is disconnected from the charge path and connected to a load, it can supply a voltage which is 95% greater than the source, E_{bb}. This system would require very critical timing. If S1 were automatically opened at t_1, C_{st} would retain its charge of 1.9 E_{bb}.

In figure 6-13A the switch is replaced by diode V1. C_{st} charges through V1 and L_{ch}.

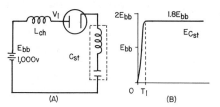

179.664

Figure 6-13.—D.c. resonance charging with diode.

When C_{st} is charged to E_{bb}, the field around L_{ch} collapses, charging C_{st} to about 1.8 E_{bb}. The charge on C_{st} is less than the charge achieved in figure 6-12B due to the voltage drop across V1. C_{st} cannot discharge due to diode V1, and will now retain a charge of 1.8 E_{bb} until a discharge path is provided.

DISCHARGING OPEN
ENDED LINES

Figure 6-14 illustrates the waveshape produced when an artificial transmission line is discharged into a load which has an impedance (Z_R) that matches the characteristic impedance (Z_O) of the line. Assume that the line (Z_O) had been charged to some value of voltage (E_{bb}) and S1 is open. At this time line voltage equals E_{bb} and load voltage equals zero. S1 is closed at time t_0 (fig. 6-14B). At this instant, line

voltage (E_{ZO}) drops to $E_{bb}/2$ and load voltage (E_{ZR}) increases to $E_{bb}/2$. The changes are equal because $Z_O = Z_R$. The discharge sequence of the line is similar to the charge sequence. An incident wave occurs from time t_0 to t_d discharging the line by half. The reflected wave occurs from time t_d to $2t_d$ completely discharging the line. When $Z_R = Z_O$, the output voltage (E_{ZR}) will be a pulse with an amplitude equal to $E_{bb}/2$ and a pulse width equal to $2t_d$.

When load impedance is greater than line impedance, several incident and reflected waves will occur before the line is completely discharged. The pulse which will appear across the load impedance is illustrated in figure 6-15. Note that the pulse width is greater than the normal time of $2t_d$.

When load impedance is less than line impedance, the load voltage is less than the line voltage and after a discharge time of $2t_d$, the line recharges in the opposite polarity. The resultant output pulse is illustrated in figure 6-16. Note that once again the total discharge time is greater than $2t_d$.

Artificial transmission lines have many applications in radar transmitters and receivers. In receivers they are used to provide controllable time delays for triggers and video signals. In transmitters they are used as pulse forming networks to store and deliver high voltage energy in the form of a pulse to the transmitter's power oscillator/power amplifier.

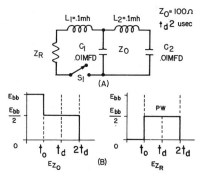

179.665

Figure 6-14.—Discharging a line when $Z_O = Z_R$.

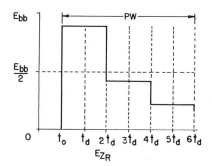

179.666

Figure 6-15.—Pulse voltage when Z_R is greater than Z_O.

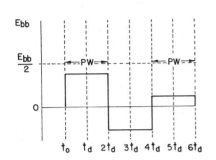

179.667

Figure 6-16. — Pulse voltage when Z_R is less than Z_0.

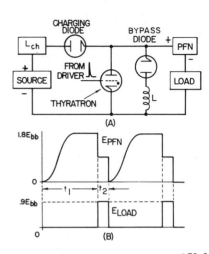

179.668

Figure 6-17. — Modulator using a thyratron switch.

SWITCHING DEVICES

After the pulse forming network has been charged, some method of switching must be provided to allow the pulse forming network to discharge through the load. The switch must be able to handle a high pulse repetition frequency and pass a current flow which may exceed 100 amperes. The switch must also provide accurate control of the on and off time interval.

One method of switching is illustrated in figure 6-17A. In this case, a hydrogen thyratron is used as the switching device. During t_1 (fig. 6-17B) the thyratron is not conducting, and the pulse forming network (line) charges through L_{ch} and the charging diode. The pulse forming network charges to approximately 1.8 times the source voltage. At the start of t_2, a trigger pulse, which corresponds in time with the master timer trigger, is applied to the grid of the thyratron. The thyratron is driven into conduction, allowing the line to discharge into the load. Since the line impedance (Z_0) of the line is closely matched to the impedance of the load Z_R, one half of the line voltage is dropped across the load.

At the end of t_2, the line voltage decreases rapidly and the thyratron de-ionizes. The width of the pulse appearing across the load is determined by the two way delay time of the line.

If there is a slight impedance mismatch (Z_R is less than Z_0) between the pulse forming network and the load, the line will tend to recharge slightly in the opposite direction (fig. 6-16). This opposite charge would increase the current through L_{ch}, causing the pulse forming network voltage to increase during the normal charge cycle. Pulse forming network voltage will keep increasing during each succeeding cycle until breakdown of either the load or thyratron occurs.

The bypass diode and inductor L are connected in shunt with the thyratron to prevent inverse charging of the line. If an inverse charge is present on the line, the bypass diode conducts providing a discharge path through the inductor. The discharge current produces a magnetic field around L1. When the discharge current stops, the field around L1 collapses causing a slight charge of the proper polarity in the pulse forming network.

PULSE TRANSFORMER

The load connected to the line or storage circuit, will be applied to the pulse transformer of a magnetron (master oscillator power amplifier). A pulse transformer is used to step up the high voltage pulse from the pulse forming network and provide impedance matching between the magnetron and the line. Pulse transformer design is critical because of the high frequency components present in the output pulse. The

111

core is composed of thin laminations of ferro-magnetic material, usually silicon steel. Close coupling between primary and secondary reduces leaking inductance to preserve the steep leading edge of the input pulse. Low interwinding capacitance is desired to prevent high frequency oscillations. Close coupling is attained between primary and secondary by winding the primary directly on the secondary and by using the same leg of the core for both windings. The secondary is usually a BIFILAR WINDING.

The bifilar winding secondary is illustrated in figure 6-18. It is made up of two insulated conductors, wound side by side so that exactly the same voltage is induced in each. The bifilar winding acts as two secondaries which have equal and inphase voltages induced in them. The bifilar winding permits the use of a filament secondary without high voltage insulation.

Bypass capacitors, C1 and C2, are often used so that pulse current will flow directly to the bifilar winding, without affecting the filament circuit.

PROTECTIVE DEVICES

Occasionally, an OVERVOLTAGE condition may exist in the pulse forming network. The line will charge to a higher voltage than normal, and an excessively high negative pulse will be applied to the magnetron. Frequent overvoltages can cause arcing and damage to the magnetron.

An overvoltage spark gap is connected across the pulse transformer secondary to prevent excessively high pulse voltages from being applied to the magnetron (fig. 6-19). The gap width is manually adjusted for the desired value of voltage which will produce arcover.

Stray capacitance and leakage inductance in the pulse transformer secondary circuit, produce a series of oscillations after the main pulse has been applied (fig. 6-19B). The negative portions of this waveshape will produce a spurious output from the magnetron which can obliterate any short range targets. These oscillations are not produced during the main pulse, due to the low impedance shunting provided by the conducting magnetron. A DAMPING DIODE is connected in parallel with the magnetron to eliminate the effects of these oscillations. When the negative main pulse is applied to the magnetron cathode, the damping diode does not conduct. The damping diode will conduct during the positive portion of the oscillations (fig. 6-19C); shunting the magnetron with a low impedance and causing the oscillations to dampen very rapidly.

DESPIKING CIRCUIT

The magnetron is a nonlinear impedance and will not be matched to the line under all con-

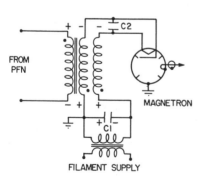

179.669
Figure 6-18.—Bifilar secondary.

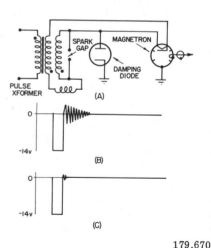

179.670
Figure 6-19.—Protective devices used with the magnetron.

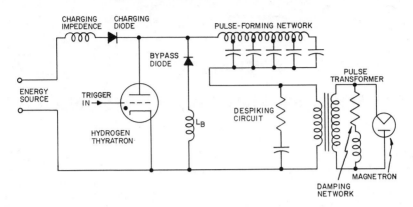

179.671
Figure 6-20. — Despiking circuit.

ditions. The mismatch can cause a spike to appear at the leading edge of the pulse. The spike can be minimized by introducing an RC circuit in parallel with the primary as shown in figure 6-20. This is called a DESPIKING CIRCUIT. The resistance is chosen to equal the impedance of the pulse forming network, and the capacitance is chosen to be small enough so as to be almost completely charged after the oscillator draws full load current.

113

CHAPTER 7

WAVEGUIDES AND CAVITY RESONATORS

The high frequencies employed by radar make possible the use of two unique, but very practical, devices—WAVEGUIDES and CAVITY RESONATORS. A waveguide is a metallic pipe which is used to transfer high frequency electromagnetic energy. A cavity resonator is a metallic cavity in which electromagnetic oscillations exist when the device is properly excited.

Fundamentally, there are two methods of transferring electromagnetic energy. One method is by means of current flow through wires. The other is by movement of electromagnetic fields. The transfer of energy by electromagnetic field motion and by current flow through conductors may appear to be unrelated. However, by considering two-wire lines as elements which guide electromagnetic fields, the current flowing through the conductors may be considered to be the result of the moving fields.

At microwave frequencies, a two-wire transmission line is a poor means of transferring electromagnetic energy because it does not confine electromagnetic fields in a direction perpendicular to the plane which contains the wires. This results in energy escaping by radiation, as illustrated in figure 7-1. Electromagnetic fields may be completely confined when one conductor is extended around the other to form a coaxial cable. Figure 7-2 illustrates this.

Energy in the form of electromagnetic fields may be transferred very efficiently through a line that does not have a center conductor. The type of line used for this purpose is a waveguide. The field configuration in a waveguide is different from that in a coaxial cable due to the missing conductor. Waveguides may be rectangular, circular, or elliptical in cross-section.

The three types of losses in RF lines are COPPER (I^2 R) LOSSES, DIELECTRIC LOSSES, and RADIATION LOSSES. The conducting area of a transmission line will determine the amount of copper loss. Dielectric losses are due to the heating of the insulation between the conductors of a transmission line. Radiation losses are due to the radiation of energy from the line.

With the above facts in mind, the advantages of waveguides over two-wire and coaxial transmission line will now be discussed. A waveguide has a large surface area. A two-wire line consists of a pair of conductors with relatively small surface area. The surface area of the outer conductor of a coaxial cable is large, however, the

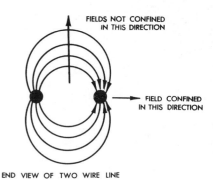

Figure 7-1.—Fields confined in two directions but not in other two.

179.672

Figure 7-2.—Fields confined in all directions.

179.673

inner conductor is relatively small. At microwave frequencies, skin effect restricts the current carrying area of a conductor to a very small layer at the conductor's surface. Although energy transfer is due to electromagnetic field motion, the magnitude of the field is limited by the current carrying area of the conductor. From the above facts, it can be seen that the waveguide will have the least copper loss of the three types of transmission line considered. Dielectric losses are very low in waveguides. This is due to there being no center conductor requiring solid dielectric supports. The dielectric in a waveguide is air, which has very low dielectric loss. Radiation losses are negligible in a waveguide, since the electromagnetic fields are contained wholly within the guide.

The power handling capability of a waveguide is greater than that of a coaxial cable of equal size. Power is a function of E^2/Z_0. E is limited by the distance between conductors. In the coaxial cable illustrated in figure 7-3, this distance is S1. In the circular waveguide illustrated (fig. 7-3) this distance is S2, which is much greater than S1. Therefore, the waveguide is able to handle greater power before the voltage exceeds the breakdown potential of the insulation.

In view of the advantages of waveguide transmission line, it would appear that the waveguide should be the only type of line used. However, waveguides have certain disadvantages which make them practical only at extremely high frequencies.

The width of a waveguide must be approximately a half-wavelength at the frequency to be propagated. A waveguide for use at one megahertz would have to be about 700-feet wide. At 200 MHz, the required width would be about four feet. At 10,000 MHz, a width of only about one inch is required. Thus, the physical dimensions required of a waveguide make the use of this type of transmission impractical below approximately 1000 MHz. If the width of the waveguide is less than a half-wavelength at a certain frequency, energy at that frequency, and all other frequencies below it, will not travel down the guide. There is also an upper limit to the frequency which a particular waveguide will transport. Therefore, the frequency range of any system utilizing waveguides is limited.

The installation of a waveguide system is carefully designed beforehand. The ideal situation would be one continuous section of waveguide between transmitting and receiving points. In practice, however, sections of waveguide must be connected together to form the complete line. This

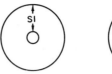

COAXIAL CABLE   CIRCULAR WAVEGUIDE

179.674

Figure 7-3.—Comparison of spacing in coaxial cable and circular waveguide.

requires special couplings to prevent discontinuities in the line and leakage of energy. Also, to reduce skin effect losses, the inside surfaces of a waveguide are often plated with gold or silver. These requirements increase the cost and decrease the practicality of a waveguide system at any but microwave frequencies.

DEVELOPMENT OF WAVEGUIDES FROM PARALLEL LINES

Figure 7-4 shows a section of two-wire transmission line supported on two insulators. The insulators may be made of plastic, porcelain, or similar material. From the view point of the line, the insulators must present a very high impedance to ground. If the insulators act as a low impedance the line would be short-circuited or bypassed to ground. At very high frequencies, the porcelain insulators are not satisfactory, because they no longer present a high impedance to ground. A superior high frequency insulator is a quarter-wave section of RF line short-circuited at one end. Such an insulator is shown in figure 7-5. The impedance of a short-circuited quarter-wave section of RF line is very high at the open end (the junction of the two-wire line). This type of insulator is known as a metallic insulator. A metallic insulator may be placed anywhere along a two-wire line. Figure 7-6 shows several metallic insulators on each side of a two-wire line. It should be noted that the insulators are a quarter-wavelength at only one frequency. This severely limits the broadband efficiency of this type of two-wire line.

The addition of each quarter-wave section increases the rigidity of the line. When more and more sections are added, until each section makes contact with the next, the result is a rectangular box. The line itself is actually part of the walls of the box. The rectangular box thus formed is a waveguide. This action is illustrated in figure 7-7.

115

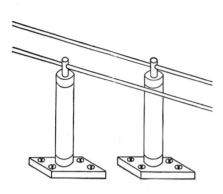

179.675
Figure 7-4.—Two-wire transmission line using ordinary insulators.

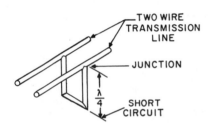

20.237(179)
Figure 7-5.—Quarter-wave section of RF line short circuited at one end.

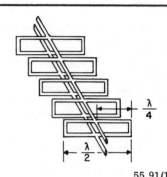

55.91(179)
Figure 7-6.—Several metallic insulators on each side of a two-wire line.

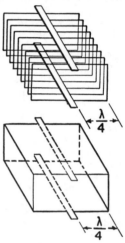

Figure 7-7.—Forming a waveguide by the addition of quarter-wave sections. 55.91.0(179)

As shown in figure 7-8, the wide dimension of a waveguide (that which contains the transmission line's conductors) is called the "a" dimension, and determines the range of operating frequencies. The narrow dimension (or short-circuiting bar) is called the "b" dimension and determines the power handling capability of the guide.

The maximum power handled by a waveguide is directly proportional to the maximum voltage which exists between the "a" walls. In a capacitor, the working voltage is dependent upon the distance between the plates and the type of dielectric material. This same condition holds true for waveguides. Thus, the narrow or "b" dimension will determine the voltage handling capability. In figure 7-8, the "b" dimension of waveguide 2

is twice that of waveguide 1. Therefore, twice as much voltage can exist across waveguide 2 than can exist across waveguide 1. Since power is a function of the square of voltage, the power handling capability of waveguide 2 will be four times that of waveguide 1.

EFFECT OF A WAVEGUIDE ON DIFFERENT FREQUENCIES

Previously, it was stated that the use of quarter-wave metallic insulators limited waveguide operation to a single frequency. However, when a solid wall of quarter-wave insulator sections is used, the guide will operate over a small range of frequencies.

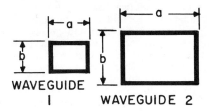

179.676

Figure 7-8.—Labeling waveguide dimensions.

A waveguide may be considered as having upper and lower sections of metallic insulators and a central section of bus bar, as in figure 7-9A. In this figure, distance ab equals cd, which in turn equals one-quarter wave-length. Distance bc is the width of the bus bar. At some higher frequency (assuming the overall dimensions of the guide to be held constant) the required length of the quarter-wave sections decreases. In turn, the width of the bus bar is in effect increased as illustrated in figure 7-9B. Theoretically, the waveguide could function at an infinite number of frequencies, as the length of the quarter-wave sections approaches zero. However, in practice, theoretical performance is prevented by certain other factors which will be discussed later.

Decreasing the frequency of a signal applied to a given waveguide will effectively lengthen the quarter-wave sections, and narrow the bus bar. Below some frequency the bus bar will cease to exist, because the quarter-wave sections converge. At a still lower frequency, the required quarter-wave sections become even longer, as shown in figure 7-9C. Now they can no longer be accommodated within the dimensions of the guide. It follows that in a waveguide there is a low frequency limit, or cutoff frequency (F_{CO}), below which the waveguide cannot transfer energy. The cutoff frequency, for a given rectangular waveguide, is the frequency at which the "a" dimension of the guide is one-half wavelength. The "a" dimension of most waveguides is made 0.7 of the desired cutoff frequency wavelength. This allows the waveguide to handle a small range of frequencies both above and below the designed center frequency. The narrow or "b" dimension is governed by the breakdown potential of the dielectric, which is usually air. Dimensions of 0.2 to 0.5 wavelengths are common for the "b" sides.

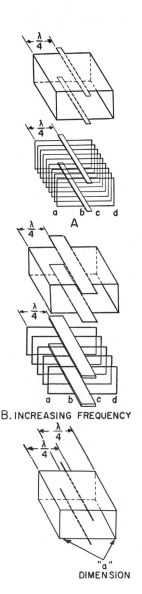

A

B. INCREASING FREQUENCY

"a"
DIMENSION

C. DECREASING FREQUENCY

179.677

Figure 7-9.—Frequency effects on waveguide "bus bar."

117

ELECTROMAGNETIC FIELDS
IN A WAVEGUIDE

A good working knowledge of the fields which exist in a waveguide is necessary, since in a waveguide energy is transferred by electromagnetic fields. The currents and voltages present are a result of the fields. In any waveguide, two inseparable fields are present. They are the MAGNETIC FIELD and the ELECTRIC FIELD.

THE ELECTRIC
FIELD

The existence of an electric field indicates a difference of potential between two points. An electric field consists of a stress in the dielectric, and is represented, in diagrams, by arrows. The simplest form of electric field is the field that forms between the plates of a capacitor, as shown in figure 7-10A. When the top plate of the capacitor is made positive with respect to the bottom plate, electrons move from the top plate to the bottom plate through the external circuit. This sets up a stress in the dielectric between the plates. This stress is represented by arrows, pointing from the more positive point toward the more negative point.

In representing electric fields, the number of arrows indicate the relative strength of the field. In the case of the capacitor of figure 7-10A, note that the arrows are evenly spaced across the area between the two plates. The reason for this is that the potential across the plates is equally distributed. This set of lines represents an electric field. Electric field is abbreviated E-field, and the lines of stress are called E-lines.

The two wire transmission line, illustrated in figure 7-10B, has an instantaneous standing wave of voltage applied to it by the generator. The line is short-circuited at one-wavelength. The instantaneous E-field strength is the same at the positive and negative voltage peaks, but the arrows, representing each field, point in opposite directions. The voltage across the line varies sinusoidally. Therefore, the density of the E-lines varies sinusoidally.

An easy way to show the development of the E-field in a waveguide is with the use of a two-wire line separated by quarter-wave metallic insulators. Figure 7-11 illustrates a two-wire line with several double quarter-wave insulators, or half-wave frames, used as insulators. The E-field on the main line is the same as that on the transmission line of figure 7-10B. The half-wave frames located at points of high voltage will have a strong E-field across them. The half-wave frames located at planes of minimum voltage will have no E-field across them.

Frame No. 1 in figure 7-11 is an example of an insulator section which has a strong E-field across it. Frame No. 2 is at zero voltage plane and has no field across it. Frame No. 3 also has a strong field, but of the opposite polarity. Frame No. 4 has a weaker field across it. Each frame is also shown below the main line for clarity. This illustration is a build up to the three dimensional aspect of the full E-field in a waveguide.

Figure 7-12A illustrates the E-field of a voltage standing wave across a one-wavelength section of waveguide which is shorted at one end. The E-field is strong at one-quarter and three-quarter distances from the shorted end. It becomes weaker, at a sine rate, toward the upper and lower ends and toward the center. It should be realized that this is an instantaneous illustration representing the instant that the standing wave is

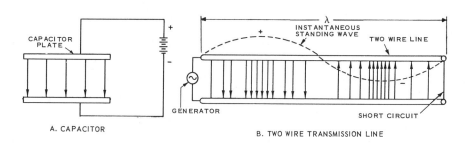

179.678
Figure 7-10.—Representing electric fields.

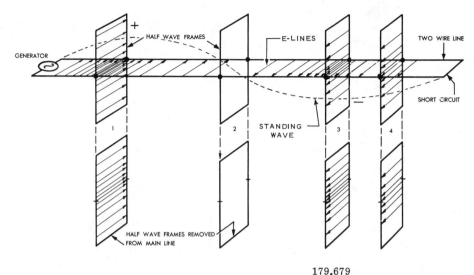

179.679
Figure 7-11.—E-fields on two-wire line with
several half-wave frames.

at its peak. At other times, the voltage and E-field vary from zero to the peak value, reversing polarity every alternation of the input. Note that the end view (fig. 7-12B) shows the field is maximum at the center and minimum near the walls of the waveguide. Figure 7-2C illustrates the three-dimensional aspect of the E-field.

THE MAGNETIC FIELD

The second field in a waveguide, is the magnetic field. The magnetic lines of force which make up the magnetic field are caused by current flow through the conducting material. The presence of the magnetic lines of force is shown by closed loops around the single wire in figure 7-13A. The line forming the loop is a magnetic line of force, or H-line. The H-line must be a continuous closed loop in order to exist.

All the lines associated with current are collectively called a magnetic field or H-field. The strength of the H-field varies directly with the amount of current. Each H-line has a certain direction. This direction may be determined by the left hand rule which states that if the conductor is gripped in the left hand with the thumb extended in the direction of current flow, the fingers will point in the direction of the magnetic lines

of force. The strength of the H-field is indicated by the number of H-lines in a given area.

Although H-lines encircle a single straight wire, they behave differently when the wire is formed into a coil, as shown in figure 7-13B. In a coil the individual H-lines tend to form around each turn of wire, but in doing so take opposite directions between adjacent turns. This results in cancellation of the field between the turns. Inside and outside the coil, the direction is the same for each H-field. Therefore the fields join and form a continuous H-line around the entire coil.

A similar action takes place in a waveguide. In figure 7-13C a two-wire line with quarter-wave sections is shown. Current flows in the main line and in the quarter-wave sections. The current direction produces the individual H-lines around each conductor as shown. At half-wave intervals on the main line, current will flow in opposite directions. This produces H-line loops having opposite directions (the individual loops on the main line are in opposite directions). When a large number of sections exist, the fields cancel between the sections, but are additive inside and outside the sections. Thus, all around the framework they join so that the long loop shown in

119

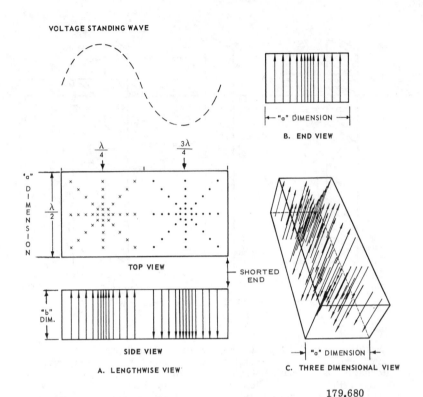

179.680
Figure 7-12.—E-field of a voltage standing wave across a
one-wavelength section of waveguide.

figure 7-13D is formed. Outside the waveguide the individual loops cannot join to form a continuous loop. Therefore, there is no magnetic field outside of a waveguide.

Figure 7-14 shows a conventional presentation of the magnetic field in a waveguide three half-wavelengths long. Note that the field is the strongest at the edges of the waveguide. This is where the current is highest. The current is minimum near the center of each set of loops because at that point the standing wave of current is zero at all times. The end view shows the field as it appears one quarter-wavelength ($\lambda/4$) from the end of the waveguide. Figure 7-14 represents an instantaneous condition. During the peak of the next half-cycle of the input, all field directions are reversed.

BOUNDARY CONDITIONS

The travel of energy down a waveguide is similar to but not identical to that of the propagation of electromagnetic waves in free space. The difference is that the energy in a waveguide is confined to the physical limits of the guide. If energy is to be propagated through a waveguide, two boundary conditions must be met.

The first condition (illustrated in fig. 7-15) is that an electric field must be perpendicular to a conductor in order to exist at the surface of that conductor. The converse of this statement is also true. Assuming a perfect conductor, if an electric field has a component parallel to the conductor, the electric field cannot exist at the surface of that conductor.

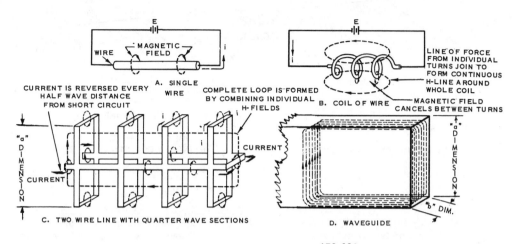

179.681

Figure 7-13.—Formation of H-lines.

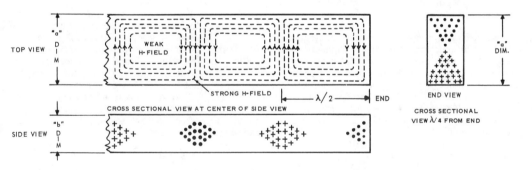

179.682

Figure 7-14.—Magnetic field in a waveguide three
half-wavelengths long.

The second boundary condition is that, at the
surface of the waveguide, there be no perpendicular component of the magnetic field. This condition is satisfied by the configuration of the
magnetic field, since all the H-lines are parallel
to the surface of the waveguide.

Electric and magnetic fields exist simultaneously in a waveguide. In fact, the E-field
causes a current flow which in turn produces an
H-field. Therefore, neither field can exist alone.

MODES OF OPERATION

Figure 7-16A illustrates the two fields which
exist in a waveguide. Details of this three dimensional figure are amplified in figures 7-16B, C,
and D. In these diagrams the number of E-lines
in a given area indicates the strength of the electric field, while the number of H-lines in a given
area indicates the strength of the magnetic field
in that area.

121

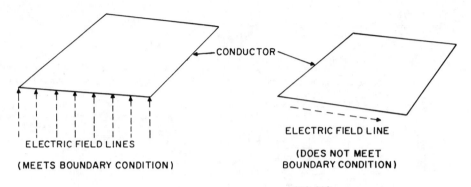

179.683
Figure 7-15.—First boundary condition.

The field configuration shown in figure 7-16, represents only one of the many ways in which fields are able to exist in a waveguide. Such a field configuration is called a MODE OF OPERATION. The DOMINANT MODE is the easiest mode to produce. Other modes (i.e. different field configurations) may occur accidentally, or may be caused deliberately.

The dominant mode is the most efficient mode (has least attenuation). Normally, waveguides are designed so that only the dominant mode will be conducted. To operate in the dominant mode, a waveguide must have an "a" dimension of at least one half-wavelength at the frequency to be propagated. To insure that only the dominant mode will exist, the "a" dimension of the guide must be maintained near the minimum allowable value. In practice this dimension is usually .7 of a wavelength.

The dominant mode has the lowest cutoff frequency. This, of course, is its low frequency limit. The high frequency limit of a waveguide is a frequency at which its "a" dimension becomes large enough to allow operation in a higher mode than that for which the system has been designed.

In some cases, waveguides may be designed to operate in a mode other than the dominant mode. An example of this is illustrated in figure 7-17. If the size of the waveguide in this figure is doubled over that of the waveguide in figure 7-16, the "a" dimension will be a full wavelength long. The two-wire line may be assumed to be a quarter-wavelength from one of the "b" walls as shown in figure 7-17A. The remaining distance to the other "b" wall is three-quarters of a wavelength. The

three-quarter wave section has the same high impedance input as the quarter-wave section, therefore, the two-wire line is properly insulated. The field configuration will show a full wave across the "a" dimension, as illustrated in figure 7-17B.

Although rectangular guides are used almost exclusively in radar systems, circular waveguides find application in specific areas. Figure 7-18 illustrates the dominant mode of a circular waveguide. The cutoff wavelength of a circular guide is 1.71 times the diameter of the guide. Thus, the diameter must be 2/1.71, or approximately 1.17, times the "a" dimension of a rectangular guide having the same cutoff frequency. Circular waveguides are often used in the rotating portion of a radar antenna system.

MODE NUMBERING
SYSTEM

Thus far, only the most basic type of E- and H-field arrangement has been shown. It is sometimes necessary to have a more complex arrangement to facilitate coupling, isolation, or types of operation. To describe the various arrangements of modes of operation, the field arrangements are first divided into two categories: TRANSVERSE ELECTRIC and TRANSVERSE MAGNETIC. These fields are abbreviated TE and TM respectively.

In a transverse electric mode, all parts of the electric field are perpendicular to the length of the guide, and no E-field is parallel to the direction of travel.

In a transverse magnetic mode, the place of the H-field is perpendicular to the length of the waveguide. No H-line is parallel to the direction

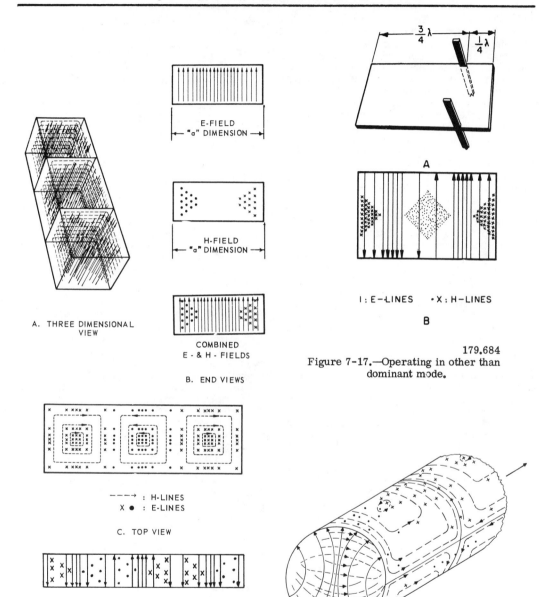

E-FIELD
"a" DIMENSION

H-FIELD
"a" DIMENSION

COMBINED
E - & H - FIELDS

A. THREE DIMENSIONAL
VIEW

B. END VIEWS

A

I : E—LINES   •X : H—LINES

B

179.684
Figure 7-17.—Operating in other than
dominant mode.

-----→ : H-LINES
X ● : E-LINES

C. TOP VIEW

↑ : E-LINES   X ● : H-LINES

D. SIDE VIEW

55.93:.94(179)
Figure 7-16.—E- and H-fields in a waveguide.

179.685
Figure 7-18.—Circular guide showing
dominant mode.

of travel. In addition to the designation TE or TM, subscript numbers are used to complete the description of the field pattern. In rectangular waveguides, the first subscript indicates the number of half-wave patterns in the "a" dimension and the second subscript indicates the number of half-wave patterns in the "b" dimension. In circular waveguides, the first subscript indicates the number of whole wavelength patterns around the circumference of the guide. The second subscript indicates the number of half-wave patterns along the diameter of the guide. The dominant mode for rectangular guides is designated TE_{10}, and the dominant mode for circular guides is TM_{01}, due to its ease in termination.

A method of numbering modes is by counting wavelengths. In the rectangular waveguide illustrated in figure 7-19A, all of the electric lines are perpendicular to the direction of movement. This makes it a TE mode. In the direction across the narrow dimension parallel to the E-line, there is no intensity change, so the first subscript is zero. Across the guide along the wide dimension, the E-field varies from zero at the top, through maximum at the center, to zero on the bottom. Since this is one-half wave, the second subscript is one. Thus, the complete description of this mode is TE_{01}.

Referring to the circular waveguide in figure 7-19B, the E-field is transverse and the letters which describe it are TE. Moving around the circumference starting at the top and going counterclockwise, the fields go from zero, through maximum positive, through zero, through maximum negative, and back to zero. This is one full wave, so the first subscript is one. Going through the diameter, the start is from zero at the top, through maximum at the center, to zero at the bottom. As this is one half-wave the second subscript is one, making this circular mode TE_{11}.

Various circular and rectangular modes are possible. You can verify the numbering system using the diagrams in figure 7-20. Note that the magnetic and electric fields are maximum in intensity in the same area. This indicates that the current and voltage are in phase. This is the condition which results when there are no reflections to cause standing waves.

BEHAVIOR OF ELECTROMAGNETIC WAVES WITHIN A WAVEGUIDE

When a small probe is inserted in a waveguide, and excited with RF energy, it will act as a vertical antenna. Positive and negative wavefronts will be radiated from the probe, as shown in figure 7-21.

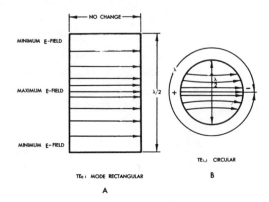

179.686
Figure 7-19.—Numbering modes by counting wavelengths.

The portion of any wavefront traveling in direction B will be rapidly attenuated because it will not fulfill the required boundary conditions. The portions of any wavefront traveling in directions A or C will resolve themselves into oblique wavefronts traveling across the guide. This action is illustrated in figure 7-22. Note that wavefronts of the same polarity cross at the center of the guide. This produces a maximum E-field along the center of the guide. Also, opposite polarity wavefronts meet at the walls of the guide, thereby cancelling. This causes the E-field to be zero at the walls of the guide. The foregoing action fulfills the boundary conditions previously stated.

The angle at the point at which the wavefront strikes the wall of the guide and the perpendicular to the surface at the point of arrival (normal) is known as the ANGLE OF INCIDENCE (θ). The angle at which the wavefront is reflected from the wall of the guide and the normal is called the ANGLE OF REFLECTION (ϕ). These angles are equal, and are depicted in figure 7-22.

The angles of incidence and reflection are a function of frequency (assuming the guide dimensions are held constant). Figure 7-23 illustrates the angles of incidence and reflection for a high frequency, a medium frequency, and a frequency just above the cutoff frequency. As the frequency being transmitted decreases, the angles decrease. Just below the cutoff frequency, the angles of incidence and reflection become zero. Therefore,

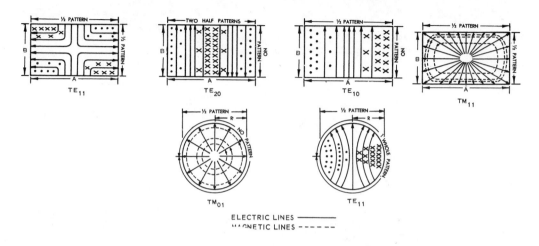

ELECTRIC LINES ————
MAGNETIC LINES — — — — —

25.149(179)
Figure 7-20.—Various modes of operation for
rectangular and circular waveguides.

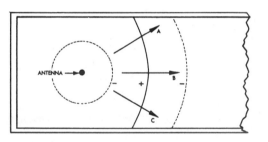

179.687
Figure 7-21.—Radiation from probe
inserted in a waveguide.

point one to point two, a distance of L, at the velocity of light (V_C). Due to this diagonal movement, the wavefront actually has moved down the guide only a distance G. The velocity with which the wavefront has moved through distance G is the group velocity. However, if an instrument were used to detect the two positions at the wall, the positions would be a distance P apart. This distance is greater than distance L or G. Therefore, the contact point between the wavefront and the wall appears to be moving with a velocity greater than the velocity of light. Since the phase of the RF has changed over distance P, this velocity is called the PHASE VELOCITY (V_p). The mathematical relationship between the three velocities is given by the equation $V_C = \sqrt{V_g V_p}$.

at any frequency lower than f_{co}, the wavefronts will be reflected back and forth across the guide, setting up standing waves and no energy will be conducted down the guide.

In figure 7-23, though the wavefront is traveling with velocity of light, it is not moving straight down the guide. Its straight line velocity or axial velocity appears to be less than the speed of light. This axial or straight line velocity is called the GROUP VELOCITY (V_g). The relationship of the group velocity to the diagonal velocity illustrates an unusual phenomenon. Referring to figure 7-24, during a given time, a wavefront will move from

When a modulated RF signal is propagated down a waveguide, the modulation envelope will appear to move forward at the group velocity, while the individual RF cycles will appear to move forward through the envelope at the phase velocity. However, group velocity decreases with a decrease in frequency, and phase velocity increases with a decrease in frequency. At the same time group velocity is always less than the velocity of light by the same amount that phase velocity is greater than the velocity of light. Therefore, the overall movement of the signal is at the velocity of light.

125

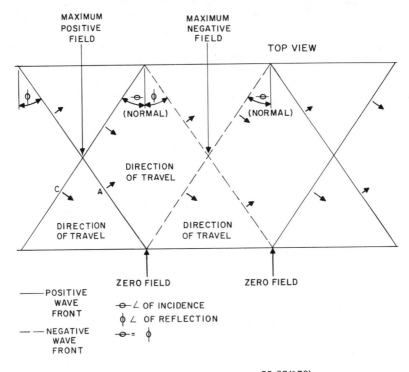

MAXIMUM POSITIVE FIELD

MAXIMUM NEGATIVE FIELD

TOP VIEW

(NORMAL)

(NORMAL)

DIRECTION OF TRAVEL

C   A

DIRECTION OF TRAVEL

DIRECTION OF TRAVEL

ZERO FIELD

ZERO FIELD

——— POSITIVE WAVE FRONT

— — NEGATIVE WAVE FRONT

∠ OF INCIDENCE

φ ∠ OF REFLECTION

= φ

55.97(179)

Figure 7-22.—Wavefront travel in a waveguide.

METHODS OF COUPLING

Fundamentally, there are three methods of coupling energy into or out of a waveguide — PROBE, LOOP, and APERATURE. Probe, or capacitive, coupling is illustrated in figure 7-25. Its action is the same as that of a quarter-wave antenna. When the probe is excited by an RF signal, an electric field is set up (fig. 7-25A). The probe should be located in the center of the "a" dimension and a quarter-wavelength, or odd multiple of a quarter-wavelength, from the short-circuited end as illustrated in figure 7-25B. This is a point of maximum E-field and, therefore, is a point of maximum coupling between the probe and the field. Usually, the probe is fed with a short length of coaxial cable. The outer conductor is connected to the waveguide wall, and the probe extends into the guide, but is insulated from it as shown in figure 7-25C. The degree of coupling may be varied by varying the length of the probe, removing it from the center of the E-field, or shielding it.

In a pulse modulated radar system there are wide sidebands on either side of the carrier frequency. In order that a probe does not discriminate too sharply against frequencies which differ from the carrier frequency, a wide band probe may be used. This type of probe is illustrated in figure 7-25D for both low and high power usage.

Figure 7-26 illustrates loop, or inductive, coupling. The loop is placed at a point of maximum H-field in the guide. As shown in figure 7-26A, the outer conductor is connected to the guide and the inner conductor forms a loop inside the guide. The current flow in the loop sets up a magnetic field in the guide. This action is illustrated in figure 7-26B. As shown in figure 7-26C, the loop may be placed in a number of locations. The degree of loop coupling may be varied by rotation of the loop.

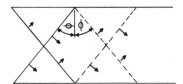

LOW FREQUENCY
(APPROACHING f_{co})

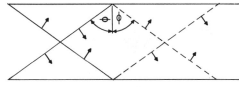

MEDIUM FREQUENCY

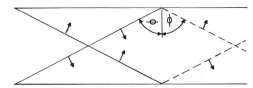

HIGH FREQUENCY

-⊖- = ∠ OF INCIDENCE

φ = ∠ OF REFLECTION

179.688
Figure 7-23.—Effects of frequency on
angles of incidence and reflection.

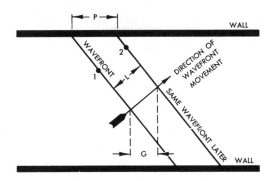

179.689
Figure 7-24.—Relationship of group velocity
to diagonal velocity.

The third method of coupling is aperture, or slot, coupling. This type of coupling is shown in figure 7-27. Slot A is at an area of maximum E-field and is a form of electric field coupling. Slot B is at an area of maximum H-field and is a form of magnetic field coupling. Slot C is at an area of maximum E- and H-field and is a form of electromagnetic coupling.

BENDS, TWISTS, AND JOINTS

Any abrupt change in the size, shape, or direction of a waveguide system will result in reflec-

tions. Usually a radar system will require bends or twists in its associated waveguides. Bends may be made if they have a radius of at least two wavelengths, as shown in figure 7-28A. The electromagnetic field may be rotated by twisting the waveguide as shown in figure 7-28B. The twists must be accomplished over a distance of at least two wavelengths.

Sometimes a sharp 90° bend is required. To avoid reflections the guide is bent 45° twice, one quarter-wave apart (fig. 7-28C). The combination of the direct reflection at one bend and the inverted reflection from the other bend will cancel and leave the fields as though no reflection had occurred. As shown, the bend can be made in either the narrow or wide dimension of the guide without changing the mode of operation.

It is practically impossible to construct a waveguide system in one piece. It is necessary to construct guide sections, which must be connected by joints. Any irregularities in the joints cause standing waves, and permit energy to be dissipated. A proper permanent joint affords a good connection between the parts of a waveguide and has very little effect on the fields. Normally, this type of joint is made at the factory. When it is used, the waveguide sections are machined within a few thousandths of an inch and then welded together. The result is a hermetically sealed and mirror smooth joint.

Where sections of waveguide must be taken apart for normal maintenance and repair, it is impractical to use a permanent joint. To permit portions of the waveguide to be separated, the sections

127

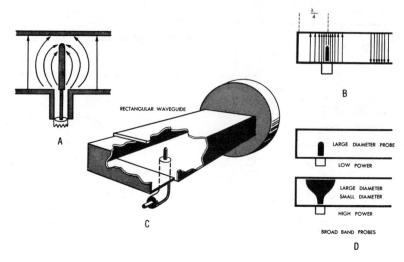

55.99(179)
Figure 7-25.—Probe, or capacitive, coupling.

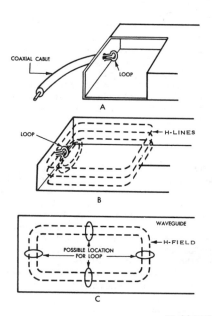

55.98(179)
Figure 7-26.—Loop, or inductive, coupling.

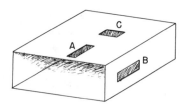

179.690
Figure 7-27.—Aperture, or slot, coupling.

are connected by semipermanent joints of which the choke joint is the most common type. A cross-sectional view of a choke joint is shown in figure 7-29A and B. It consists of two flanges which are connected to the waveguide at the center (fig. 7-29C). The right flange is flat, and the one at the left is slotted one-quarter wavelength deep from the inner surface of the waveguide, at a distance of one-quarter wavelength from the point where the flanges are joined. The quarter-wavelength slot is short-circuited at the end. The two quarter-wavelength sections form a half-wavelength section and reflect a short-circuit at the place where the walls are joined together.

Electrically, this creates a short-circuit at the junction of the two waveguides. The two sections may actually be separated as much as a tenth of a wavelength without excessive loss of energy at the joint. This separation allows room to seal the interior of the waveguide with a rubber gasket for pressurization. The quarter-wave distance from the walls to the slot is modified slightly to compensate for the slight reactance introduced by the short space and the open circuit from the slot to the periphery of the flange.

The loss introduced by a well designed choke is less than 0.03 db, while an unsoldered permanent joint, well machined, has a loss of 0.05 db, or more.

Rotating joints are usually required in a radar system where the transmitter is stationary and the antenna is rotatable. A simple method for rotating part of a waveguide system uses a mode of operation that is symmetrical about the axis, as shown in figure 7-30A and B. This requirement is met by using a circular waveguide and a mode such as TM_{01}. In this method, a choke joint may be used to separate the sections mechanically and to join them electrically as shown in figure 7-30C. The waveguide rotates, but not the field. The fixed field minimizes reflections. This is basically the method used for all rotating joints. Since most radars use rectangular waveguides,

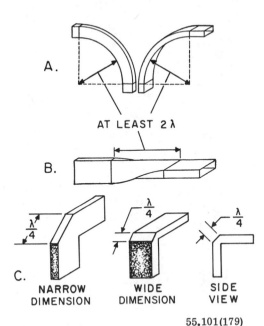

A.

AT LEAST 2λ

B.

$\frac{\lambda}{4}$ $\frac{\lambda}{4}$

$\frac{\lambda}{4}$ $\frac{\lambda}{4}$

C.

NARROW
DIMENSION

WIDE
DIMENSION

SIDE
VIEW

55.101(179)
Figure 7-28.—Waveguide bends and twists.

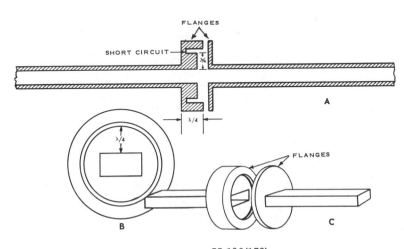

FLANGES

SHORT CIRCUIT

$\lambda/4$

A

$\lambda/4$

$\lambda/4$

FLANGES

B

C

55.102(179)
Figure 7-29.—Choke joint.

the circular rotating joint must be inserted between two rectangular sections. A typical rotating
joint is shown in figure 7-31. The joint consists of
two sections of circular guide, one stationary, the
other rotating. At the ends of each section of circular guide a transition, from the circular to the
rectangular waveguide, occurs.

Note that the rectangular guide is operating in
the TE_{10} mode (the lines in the illustration of the
guide represent the E-lines). The E-lines couple
from the rectangular guide into the circular guide
and excite the circular guide in the TM_{01} mode.
This is the mode that provides the required axial
symmetry for rotating joints. At the top of the
joint, the E-lines couple the energy into the rectangular guide which leads to the antenna. Here
the guide is operating in the TE_{10} mode.

IMPEDANCE CHANGING DEVICES

Often, RF transmission systems are not perfectly matched by their load devices. The standing
waves that result are undesirable because of the
associated power loss, reduction of power handling
capability, and increase in frequency sensitivity.
Impedance changing devices are therefore introduced near the sources of reflected waves. These
devices provide a standing wave ratio (SWR) nearly
equal to unity. Impedance changing devices are
also employed near the source to provide a load
impedance of optimum value for the source.

The principles used to achieve impedance
matching in waveguides are the same as those used
in conventional circuits. However, the physical
appearance of the matching elements is different.

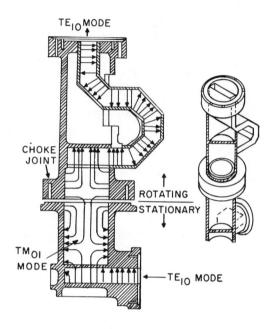

179.692

Figure 7-31.—Representative rotating
joint.

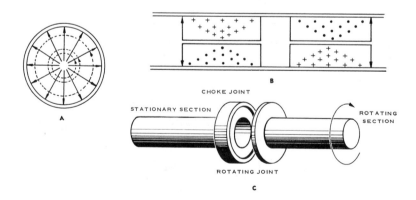

179.691

Figure 7-30.—Basic rotating joint.

Figure 7-32 illustrates three irises used to introduce inductance or capacitance into a waveguide, their placement in the waveguide, and their equivalent impedance circuits. An iris is a metal diaphragm located in a transverse plane of a waveguide. It contains openings (called windows or apertures) through which the waves may pass.

Figure 7-32A illustrates an inductive iris. It places a shunt inductive reactance across the line. The wider the opening, the higher is the reactance. A capacitive iris is illustrated in figure 7-32B. It places a shunt capacitive reactance across the line. Again, the wider the opening, the higher is the reactance. The iris illustrated in figure 7-32C forms an equivalent parallel LC circuit across the line. At resonance, the iris would place a high shunt resistance across the guide. Above or below resonance, the iris will act as a capacitive or inductive reactance.

Figure 7-33 illustrates posts and screws used for impedance changing. Conducting posts or screws which penetrate only part way into the guide, as illustrated in figure 7-33A, form shunt capacitive reactances.

When the conducting post or screw extends completely through the guide, making contact with the top and bottom walls, it will act as an inductive reactance as shown in figure 7-33B. The post is not a perfect short-circuit because it is encircled by magnetic fields and possesses an inductive reactance. The difference in screws and posts is that the screw can be adjusted to vary the reactance.

TERMINATING A WAVEGUIDE

Since a waveguide is a single conductor, it is not easy to define its characteristic impedance (Z_0). Nevertheless, you may think of the characteristic impedance of a waveguide as being approximately equal to the ratio of the strength of the electric field to the strength of the magnetic field for energy traveling in one direction. This ratio is equivalent to the voltage-to-current ratio in coaxial lines on which there are no standing waves.

On a waveguide there is no place to connect a fixed resistor with which to terminate the guide in its characteristic impedance. However, there are a number of special arrangements that accomplish the same thing. One arrangement consists of filling the end of the waveguide with graphited-sand as shown in figure 7-34A. As the fields enter the sand, they cause current flow through the sand. This current flow creates heat, thereby dissipating energy. None of the energy thus dissipated is reflected back into the guide. Another arrangement (fig. 7-34B) uses a high resistance rod placed at the center of the E-field. The E-field causes current to flow through the rod. The high resistance of the rod dissipates the energy as an I^2R loss.

Still another method for terminating a waveguide is the use of a wedge of high resistance material (fig. 7-34C). The plane of the wedge is placed perpendicular to the magnetic lines of force. When the H-lines cut the wedge a voltage is induced in it. The current produced by this induced voltage creates an I^2R loss. This loss is dissipated in the form of heat. Therefore, very little energy is permitted to reach the closed end of the guide, thus preventing reflections.

Each of the preceding terminations is designed to match the impedance of the guide in order to insure a minimum of reflection. On the other hand,

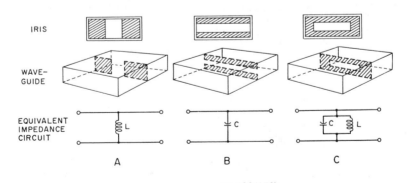

55.103(179)

Figure 7-32.—Waveguide irises.

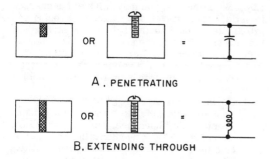

A. PENETRATING

B. EXTENDING THROUGH

179.693
Figure 7-33.—Conducting posts and screws.

there are many instances where it is desirable for all the energy to be reflected from the end of the guide. The best way to accomplish this is to permanently weld a metal plate at the end of the waveguide, as shown in figure 7-34D.

When it is necessary that the end of the guide be removable, a removable plate is attached to the end of the guide. The contact between the guide and the end plate must be exceptionally good. This insures that current flow will not attenuate the H-field. Perfect contact is not required when the connection is made at a point of minimum current. Such a point is located one-quarter wavelength from the end of the guide. When connection is made at this point, a cup is used in place of the end plate (fig. 7-34E). This cup is a quarter-wavelength long and large enough to fit over the end of the guide. The voltage across opposite sides of the cup opening is high, but as the reflected H-field cancels the incident H-field, the resulting current is very small and reflection is at a minimum.

When the end of the guide must be adjustable, an arrangement similar to the choke joint previously explained is employed. It consists, principally, of an adjustable plunger that fits into the guide, as shown in figure 7-34F. The walls of the waveguide and of the plunger form a half-wave channel. The half-wave channel is closed at one end and reflects a short-circuit across the other end. This results in a perfect connection between the wall and the plunger. The actual physical contact is made at a quarter-wavelength from the short circuit, where the current is minimum due to the standing waves. This makes it possible for the plunger to slide loosely in the guide at a point where the contact resistance to current flow is very low.

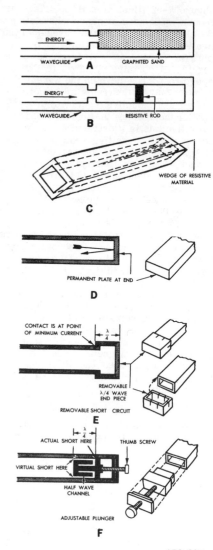

179.694
Figure 7-34.—Terminating waveguides.

CAVITY RESONATORS

Circuits composed of lumped inductance and capacitance elements may be made to resonate at any frequency from less than one hertz to many thousand megahertz. At extremely high frequencies, however, the physical size of the inductors and capacitors becomes extremely small. Also, losses in the circuit become extremely great. Resonant devices of different construction are therefore preferred at extremely high frequencies. In the UHF range, secions of parallel wire or coaxial transmission line are commonly employed in place of lumped constant resonant circuits. In the microwave region, cavity resonators are used. Cavity resonators are metal walled chambers fitted with devices for admitting and extracting electromagnetic energy. The Q of these devices may be much greater than that of conventional LC tank circuits.

Although cavity resonators, built for different frequency ranges and applications, have a variety of physical forms, the basic principles of operation are essentially the same for all. Operating principles of cavity resonators are explained in this chapter. These principles are applied, in subsequent chapters, to the study of important microwave components employing cavity resonators.

Resonant cavity walls are made of highly conductive material and enclose a good dielectric, usually air. One example of a cavity resonator is the rectangular box of figure 7-35A. It may be though of as a section of rectangular waveguide closed at both ends by conducting plates. Because the end plates are short-circuits for waves traveling in the Z direction, the cavity is analogous to a transmission line section with short-circuits at both ends. Resonant modes occur at frequencies for which the distance between end plates is a half-wavelength or multiple of a half-wavelength.

Cavity modes are designated by the same numbering system that is used with waveguides, except that a third subscript is used to indicate the number of half-wave patterns of the transverse field along the axis of the cavity (perpendicular to the transverse field). For example, the mode illustrated in figure 7-35B is TE_{101}. This field configuration is obtained by addition of the fields of the component traveling waves. Observe that the pattern is similar to one segment of the pattern of the TE_{10} mode of a waveguide. The only difference is that the loops formed by the magnetic field lines are shifted one-quarter wavelength with respect to the electric field lines. This shift is the result of adding the two oppositely directed traveling waves.

The rectangular cavity is only one of many cavity devices useful as high frequency resonators. By appropriate choice of cavity shape advantages such as compactness, ease of tuning,

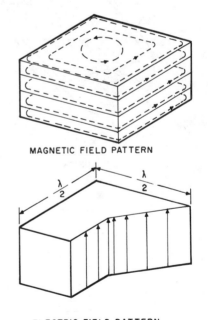

A. RESONATOR SHAPE

B. FIELD PATTERNS OF A SIMPLE MODE

MAGNETIC FIELD PATTERN

ELECTRIC FIELD PATTERN

179.695

Figure 7-35.—Rectangular waveguide resonator.

simple mode spectrum, and high Q may be se-
cured as required for special applications.

Figure 7-36 illustrates some other cavity
types and possible field configurations. Part A
is a cylindrical cavity. Several examples of a
type of cavity known as a reentrant cavity are
shown in part B. Part C depicts a hole and slot
type cavity commonly used in magnetron oscillators.

CAVITY TUNING

The resonant frequency of a cavity may be
varied by changing any of three parameters:

CAVITY VOLUME, CAVITY INDUCTANCE, or
CAVITY CAPACITANCE. Although the mechanical methods for tuning cavity resonators may
vary, they all utilize the electrical principles
explained below.

Figure 7-37 illustrates a method of tuning a
cylindrical type cavity by varying its volume.
Varying the distance "d" will result in a new
resonant frequency. Increasing distance "d" will
lower the resonant frequency, while decreasing
"d" will cause an increase in resonant frequency.
The movement of the disk may be calibrated in

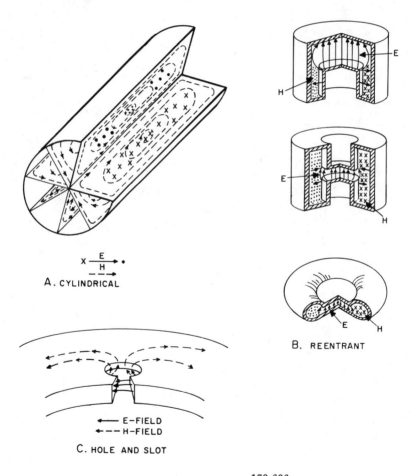

A. CYLINDRICAL

B. REENTRANT

C. HOLE AND SLOT

E-FIELD
H-FIELD

179.696
Figure 7-36.—Types of cavities.

terms of frequency. A micrometer scale is usually used to indicate the position of the disk, and a calibration chart is used to determine frequency.

A second method for tuning a cavity resonator is illustrated in figure 7-38A. Here, a nonferrous metallic screw is inserted into a cavity at a point of maximum H-field. This decreases the permeability of the cavity and decreases its effective inductance, which raises its resonant frequency. The farther the screw penetrates into the cavity the higher is the resonant frequency. A paddle (fig. 7-38B) can be used in place of the screw. Turning the paddle to a position more nearly perpendicular to the H-field increases resonant frequency.

Tuning a reentrant cavity by varying its capacitance is illustrated in figure 7-39. The E-field in this type of cavity may be assumed to be uniform and contained wholly within the gap between center plates. Adjusting the screw expands or contracts the lever which varies the distance between the center plates. Varying the distance between these center plates varies the effective cavity capacitance, which, in turn, varies the cavity resonant frequency. Increasing this distance decreases capacitance, and there, increases resonant frequency. The wall of this type cavity has to be flexible to allow the movement necessary for tuning.

CAVITY COUPLING
DEVICES

As with conventional LC tank circuits, a method of introducing and removing energy is required.

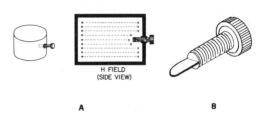

179.698

Figure 7-38.—Cavity tuning by inductance.

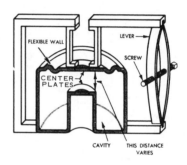

179.699

Figure 7-39.—Cavity tuning by capacitance.

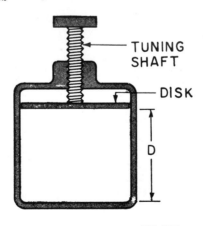

179.697

Figure 7-37.—Cavity tuning by volume.

135

There are three principal methods of accomplishing this: PROBE COUPLING, LOOP COUPLING, and APERTURE, or SLOT, COUPLING.

Figure 7-40A illustrates probe coupling. The probe is inserted into the cavity at a point of maximum E-field and sets up E-lines, thus starting and maintaining oscillation. This is a form of capacitive coupling. Part B illustrates loop coupling. Here, a loop is placed in the cavity at a point of maximum H-field, and introduces an H-field into the cavity. This is a form of inductive coupling. Energy may be introduced or removed through an aperture or slot cut in a wall that is common to the guide and resonator. This is illustrated in part C of figure 7-40. Coupling results from the penetration of electromagnetic fields through the opening.

In reentrant cavities, energy may be introduced into the cavity by pulses of electron energy passed through holes in the center of perforated plates. This is illustrated in figure 7-41.

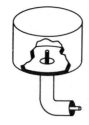

A. PROBE COUPLING

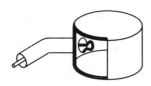

B. LOOP COUPLING

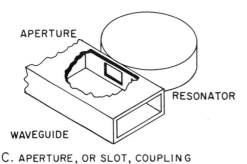

C. APERTURE, OR SLOT, COUPLING

179.700
Figure 7-40.—Cavity coupling.

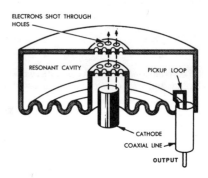

179.701
Figure 7-41.—Reentrant cavity coupling.

CHAPTER 8

MICROWAVE TRANSMITTING DEVICES

Microwave transmitters require the same stages as do the lower frequency transmitters discussed in previous chapters. However, due to their high frequencies special components have been developed. Among these are magnetrons, klystrons, traveling wave tubes, and gallium arsinide (Gunn) oscillators.

MAGNETRONS

The magnetron is a power oscillator. It is used because conventional tubes are not practical at the frequencies and power levels required for radar (microwave) applications.

The magnetron is an oscillator unlike any other that has previously been discussed in this text. The magnetron is a self-contained unit. That is, it produces a microwave frequency output within its enclosure without the use of external components such as crystals, inductors, capacitors, etc.

MAGNETRON CONSTRUCTION

Basically, the magnetron is a diode and has no grid. A magnetic field in the space between the plate (anode) and the cathode serves as a grid. The plate of a magnetron does not have the same physical appearance as the plate of an ordinary electron tube. Since conventional LC networks become impractical at microwave frequencies, the plate is fabricated into a cylindrical copper block containing resonant cavities which serve as tuned circuits. The magnetron base differs greatly from the conventional base. It has short, large diameter leads that are carefully sealed into the tube and shielded, as shown in figure 8-1.

The cathode and filament are at the center of the tube. It is supported by the filament leads which are large and rigid enough to keep the cathode and filament structure fixed in position.

The output lead is usually a probe or loop extending into one of the tuned cavities and coupled into a waveguide or coaxial line. The plate structure, as shown in figure 8-2, is a solid block of copper. The cylindrical holes around its circumference are resonant cavities. A narrow slot runs from each cavity into the central portion of the tube and divides the inner structure into as many segments as there are cavities. Alternate segments are strapped together to put the cavities in parallel with regard to the output. These cavities control the output frequency. The straps are circular metal bands that are placed across the top of the block at the entrance slots to the cavities. Since the cathode must operate at high power, it must be fairly large and must be able to withstand high operating temperatures. It must also have good emission characteristics, particularly under back bombardment, because much of the output power is derived from the large number of electrons emitted when high velocity electrons return to strike the cathode. The cathode is indirectly heated, and is constructed of a high emission material. The open space between the plate and the cathode is called the INTERACTION SPACE because it is in this space that the electric and magnetic fields interact to exert force upon the electrons.

The magnetic field is usually provided by a strong permanent magnet mounted around the magnetron so that the magnetic field is parallel with the axis of the cathode. The cathode is mounted in the center of the interaction space.

BASIC MAGNETRON OPERATION

The theory of operation of the magnetron is based on the motion of electrons under the influence of combined electric and magnetic fields. The following laws govern this motion.

The direction of an electric field is from the positive electrode to the negative electrode. The law governing the motion of an electron in an

137

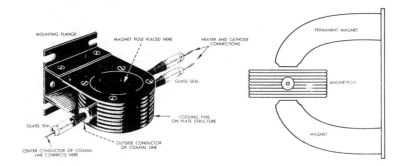

20.290(179)
Figure 8-1. — Magnetron.

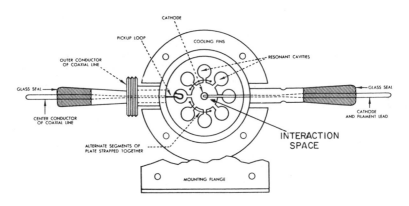

179.702
Figure 8-2. — Cutaway view of a magnetron.

electric, or E, field states that the force exerted by an electric field on an electron is proportional to the strength of the field. Electrons tend to move from a point of negative potential toward a positive potential as shown in figure 8-3. In other words, electrons tend to move against the E-field.

When an electron is being accelerated by an E-field, as shown in figure 8-3, energy is taken from the field by the electrons.

The law of motion of an electron in a magnetic, or H, field states that the force exerted on an electron in a magnetic field is at right angles to both the field and the path of the electron. The direction of the force is such that the electron trajectories are clockwise when

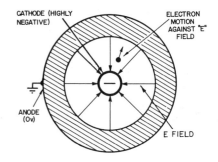

179.703
Figure 8-3.—Electron motion in an electric field.

ewed in the direction of the magnetic field s shown in figure 8-4.

In figure 8-4, it is assumed that a south pole below the paper and a north pole is above the uper so that the magnetic field is going into e paper. When an electron is moving in space magnetic field is built around the electron st as there would be a magnetic field around wire when electrons are flowing through a wire. ote in figure 8-4 that the magnetic field around e moving electron adds to the permanent magtic field on the left side of the electron path d subtracts from the permanent magnetic eld on the right side of the electron path, thus eakening the field on that side. Therefore, the ectron path bends to the right (clockwise). the permanent magnetic field strength is creased, the electron path will bend sharper. ikewise, if the velocity of the electron inreases, the field around it increases and its ath will bend more sharply.

A schematic diagram of a basic magnetron shown in figure 8-5A. The tube consists of cylindrical plate with a cathode placed coially with it. The tuned circuit (not shown) which oscillations take place are cavities aysically located in the plate.

When no magnetic field exists, heating the athode results in a uniform and direct movement in the field from the cathode to the plate, s illustrated in figure 8-5B. However, as the agnetic field surrounding the tube is inreased, a single electron is affected as shown figure 8-6. In figure 8-6A, the magnetic field as been increased to a point where the electron

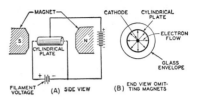

179.705

Figure 8-5. — Basic magnetron.

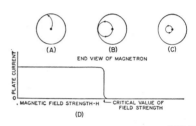

179.706

Figure 8-6. — Effect of magnetic field on single electron.

proceeds to the plate in a curve rather than a direct path.

In figure 8-6B, the magnetic field has reached a value great enough to cause the electron to just miss the plate and return to the filament in a circular orbit. This value is the CRITICAL VALUE of field strength. In figure 8-6C, the value of the field strength has been increased to a point beyond the critical value, and the electron is made to travel to the cathode in a circular path of smaller diameter.

Figure 8-6D shows how the magnetron plate current varies under the influence of the varying magnetic field. In figure 8-6A, the electron flow reaches the plate, so that there is a large amount of plate current flowing. However, when the critical field value is reached, as shown in figure 8-6B, the electrons are deflected away from the plate; and the plate current drops abruptly to a very small value. When the field strength is made still larger, figure 8-6C, the plate current drops to zero.

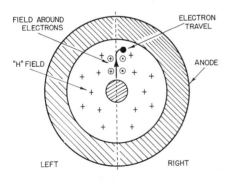

179.704

igure 8-4.—Electron motion in a magnetic field.

139

When the magnetron is adjusted to the plate current cutoff or critical value, and the electrons just fail to reach the plate in their circular motion, the magnetron can produce oscillations at microwave frequency by virtue of the currents induced electrostatically by the moving electrons. This frequency is determined by the time it takes the electrons to travel from the cathode toward the plate and back again. A transfer of microwave frequency energy to a load is made possible by connecting an external circuit between the cathode and plate of the magnetron. Magnetron oscillators are divided into two classes, NEGATIVE RESISTANCE and ELECTRON RESONANCE magnetron oscillators.

A negative resistance magnetron oscillator operates by reason of a static negative resistance between its electrodes and has a frequency equal to the natural period of the tuned circuit connected to the tube.

An electron resonance magnetron oscillator operates by reason of the electron transit time characteristics of an electron tube, that is, the time it takes electrons to travel from cathode to plate. This oscillator is capable of generating very large peak power outputs at frequencies in the thousands of megahertz. Although its average power output over a period of time is low, it can put out very high power oscillations in short bursts of pulses.

NEGATIVE RESISTANCE
MAGNETRON

The split-anode negative resistance magnetron is a variation of the basic magnetron which operates at a higher frequency, and is capable of more output. Its general construction is similar to the basic magnetron, except that it has a split plate as shown in figure 8-7. These half plates are operated at different potentials to provide an electron motion as shown in figure 8-8. The electron leaving the cathode and progressing toward the high potential plate

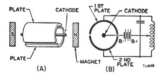

179.707
Figure 8-7. — Split-anode magnetron and its output.

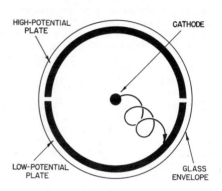

55.77
Figure 8-8. — Movement of electron in a split-anode magnetron.

is deflected by the magnetic field at a certain radius of curvature. After passing the split between the two plates, the electron enters the electrostatic field set up by the lower potential plate.

Here the magnetic field has more effect on the electron, which is deflected at a smaller radius of curvature. The electron then continues to make a series of loops through the magnetic field and electric field until it finally falls on the low potential plate.

Oscillations can be started by applying the proper value of magnetic field to the tube. The value of field required is somewhat beyond the critical value which, for the split-anode tube, is the field required to cause all the electrons to miss the plate when its halves are operating at the same potential. However, the alternating voltages impressed on the plates as a result of the oscillation generated in the tank circuit will cause electron motion such as that shown in figure 8-8, and current will flow. Since a very concentrated magnetic field is required for the negative resistance magnetron oscillator, the length of the tube plate is limited to a few centimeters for a magnet of reasonable dimensions. In addition a small diameter tube is required to make the magnetron operate efficiently at microwave frequencies. For this reason, a heavy-walled plate is used to increase the radiating properties of the tube. To obtain still greater dissipation, tubes with high outputs use an artificial cooling method such as forced air or water cooling.

140

The output of magnetrons is somewhat reduced by the bombardment of the filament by electrons which travel in loops (figs. 8-6B and 8). This effect causes an increase of filament temperature under certain conditions of high magnetic field and high plate voltage; and sometimes results in unstable operation of the tube. The effects of filament bombardment can be compensated for by operating the filament at reduced voltage. In some cases, the plate voltage and field strength also are reduced to prevent destructive filament bombardment.

ELECTRON RESONANCE MAGNETRON

In the electron resonance type of magnetron, the plate itself may be so constructed as to resonate and function as a tank circuit. Thus, there are no external tuned circuits; power is delivered directly from the tube to a transmission line as shown in figure 8-9. The tube constants and operating conditions are such that the electron paths are somewhat different from those in figure 8-8. Instead of having closed spirals or loops, the path is a curve having a series of abrupt points as illustrated in figure 8-10. Ordinarily, this type of magnetron also has more than two segments in the plate. For example, figure 8-10 illustrates an eight-segment plate.

This type of magnetron is the most widely used at present for microwave frequencies. Modern designs have a reasonably high efficiency and relatively high output. However, one disadvantage of the electron resonance magnetron is that its average power is limited by the cathode emission. Furthermore, the peak power is limited by the maximum voltage which it can withstand without injury. Three common types of anode blocks used in electron resonance magnetrons are shown in figure 8-11.

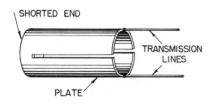

Figure 8-9. — Plate tank circuit of magnetron.

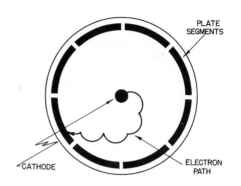

179.709
Figure 8-10. — Electron path in electron resonance magnetron.

The first type shown in figure 8-11 has cylindrical cavities. This type is called a HOLE-AND-SLOT ANODE. The second type is called the VANE ANODE which has trapezoidal cavities. These first two anode blocks operate in such a way that alternate segments must be connected, or strapped, to ensure that each segment is opposite in polarity to its neighboring segment on either side as shown in figure 8-12. This also requires an even number of cavities.

The third type, illustrated in figure 8-11, is called a RISING SUN BLOCK because of its appearance. The alternate large and small trapezoidal cavities in this block results in a stable frequency between the resonant frequencies of the large and the small cavities.

Figure 8-13A shows the physical appearance of the resonant cavities contained in the hole-and-slot anode which we will use when analyzing the operation of the electron resonance magnetron.

Notice that the cavity consists of a cylindrical hole in the copper anode, and a slot which connects the cavity to the interaction space.

The electrical equivalent circuit of the cavity and slot is shown in figure 8-13B. The parallel sides of the slot form the plates of a capacitor, while the walls of the hole act as an inductor. The hole and slot thus form a high Q resonant LC circuit. As shown in figure 8-11, the anode of a magnetron contains a number of these cavities.

An analysis of the anode in figure 8-11 would reveal that the LC tank of each cavity are in

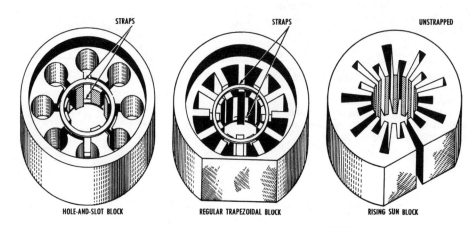

55.79

Figure 8-11.—Common types of anode blocks.

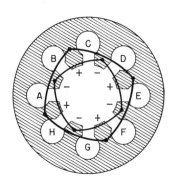

179.710

Figure 8-12.—Strapping alternate segments.

series as shown in figure 8-14. This is assum
ing the straps has been removed. Howeve.
an analysis of the anode block after alterna
segments had been strapped (fig. 8-12) wi
reveal that the cavities are now connected
parallel. This is due to the strapping. The re
sult is shown in figure 8-15.

Operation

The electric field in the electron resona
oscillator is a product of an a.c. and a d.
field. The d.c. field extends radially betwee
adjacent anode segments by the RF oscillatior
induced in the cavity tank circuits of the anoc
block.

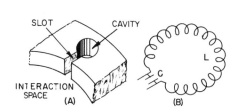

Figure 8-13.— Equivalent circuit of a
hole-and-slot cavity.

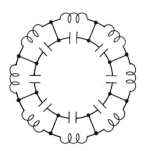

179.712

Figure 8-14.—Cavities connected in series.

142

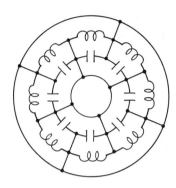

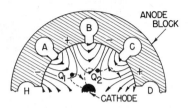

179.714

Figure 8-16. — Probable electron paths in an electron resonant magnetron oscillator — RF oscillations occurring.

179.713

Figure 8-15. — Cavities in parallel due to strapping.

Figure 8-16 shows the a.c. fields between adjacent segments at an instant of maximum magnitude of one alternation in the RF oscillations occurring in the cavities.

The strong d.c. field going from anode to cathode, due to a large negative d.c. voltage pulse applied to the cathode is not shown in figure 8-16, but is assumed to be present (it was shown in fig. 8-3). It is actually this strong d.c. field which causes electrons to accelerate toward the plate after they have been emitted from the cathode. Earlier in this chapter it was pointed out that an electric field went from a positive electrode to a negative electrode. Also, an electron moving against an E-field would be accelerated by the field thus taking energy from the field. An electron would give up energy to a field and slow down if it were moving in the same direction as the field (positive to negative). Oscillations are sustained in a magnetron because electrons gain energy from the d.c. field, and give up this energy to the a.c. fields as they pass through these fields. These electrons are sometimes referred to as WORKING ELECTRONS. However, not all of the electrons give up energy to the a.c. fields. Some electrons may actually take energy from the a.c. fields. This action is undesirable.

In figure 8-16 consider electron 1, which is shown entering the field around the slot entrance to cavity A. The clockwise rotation of the electron path is due to the interaction of the magnetic field around the moving electron with the permanent magnetic field which is assumed to be going into the paper in figure 8-16. The action

of an electron moving in an H-field was explained earlier and illustrated in figure 8-4. Notice that electron 1 which has entered the a.c. field around cavity A is going against this a.c. field. Thus, it will take energy from the a.c. field and be accelerated. The electron will turn more sharply when its velocity increases as was explained earlier. Thus, electron 1 will turn back toward the cathode. When it strikes the cathode, it will give up the energy it received from the a.c. field in the form of heat. This will also force more electrons to leave the cathode and accelerate toward the anode. Electron 2 is, therefore, slowed down by the field and gives up some of its energy to the a.c. field. Since electron 2 looses velocity, the deflective force exerted by the H-field is reduced and the electron path deviates to the left in the direction of the anode, rather than return to the cathode as did electron 1.

The cathode to anode potential and the magnetic field strength (E-field to H-field relationship) determines the time taken by electron 2 to travel from a position in front of cavity B to a position in front of cavity C. Cavity C is equal to approximately one-half cycle of the RF oscillation of the cavities. When electron 2 reaches a position in front of cavity C, the a.c. field of cavity C will be reversed from that shown.

Therefore, electron 2 will give up energy to the a.c. field of cavity C and will slow down more. Electron 2 will actually give up energy to each cavity as it passes and will eventually reach the anode when its energy is expended. Thus electron 2 will have helped sustain oscillations because it has taken energy from the d.c. field and given it to the a.c. fields. Electron 1 which took energy from the a.c. field around cavity A did little harm because it immediately returned to the cathode.

Electrons such as electron 2 which give up energy to the a.c. field as they rotate clockwise from one a.c. field to the next are called working electrons, and stay in the interaction space for a considerable time before striking the anode.

The cummulative action of many electrons being returned to the cathode and directed toward the anode forms a pattern resembling the spokes of a wheel as indicated in figure 8-17. Electrons in the spokes of the wheel are the working electrons.

This overall space charge wheel rotates about the cathode at an angular velocity of two poles (anode segments) per cycle of the a.c. field, and of a phase that enables the concentration to continuously deliver energy to sustain the RF oscillations. Electrons emitted from the area of the cathode between the spokes are, as previously explained, quickly returned to the cathode.

In figure 8-17 it is assumed that alternate segments between cavities are at the same potential at the same instant and that there is an a.c. field existing across each individual cavity. This type of mode of operation is called the PI MODE, since adjacent segments of the anode have a phase difference of 180° or one-pi radian. There are several other possible modes of oscillation.

A magnetron operated in the pi mode will have greater power output. Therefore, the pi mode is the most commonly used mode.

In order to ensure that alternate segments have identical polarities, an even number of cavities, usually six or eight, are used and alternate segments are strapped as pointed out earlier. The frequency of the pi mode is separated from the frequency of the other modes by strapping. Operation in the pi mode is ensured as follows:

For the pi mode, all parts of each strapping ring are at the same potential; but the two rings have alternately opposing potentials as shown in figure 8-18. Stray capacitance between the rings adds capacitive loading to the resonant mode. For other modes, however, there is a phase difference between the successive segments connected to a given strapping ring which causes current to flow in the straps.

The straps contain inductance and an inductive shunt is placed in parallel with the equivalent circuit thereby lowering the inductance and increasing the frequency at modes other than the pi mode.

Coupling Methods

RF energy can be removed from a magnetron by means of a coupling loop. At frequencies lower than 10,000 MHz the coupling loop is made by bending the inner conductor of a coaxial cable into a loop and soldering the end to the outer conductor, so that the loop projects into the cavity as shown in figure 8-19A. To obtain sufficient pickup at higher frequencies the loop is located at the end of the cavity as shown in figure 8-19B.

The segment-fed loop method is shown in figure 8-19C. The loop intercepts the flux passing between cavities. The strap-fed loop method (fig. 8-19D) intercepts the energy between the strap and the segment. On the output side, the coaxial line feeds another coaxial line directly, or it feeds a waveguide through a choke joint, with the

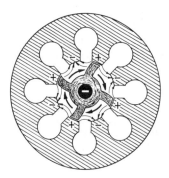

179.715
Figure 8-17. — Rotating space charge wheel in eight cavity magnetron.

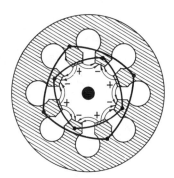

179.716
Figure 8-18. — Alternate segments connected by strapping rings.

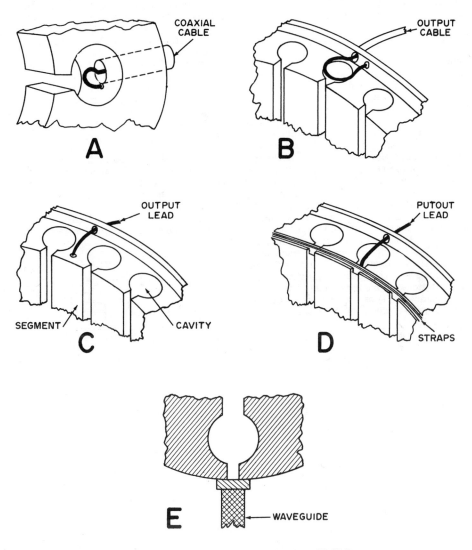

179.717
Figure 8-19.—Magnetron coupling methods.

vacuum seal at the inner conductor helping to support the line.

Aperture, or slot, coupling is illustrated in figure 8-19E. Here, energy is coupled directly to a waveguide through an iris feeding into the waveguide connector through a window.

Tuning

A tunable magnetron permits the system to be operated at a precise frequency anywhere within a band of frequencies, as determined by the magnetrons characteristics.

The resonant frequency of a magnetron may be varied by varying the inductance or capacitance of the resonant cavities. In figure 8-20, an inductive tuning element is inserted into the hole portion of the hole-and-slot cavities. It changes the inductance of the resonant circuits by altering the surface to volume ratio in a high current region. The type of tuner illustrated in figure 8-20 is called a SPROCKET TUNER or CROWN OF THORNS TUNER. All of its tuning elements are attached to a frame which is positioned by means of a flexible bellows arrangement. The insertion of the tuning elements into each anode hole decreases the inductance of the cavity and therefore increases the resonant frequency. One of the limitations of inductive tuning is that it lowers the unloaded Q of the cavities and therefore reduces the efficiency of the tube.

The insertion of an element (ring) into the cavity slot as shown in figure 8-21 increases the slot capacitance and decreases the resonant frequency. Because the gap is narrowed in width the breakdown voltage will be lowered, and capacity tuned magnetrons must be operated with low voltages and hence low powers. The type of capacity tuner illustrated in figure 8-21 is called a COOKIE CUTTER TUNER. It consists of a metal ring inserted between the two rings of a double strapped magnetron, thereby increasing the strap capacitance. Because of the mechanical and voltage breakdown problems associated with the cookie cutter tuner, it is more suited for use at longer wavelengths.

Both the capacitance and inductance tuners described above are symmetrical. Each cavity is affected in the same manner, and the angular symmetry of the pi mode is preserved.

A 10 percent frequency range may be obtained with either of the two tuning methods described above. There is some indication that the cookie cutter is more restricted than the crown of thorns tuner. The two tuning methods may be used in combination to cover a larger

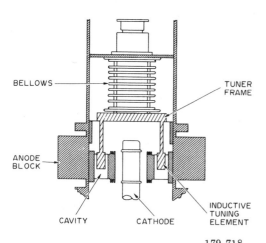

179.718
Figure 8-20. — Inductive magnetron tuning.

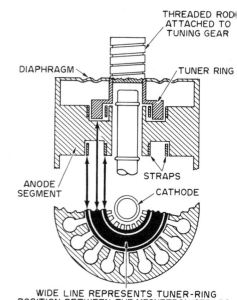

179.719
Figure 8-21. — Capacitive magnetron tuning.

tuning range than is possible with either one alone.

Seasoning

During initial operation a high power magnetron arcs from cathode to plate, and must be properly BROKEN IN or BAKED IN. Actually, arcing in magnetrons is very common. It occurs with a new tube or following long periods of idleness.

One of the prime causes of arcing is the liberation of gas from tube elements during idle periods. Arcing may also be caused by the presence of sharp surfaces within the tube, mode shifting, and by drawing excessive current. While the cathode can withstand considerable arcing for short periods of time, continued arcing will shorten the life of the magnetron and may destroy it entirely. Hence, each time excessive arcing occurs, the tube must be baked in again until the arcing ceases and the tube is stabilized.

The baking in procedure is relatively simple. Magnetron voltage is raised from a low value until arcing occurs several times a second. The voltage is left at that value until arcing dies out. Then the voltage is raised further until arcing again occurs, and is left at that value until the arcing again dies out. Whenever the arcing becomes very violent and resembles a continuous arc, the applied voltage is excessive and should be reduced to permit the magnetron to recover. When normal rated voltage is reached and the magnetron remains stable at the rated current, the baking in is complete. It is good maintenance practice to bake in magnetrons left idle in the equipment, or those used as spares, when long periods of nonoperating time have accumulated.

The preceding information is general in nature. The equipment technical manuals recommended times and procedures should be followed when baking in a specific type magnetron.

KLYSTRONS

The klystron tube is a stable microwave power amplifier which provides high gain at medium efficiency. Depending on the type of tube, klystron power outputs range from a few milliwatts to several megawatts peak power, and over 100 kilowatts average power. Power gains vary from 3 to 90 db.

Klystron amplifiers are somewhat noisy and are therefore used mainly as power amplifiers.

However, they have applications in many facets of microwave transmissions.

KLYSTRON AMPLIFIER OPERATION

The basic theory of a klystron amplifier is quite simple. In fact, the klystron amplification principle may be readily explained with an analogy to a simple triode RF amplifier. Obviously there are some differences (which will be explained), and these differences allow the klystron to amplify at microwave frequencies, whereas a triode will not.

Figure 8-22 shows a simplified schematic of a triode amplifier with resonant circuits at both the input and output. Such resonant circuits restrict the bandwidth of the amplifier and increase the gain. Reviewing, a triode tube consists of three elements: a cathode which emits a stream of electrons, a grid which controls the electron stream, and a plate which attracts the electrons and catches them after they pass through the grid. The grid acts as a valve, opening or closing the current path according to the voltage applied to it. The RF input signal comes to the grid as a weak alternating voltage. This voltage modulates the electron flow through the tube, at the radio frequency. The electron stream then delivers, at the plate, an alternating current which is an amplified reproduction of the input signal.

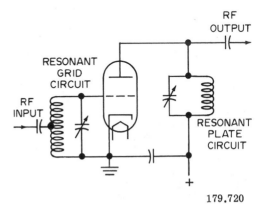

179.720

Figure 8-22. — Simplified schematic of a triode amplifier with resonant circuits at both input and output.

This alternating current flows through the resonant plate circuit and excites alternating voltages across it. These voltages constitute the RF output from the amplifier.

The time it takes electrons to cross the tube is on the order of a billionth of a second. This transit time is short compared to the period of a cycle of a radio wave below the microwave range (approximately a millionth of a second). Hence, the electrons are slowed down or speeded up by the voltage on the grid at a given instant. The flow of electrons, therefore, can follow the voltage fluctuation on the grid. In the case of microwaves, however, the oscillations are so rapid (i.e., the period of a cycle is so short) that the voltage on the grid may go through several complete oscillations while a particular quantity of electrons travel across the tube. In other words, the grid voltage changes too rapidly for the electrons to follow the fluctuation. There are other reasons why the conventional triode tube fails at microwave frequencies, but the most fundamental reason is that the transit time of the electrons is long compared to the period of one cycle of the microwave signal.

The klystron tube makes a virtue of the very thing that defeats the triode—the transit time of the electrons. The klystron modulates the velocity of the electrons, so that as the electrons travel through the tube, electron bunches are formed. These bunches deliver an oscillating current to the output resonant circuit of the klystron. Figure 8-23 shows a cutaway representation of a basic klystron amplifier. The klystron amplifier consists of three separate sections: the electron gun, the RF section, and the collector.

Consider first the electron gun structure. It consists of a heater, cathode, control grid, and anode. Electrons are emitted by the cathode and drawn toward the anode, which is operated at a positive potential with respect to the cathode. The electrons are formed into a narrow beam by either electrostatic or magnetic focusing techniques (not shown). The control grid is used to control the number of electrons which reach the anode region. It may also be used to turn the tube completely on or off in certain pulsed amplifier applications.

The electron beam is well formed by the time it reaches the anode. The beam passes through a hole in the anode and on to the RF section of the tube, and eventually strikes the collector. The electrons are returned to the cathode through an external power supply (not shown in fig. 8-23). It is evident that the collector of a klystron acts

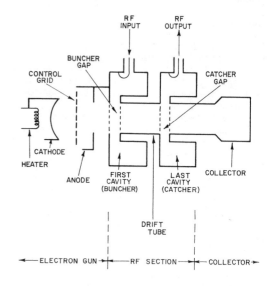

32.82(179)A

Figure 8-23.—Cutaway representation of a basic klystron amplifier.

much like the plate of a triode insofar as the collecting of electrons is concerned. However, there is one important difference. The plate of a triode is normally connected, in some fashion, to the output RF circuit, whereas, in a klystron amplifier, the collector has no connection to the RF circuitry at all.

Consider the RF section of a basic klystron amplifier. This part of the tube is quite different from a conventional triode amplifier. The resonant circuits used in a klystron amplifier are reentrant cavities. The characteristics of this type of cavity have previously been discussed in chapter 7 and will be reviewed only briefly.

Referring to figure 8-23, electrons pass through the cavity gaps in each of the resonators as well as the cylindrical metal tube between the gaps. These metal tubes are called DRIFT TUBES. In a klystron amplifier the low level RF input signal is coupled to the first resonator, which is called the BUNCHER CAVITY. The signal may be coupled through either a waveguide or a coaxial connection. If the cavity is tuned to the frequency of the RF input it will be excited into oscillation. An electric field will exist across the buncher gap, alternating at the input frequency. For half a cycle, the electric field will be in a direction which will cause the field to increase the velocity of electrons flowing through the gap. On the other half-cycle, the field will be in a direction which will cause the field to decrease electron velocity. This effect is called VELOCITY MODULATION, and is illustrated in figure 8-24. (Note that when the voltage across the cavity gap is negative, electrons will decelerate; when the voltage is zero, the electrons will be unaffected; and when the voltage is positive, the electrons will accelerate.)

After leaving the buncher gap (fig. 8-23) the electrons proceed through the drift tube region, toward the collector. In the drift tube region, electrons which have been speeded up by the electric field in the buncher gap will tend to overtake electrons which have been slowed down. Due to this action, bunches of electrons will begin to form in the drift tube region and will be completely formed by the time they reach the gap of the last cavity. The last cavity is called the catcher cavity. Bunches of electrons periodically flow through the gap of this catcher cavity, and during the time between bunches relatively few electrons flow through the gap. The time between arrival of electron bunches is equal to the period of one cycle of the RF input signal.

The initial bunch of electrons flowing through the catcher cavity will cause the cavity to oscillate at its resonant frequency. This sets up an alternating electric field across the catcher cavity gap, as illustrated in figure 8-25.

With proper design and operating potentials, a bunch of electrons will arrive in the catcher cavity gap at the proper time to be retarded by the RF field. Thus, energy will be given up to the catcher cavity. This action is illustrated in figure 8-26.

The RF power in the catcher cavity will be much greater than that applied in the buncher cavity. This is due to the ability of the concentrated bunches of electrons to deliver great amounts of energy to the catcher cavity. Since the electron beam delivers some of its energy to the output cavity, it arrives at the collector with less total energy than it had when it passed through the input cavity. This difference in beam energy is approximately equal to the energy delivered to the output cavity.

It is appropriate to mention here that velocity modulation does not form perfect bunches of electrons. There are some electrons which come through out of phase. These electrons show up in the output cavity gap between bunches. The electric field across the gap at the time these out of phase electrons come through is in a direction to accelerate them. This causes some energy to be taken from the cavity. However, much more energy will be contributed to the output cavity by the concentrated bunches of electrons than will be withdrawn from it by the small number of out of phase electrons.

Multicavity Power Klystron Amplifier

In the above discussion, only a basic two-cavity klystron has been considered. This simple type of klystron amplifier is not capable of high gain, high output power, or suitable efficiency. With the addition of intermediate cavities and other physical modifications the basic two-cavity klystron may be converted to a multicavity power klystron. This amplifier is capable of high gain, high power output, and satisfactory efficiency. Figure 8-27 illustrates a multicavity power klystron amplifier.

In addition to the intermediate cavities, there are several physical differences between the basic and the multicavity klystron. The cathode of the multicavity power klystron must be larger, in order to be capable of emitting large numbers of electrons. The shape of the cathode is usually concave, to aid in focusing the electron beam. The collector must also be larger to allow for greater heat dissipation. In a high power klystron, the electron beam may strike the collector with sufficient energy to cause the emission of x-rays from the collector. Many klystrons have a lead shield around the collector as protection against these x-rays. Most high power klystrons are liquid cooled and must be constructed to facilitate the cooling system. Klystron cooling will be discussed in detail in a subsequent section.

Klystron amplifiers have been built (to the present time) with as many as seven cavities (i.e., with five intermediate cavities). The effect of the intermediate cavities is to improve the bunching process. This results in increased amplifier gain, and to a lesser extent increased efficiency. Adding more intermediate cavities is roughly analogous to adding more stages to an IF amplifier. That is, the overall amplifier gain is increased and the overall bandwidth is reduced, if all the stages are tuned to the same frequency. The same effect occurs with klystron amplifier tuning. A given klystron amplifier tube will deliver high gain and narrow bandwidth if all the cavities are tuned to the same

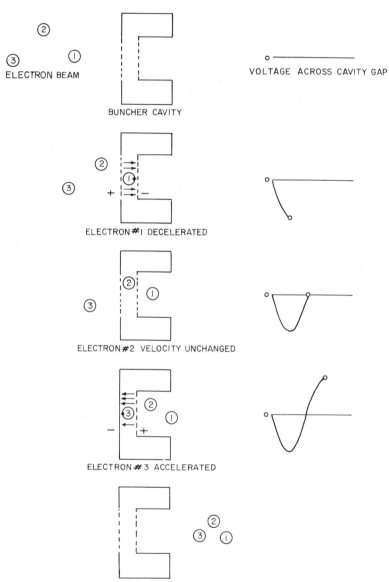

VOLTAGE ACROSS CAVITY GAP

BUNCHER CAVITY

ELECTRON BEAM

ELECTRON #1 DECELERATED

ELECTRON #2 VELOCITY UNCHANGED

ELECTRON #3 ACCELERATED

ELECTRONS BEGINNING TO BUNCH, DUE TO VELOCITY DIFFERENCES

179.721
Figure 8-24. — Velocity modulation.

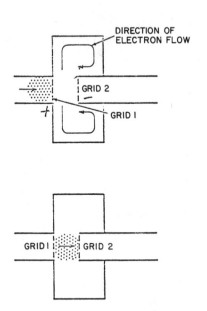

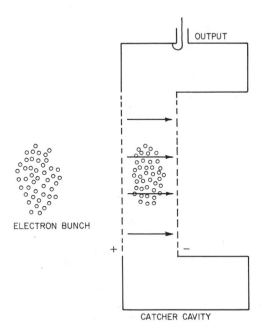

179.723
Figure 8-26. — Electron bunch arriving in catcher cavity gap at the proper time to be retarded in the RF field.

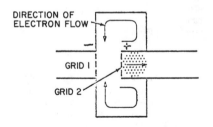

179.722
Figure 8-25. — Catcher cavity gap alternating electric field.

frequency. This is called SYNCHRONOUS TUNING. If the cavities are tuned to slightly different frequencies the gain of the klystron amplifier will be reduced and the bandwidth may be appreciably increased. This is called STAGGER TUNING. Most klystron amplifiers which feature relatively wide bandwidths are stagger tuned.

The klystron is not a perfectly linear amplifier. That is, the RF power output is not linearly related to the RF power input at all operating levels. Another way of stating this is that the

klystron amplifier will saturate, just as a triode amplifier will limit if the input signal becomes too large. In fact, if the RF input is increased to levels above saturation the RF power output will actually decrease. Figure 8-28 shows a plot of typical klystron amplifier performance for various tuning conditions. The RF output is plotted as a function of the RF input. Curve A of figure 8-28 shows typical performance for synchronous tuning. Under these conditions the tube has maximum gain. The power output is almost perfectly linear with respect to the power input, up to about 70 percent of saturation. However, as the RF input is increased beyond that point, the gain decreases and the tube saturates. As the RF input is increased beyond saturation the RF output decreases.

To better understand the reason for this decrease, recall that in the previous discussion, electron bunches were formed by the action of the RF voltage across the buncher cavity gap.

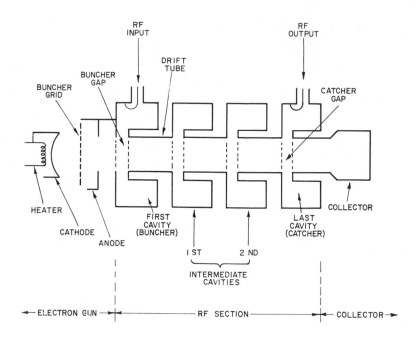

179.724
Figure 8-27. — Multicavity power klystron amplifier.

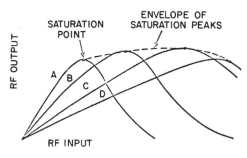

179.725
Figure 8-28. — Plot of typical klystron amplifier performance for various tuning conditions.

This RF voltage accelerated some electrons and slowed down other electrons, resulting in formation of bunches in the drift tube region. Obviously, this speeding up and slowing down effect will be increased as the RF drive power is increased. The saturation point on figure 8-28 is reached when the bunches are most perfectly formed at the instant they reach the output cavity gap. This results in the maximum power output condition. When the RF input is increased beyond this point, the bunches are perfectly formed before they reach the output gap. That is, they form too soon. By the time the bunches have reached the output gap, they tend to debunch because of the mutual repulsion of electrons and because the faster electrons have overtaken and passed slower electrons. This causes the power output to decrease.

If a multicavity klystron power amplifier is synchronously tuned, and the next to last cavity is then tuned to a higher frequency, the gain of the amplifier is reduced, but the saturation power output level may be increased. This effect is shown by curves B and C of figure 8-28. Curve B represents a small amount of detuning of the next to last cavity, and curve C represents even more detuning. Note that the gain of the tube has been reduced, and that the saturation output power is higher than that obtained with synchronous tuning (curve A). Many klystron amplifiers are operated stagger tuned because of the resulting higher output power capability with the same beam power input. This increases the efficiency, provided, of course, that enough RF drive power is available to operate under the stagger tuned condition. Also, as mentioned previously, stagger tuning results in a wider amplifier bandwidth.

As might be expected, stagger tuning may be carried too far, at which point the saturation output power will drop. This is illustrated by curve D of figure 8-28.

Focusing

One very important item which is required for high power klystron amplifier operation is an axial magnetic field (i.e., a magnetic field parallel to the axis of the klystron). In klystron amplifiers which are physically long, it is quite difficult to keep the electron beam properly formed during its travel through the RF section. The mutual repulsion between electrons causes the beam to spread in a direction perpendicular to the axis of the tube. If this is allowed to occur electrons will strike the drift tube and be collected there, rather than passing through the drift tube to the collector.

To overcome beam spreading, an axial magnetic field is used. The action of the magnetic field is to exert a force on the electrons which keeps them focused into a narrow beam. The magnetic field may be developed by a permanent magnet or by one or more electromagnets. A permanent magnet is used on tubes which are physically small or of medium power rating. Unfortunately, the size and weight of a permanent magnet are excessive for long or high power tubes, making it necessary to use electromagnets. In some large tubes, several separate electromagnets are used. The current in each coil is individually adjustable to optimize the magnetic field shape. The magnetic field is normally terminated a short distance beyond the output cavity so that the beam may spread before it hits the collector. This tends to spread the electron beam interception over a large surface

on the collector, minimizing collector cooling problems which would result from the beam remaining concentrated at the time of interception.

Even with an axial magnetic field, some electrons will go astray and not remain in the main electron beam. These electrons will be intercepted by the anode or klystron drift tubes. In high power tubes it is particularly important to minimize the number of stray electrons, because of the heat generated when they strike the drift tubes. In a high power klystron this heating may be a very severe problem because drift tubes are very difficult to cool. Temperatures may become high enough to melt the drift tubes and destroy the tube.

The collector is normally insulated from the RF section of large klystron amplifiers to permit separate metering of the electrons intercepted by the drift tubes and those intercepted by the collector. The electrons intercepted by the RF section are normally called BODY CURRENT, while electrons intercepted by the collector are normally referred to as COLLECTOR CURRENT. Obviously, the sum of body current and collector current is equal to the total current in the electron beam, which is called BEAM CURRENT. Klystron amplifier specifications will quite often place a maximum limit on allowable body current.

In the previous discussion of klystron operation, it was implied that klystron amplifiers normally have actual metal grid structures across the gaps in the resonant cavities. Many low power klystrons do indeed have wire mesh grids. However, most high power klystrons do not have actual grids across the gaps. Such grids would intercept sizable quantities of electrons.

It is very difficult to cool grid structures, and a large amount of beam interception would melt the grids, thus destroying the tube. Fortunately, by proper design the klystron may be made to operate efficiently without actual grid wires across the cavity gaps. The absence of these grids does not change the operating principles discussed previously, but it does have a secondary effect on klystron performance. If the electron beam has a small diameter compared to the diameter of the drift tube, the beam does not couple energy to the cavities very well. Therefore, the performance of a klystron amplifier, which does not have gridded gaps, may sometimes be improved by permitting the electron beam to be as wide as possible, while keeping the body current down to the maximum specified for the tube. The width of the beam may be somewhat controlled by the magnetic field strength.

Body current usually increases with RF input level because it is the RF input which causes the bunches to form. The dense electron concentration in the bunch causes mutual repulsion of electrons, and the diameter of the bunch may become larger than the diameter of the beam with no bunches present. Consequently, some of the electrons in the bunch may be lost to the drift tubes, and the body current may increase.

ASSOCIATED EQUIPMENT

In the preceding sections the basic theory of klystron operation has been discussed. Considerable additional equipment is required for a complete amplifier system. Various power supplies are necessary, to deliver the required voltages and currents. In high power systems, a method of cooling is required. Various RF circuit components are required to control and measure the RF input to the klystron tube and to measure the RF output from the tube. A large collection of meters and protective devices are needed to monitor performance and protect operating personnel and equipment, in the event of a malfunction or operator error. Figure 8-29 is a simplified diagram illustrating some of the power supplies, monitoring devices, and protective devices used in a typical power klystron amplifier.

In most klystron tubes the anode and RF section are connected together inside the vacuum envelope. These parts are normally called the TUBE BODY, and are generally operated at ground potential. It is convenient to operate the tube body at ground potential because the input and output connections (either waveguide or coaxial) are then also at ground potential. This makes it easier to connect the klystron into the rest of the system. In addition, the cavity tuners are at ground potential, eliminating any danger to personnel tuning the tube.

The beam supply shown in figure 8-29 supplies the voltage required to accelerate the electrons and form the beam. It must also deliver the required beam current. The crowbar system quickly discharges the beam supply in the event of an internal klystron arc, or other high voltage fault condition.

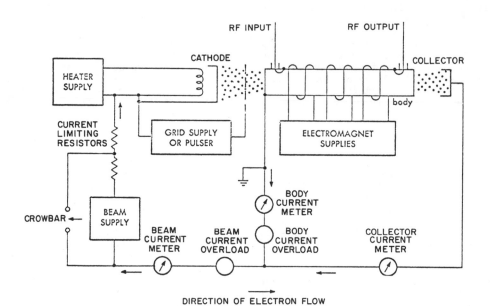

179.726

Figure 8-29. — Associated klystron amplifier equipment.

For high power systems, it is normal to have some value of series resistance between the beam supply and the klystron cathode. This limits tube current to a finite value if the tube should arc from cathode to ground.

Some klystrons have a grid or modulating anode which is used to control the number of electrons in the electron beam. Such grids are often used in pulsed systems to turn the tube full-on or full-off. A few systems employ grid modulation for the transmission of intelligence. In most gridded klystron tubes the grid is never allowed to go positive with respect to the cathode. This might cause undue grid interception of the beam and result in burnout of the grid element. A grid power supply is required in those tubes which have grids. These power supplies and pulsers may take many forms, depending on the system application, and will not be discussed here.

The collector of most high power klystrons is insulated from the body of the tube. This allows separate metering and overload protection for the body current and collector current. In most systems the collector and body operate at nearly the same potential. Any potential difference is normally only the difference in voltage drop across the various metering circuits.

Figure 8-29 shows three electromagnets wrapped around the body of the klystron. Some klystrons are made with the electromagnets physically part of the tube itself. However, in most systems the electromagnets are separate from the tube, and the klystron is inserted into the electromagnet structure. Many modern klystrons have only one electromagnet and therefore require only one power supply. Others may have as many as six separate coils requiring one power supply for each coil. Voltage and current metering is usually supplied for each of the electromagnet power supplies. If an electromagnet power supply should fail, the electron beam would almost certainly spread, and most of the beam current would be intercepted on a small section of the drift tube. In most cases this would cause the drift tube to melt and permanently destroy the tube. Therefore, klystron amplifier equipment normally has under-current protection in each of the electromagnet circuits. When the magnet current falls below a predetermined level the beam supply is automatically turned off to prevent damage to the klystron. Redundant protection is provided by the body current overload circuits which also turn off the beam supply in the event of magnet current failure or misadjustment.

Also shown in figure 8-29 is a method used to monitor body current, collector current, and beam current separately. In many systems separate monitoring of collector current is not done since the collector current and total beam current are almost equal. It is quite unusual, in a relatively high power klystron amplifier system, to allow the body current to exceed 10 percent of the beam current. High body current usually means low efficiency and increases the danger of burning out drift tubes. In very high power klystrons the body current is often limited to 1 or 2 percent of the total beam current. Overcurrent protection is almost always supplied for both body current and beam current. If a klystron arcs internally, the arc will always occur between cathode and anode. When this occurs, the body current immediately becomes excessive, tripping out the body current overload relay. An arc also causes beam current to be much higher than normal, and the beam current overload will also trip out. In fact, almost any high voltage system fault (such as an insulation breakdown) will cause excessive current through the body current meter and overload relay.

Because of the possibility of extremely high currents flowing under fault conditions, the protection of body current and beam current meters presents a somewhat difficult problem. This problem is usually solved by using very high current solid-state rectifiers, back to back, across the meters. In some cases it is necessary to add a small resistance or inductance in series with the meter. Surge capacitors are normally placed across the combination. It is necessary to connect the rectifiers back to back because fault conditions often cause oscillating currents to flow through the meters.

Cooling

Most low power klystron amplifiers are air cooled, while all high power klystrons are liquid cooled. At the present state of the art, air cooling is usable up to about one kilowatt, CW. However there are a few applications where liquid cooling is employed with tubes having an output as low as 10 watts.

The main source of power (and therefore heat) in a klystron amplifier package is the beam power supply. The power generated by the beam supply must go somewhere. Part of it is converted to RF power, while the remainder eventually shows up as heat somewhere in the klystron. Klystron cooling is required to be able to dissipate the entire beam power. This is necessary because, if no RF output is being generated (either due to low RF input power or detuning of the klystron tube) all of the beam power is dissipated as heat somewhere within the tube. As discussed previously, most of the electrons in the beam eventually strike the collector. When they strike the collector their energy is dissipated as heat. The small fraction of the beam lost to the drift tube also generates heat.

Klystron amplifiers are normally between 30 and 50 percent efficient. Therefore, a tube rated at 10-kilowatts output must be designed to dissipate between 23 and 10 kilowatts respectively, depending on its efficiency. A tube rated at 100 kilowatts must be capable of dissipating between 100 and 230 kilowatts depending on efficiency. As may be seen from the above discussion, very advanced cooling techniques are necessary. The power levels involved may melt a hole in the drift tube or collector in a small fraction of a second, if the cooling system fails and adequate protective devices are not provided.

There are other smaller sources of heat in a klystron amplifier system. The heat produced by the heater will be conducted and radiated to the exterior surfaces of the electron gun assembly, and must be dissipated. Large tubes require a blower on the electron gun assembly in order to dissipate this heat. The power generated by the focus coil power supply is all dissipated in the electromagnet(s). Large electromagnets are usually liquid cooled. Should the electromagnet cooling fail, for any reason, the focus coil power supply must be shut off very quickly, or the magnet will burn out. The beam voltage must also be removed (preferably before turning off the focus coil supply) to protect the tube from excessive body current.

During operation, the walls of the resonant cavities have oscillating currents flowing in them. Although these cavities are made of very high conductivity metal, they still present a finite resistance to these oscillating currents. Therefore, heat will be generated in the cavity walls. The amount of heat generated may be quite sizable in high power high frequency tubes. For example, in a 20 kilowatt, CW, X-band klystron amplifier, approximately 1 killowatt of heat is generated by the circulating RF currents in the output cavity. Since the cavity is approximately a one-inch cube, it is apparent that removing the heat is a formidable problem.

Another problem associated with cavity heating is not immediately apparent. The resonant frequency of a cavity depends on its physical size. The cavities are made of metal, which expands as its temperature increases. This effect tends to change the resonant frequency of the cavities and, thereby, detune the tube. As the tube detunes, the power output drops. This, in turn, results in a reduction of the RF heating, allowing the tube to come back into tune. If this problem was not considered in the initial tube design the resulting tube would be unstable in its operation. This situation indeed exists in some tubes which are external cavities. These external cavities are cooled by air rather than by liquid, and the cavity tuning is seriously affected by the ambient air temperature. All high power klystrons are liquid cooled, including the cavities and tuners. The cavities are maintained at a stable temperature by controlling the temperature of the cooling liquid, therefore, thermal detuning is no longer a problem.

Drift tube heating is a serious problem in very high power and in medium power high frequency klystrons. The drift tubes, which are inside the vacuum envelope are physically small, and it is difficult to conduct the drift tube heat into the region outside the vacuum envelope. In some high power tubes it is necessary to bring the cooling liquid inside the vacuum envelope, and around the drift tubes, in order to remove the heat from them.

In some high power high frequency systems it is necessary to cool the output waveguide. An X-band waveguide carrying 5-kilowatts CW, becomes too hot to touch in normal ambient air. Fortunately, waveguides may be cooled easily by soldering copper tubing along the sides of the guide and running cooling liquid through the tubing.

Systems which use blowers for cooling will usually have an air flow switch. If the blower fails, the switch will open and remove power from appropriate power supplies. Systems employing liquid cooling normally distribute the liquid into a large number of paths, since the flow requirements are quite dissimilar. Each of the various paths will have a low flow interlock. If one of the liquid cooling circuits becomes plugged, the low flow interlock will open and remove power from the system. Liquid cooling systems also include pressure gauges and switches, temperature gauges, and over-temperature switches. Many systems have pressure or flow regulators. Some systems include devices which will sound an alarm before trouble actually occurs. In some cases this allows the problem to be corrected without shutting down the equipment.

Some discussion of cooling liquids is appropriate here. Distilled water is the best medium for cooling klystron amplifiers. Some very high power amplifiers specify that only distilled water may be used. Unfortunately, water freezes at an inconveniently high temperature. Many low and medium power klystrons permit the use of ethylene-glycol and water as the cooling liquid. Ethylene-glycol reacts with certain types of metals and hoses which might be used in the system. Therefore, special care must be taken in designing a system which is to use ethylene-glycol. Only non-ferrous metals should be used in a cooling system for a klystron amplifier.

Figure 8-30 illustrates a typical power klystron cooling system. Shown are the many protective devices associated with the complex cooling system. It should be stressed that the technician responsible for maintaining the klystron transmitter must be familiar with the cooling system of his equipment. Figure 8-31 illustrates a pictorial diagram of a high power klystron amplifier, and some of its associated components.

The above discussion should impress the reader with the fact that an expensive klystron amplifier system may be destroyed in a matter of seconds if the cooling system fails. A well designed system uses many protective devices to prevent this from happening.

NOISE IN KLYSTRON AMPLIFIERS

Volumes have been written about noise in microwave systems. Here, only the high points will be covered.

The output of a klystron amplifier contains harmonics. This is primarily because the output

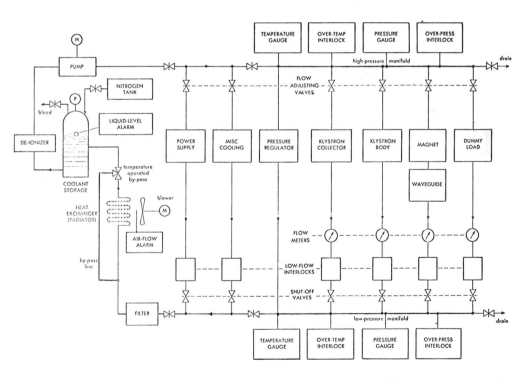

179.727

Figure 8-30. — Typical power klystron cooling system.

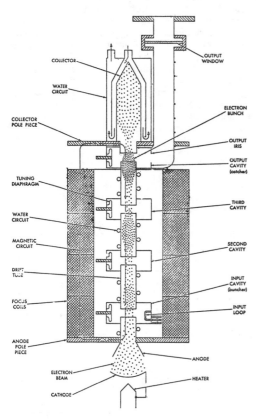

COLLECTOR

WATER
CIRCUIT

COLLECTOR
POLE PIECE

OUTPUT
WINDOW

ELECTRON
BUNCH

OUTPUT
IRIS

OUTPUT
CAVITY
(catcher)

TUNING
DIAPHRAGM

THIRD
CAVITY

WATER
CIRCUIT

MAGNETIC
CIRCUIT

SECOND
CAVITY

DRIFT
TUBE

INPUT
CAVITY
(buncher)

FOCUS
COILS

INPUT
LOOP

ANODE
POLE
PIECE

ANODE

ELECTRON
BEAM

HEATER

CATHODE

32.82(179)B

Figure 8-31.— Pictorial drawing of a high power
klystron amplifier.

cavity is excited by bunches of electrons which
pass through the output gap once every cycle.
Since the driving energy supplied to the output
cavity is not continuous, but rather occurs in
quick pulses, it is evident that the output current
may not be purely sinusoidal. Therefore, the
output will contain harmonic components. This
situation is analogous to that of a class C triode
amplifier in which the plate current flows in
pulses and sets up oscillating currents in the
resonant plate circuit. In general, the harmonic
output of a klystron amplifier is largest when
the tube is operated at or above saturation.
Harmonic content decreases when the tube is
operated below saturation. Harmonics in the

output may be reduced also by the use of harmonic
filters.

Another source of distortion is the non-
linearity of the klystron. If the RF signal is
amplitude modulated, the RF output may not
perfectly follow the RF input. This may result in
distortion, increasing as the tube is driven closer
to saturation on the peaks of the RF input signal.
In general, klystron amplifiers should not be
used to amplify amplitude modulated signals if
the RF output is driven higher than 70 percent
of the saturation level. Between 70 and 100 percent of saturation, considerable distortion may
occur.

A klystron amplifier will generate a certain
amount of white noise, just as in any other
electron tube. White noise occurs primarily
because an electron beam is never perfectly
homogeneous. The amount of electrons will vary
slightly with time, primarily due to shot noise
at the cathode surface. This variation shows up
as random noise in the RF output. A certain
amount of noise may also be generated by electrons striking the drift tubes.

From the above discussion it may be seen
that a klystron amplifier is relatively noisy.
Therefore, power klystrons are not usually used
to amplify weak microwave signals.

KLYSTRON MIXER

In addition to its use as a power amplifier,
a klystron also may be used as a mixer. Heterodyning is accomplished in the electron beam.
Usually the original input frequencies consist
of a high frequency signal and low frequency
signal. The high frequency signal is applied to
the buncher cavity, while the low frequency
signal may be applied either to the cathode
or modulating grid. The buncher cavity is tuned
so as to be resonant to the original high frequency signal and the intermediate and catcher
cavities are tuned to the difference of the two
original input frequencies.

KLYSTRON MODULATION

In a pulsed microwave system using a power
klystron, one of three methods may be used to
accomplish modulation. The first method is to
switch the beam accelerating voltage on and off.
The second is periodic interruption of the RF
input signal. The last method is to turn the
klystron beam current on and off.

When a klystron is pulsed by turning the
accelerating voltage on and off, the entire beam

current must be pulsed as well. This action is similar to modulating a magnetron, and requires a modulator capable of handling the full power of the beam.

When modulation is accomplished by switching of the RF input signal, the beam current must also be pulsed. If this is not done, beam power will be dissipated, to no useful purpose, in the interval between RF input pulses. This reduces the efficiency of the tube.

Of the three methods, pulsing the modulating grid or anode is the most commonly used. For communications use, the klystron is usually modulated by applying the intelligence to the modulating grid.

KLYSTRON ADVANTAGES AND DISADVANTAGES

The chief advantage of a klystron amplifier is that it is capable of high power output along with good stability, efficiency, and gain. Since a klystron is basically a power amplifier it may be driven by a stable oscillator, operating at a lower frequency, followed by a frequency multiplier chain. This arrangement results in more stable operation than is possible with a self-excited power oscillator.

Another advantage of a klystron is that its d.c. and RF sections are separate. This allows the cathode and collector regions to be designed for optimum performance, without concern for their effect on RF fields. As a result, the life of a klystron is increased over other types of microwave power generators.

The chief limitations of klystron amplifiers are their large size, high operating voltages, and the complexity of their associated equipment.

TRAVELING WAVE TUBE (TWT)

The traveling wave tube is a high gain, low noise, wide bandwidth microwave amplifier. TWT's are capable of gains of 40 db or more, with bandwidths of over an octave. (A bandwidth of one octave is one in which the upper frequency is twice the lower frequency.) TWT's have been designed for frequencies as low as 300 MHz and as high as 50 GHz.

The primary use for the TWT is voltage amplification (although high power TWT's, with characteristics similar to those of a power klystron, have been developed). Their wide bandwidth and low noise characteristics make them ideal for use as RF amplifiers in microwave and electronic countermeasures equipment.

CONSTRUCTION

Figure 8-32 is a pictorial diagram of a traveling wave tube. The electron gun produces a stream of electrons which are focused into a narrow beam by an axial magnetic field, much the same as in a klystron tube. The field is produced by a permanent magnet or electromagnet (not shown) which surrounds the helix portion of the tube. The narrow beam is accelerated, as it passes through the helix, by a high potential on the helix and collector. The function of the helix will be discussed later.

OPERATION

Whereas the electron beam in a klystron travels, for the most part, in regions free from RF electric fields, the beam in a TWT is continually interacting with an RF electric field

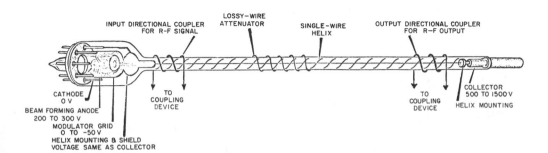

179.728

Figure 8-32. — Pictorial diagram of a traveling wave tube.

propagating along an external circuit surrounding the beam.

To obtain amplification, the TWT must propagate a wave whose phase velocity is nearly synchronous with the d.c. velocity of the electron beam. It is difficult to accelerate the beam to greater than about one-fifth the velocity of light. Therefore, the forward velocity of the RF field propagating along the helix must be reduced to nearly that of the beam.

The phase velocity in a waveguide which is uniform in the direction of propagation is always greater than the velocity of light. However, this velocity may be reduced below the velocity of light by introducing a periodic variation of the circuit in the direction of propagation. The simplest form of variation is obtained by wrapping the circuit in the form of a helix whose pitch is equal to the desired slowing factor.

As previously explained, the electron beam is focused and constrained to flow along the axis of the helix. The longitudinal components of the input signal's RF electric field, along the axis of the helix or slow wave structure, continually interact with the electron beam to provide the gain mechanism of TWT's. This interaction mechanism is pictured in figure 8-33. This figure illustrates the RF electric field of the input signal, propagating along the helix, infringing into the region occupied by the electron beam.

Consider first the case where the electron velocity is exactly synchronous with the circuit phase velocity. Here, the electrons experience a steady d.c. electric force which tends to bunch them around position A, and debunch them around position B. This action is due to the accelerating and decelerating electric fields, and is similar to velocity and density modulation previously discussed. In this case, as many electrons are accelerated as are decelerated; hence, there is no net energy transfer between the beam and the RF electric field. To achieve amplification, the electron beam is adjusted to travel slightly faster than the RF electric field propagating along the helix. The bunching and debunching mechanisms just discussed are still at work, but the bunches now move slightly ahead of the fields on the helix. Under these conditions more electrons are in the decelerating field to the right of A than are in the accelerating field to the right of B. Since more electrons are decelerated than are accelerated, the energy balance is no longer maintained. Thus, energy transfers from the beam to the RF field, and the field grows.

Fields may propagate in either direction along the helix. This leads to the possibility of oscillation due to reflections back along the helix. This tendency is minimized by placing resistive material near the input end of the slow wave structure. This resistance may take the form of a lossy wire attenuator (fig. 8-32) or a graphite coating placed on insulators adjacent to the helix. Such lossy sections completely absorb any backward

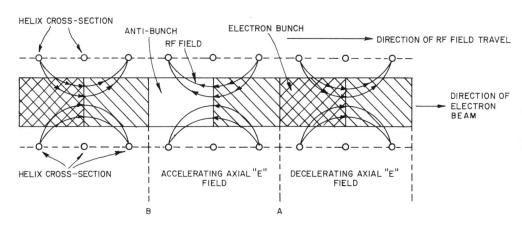

179.729
Figure 8-33.—Helix field interaction.

traveling wave. The forward wave is also absorbed to a great extent, but the signal is carried past the attenuator by the bunches of electrons. These bunches are not affected by the attenuator, and, therefore, reinstitute the signal on the helix, after they have passed the attenuator.

Coupling

Some means must be provided to apply RF energy to one end of the helix and remove it from the other. Four methods of coupling are illustrated in figure 8-34.

Figure 8-34A illustrates waveguide matching. The waveguide is terminated in a nonreflecting impedance, and the helix is inserted into the waveguide, as shown. The efficiency of the system is rather good, but the waveguide has a far higher Q than the traveling wave tube. This means that the broad band characteristics of the TWT suffer

in that the entire bandwidth is not available for amplification. The waveguide will not respond over such a wide spectrum.

The cavity match, illustrated in figure 8-34B, is very similar to the waveguide match. Cavities may be made to resonate over wider ranges than waveguides, but they still have a high Q compared to the TWT. The helix is placed at the mouth of the cavity, thereby absorbing any E-field produced. The RF is fed into the cavity by a coaxial cable.

Figure 8-34C illustrates a direct coax-helix match. It is the simplest system of all. The center conductor of the input coaxial cable is connected directly to the helix. Although this method is used quite frequently, it has a disadvantage. A high VSWR is set up by this match, and this causes heating around the input connection. Since this connection passes through the glass envelope, the envelope is subject to heating

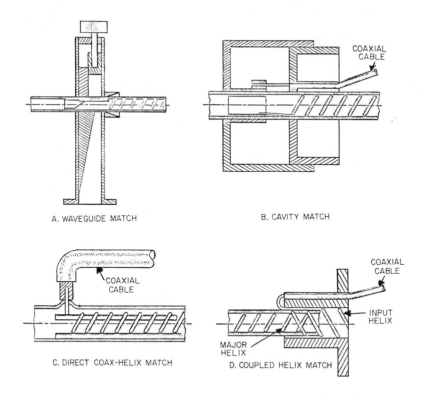

A. WAVEGUIDE MATCH

B. CAVITY MATCH

C. DIRECT COAX-HELIX MATCH

D. COUPLED HELIX MATCH

179.730
Figure 8-34. — TWT RF coupling.

and breaking at this point. However, this is a major problem only in higher power TWT's.

Figure 8-34D illustrates the coupled helix match. In this system, the coaxial center conductor is attached to a small helix. The major helix is inserted within this input helix where it acts as the secondary of a transformer. This system has a good VSWR and is broader in bandwidth than cavities or waveguides, although it is unable to handle large amounts of power.

It should be noted, that any of the above coupling methods may be used for input as well as output coupling.

TWT MIXER

The traveling wave tube has also found application as a microwave mixer. By virtue of its wide bandwidth, the TWT can accommodate the frequencies generated by the heterodyning process (provided of course, that the frequencies have been chosen to be within the range of the tube). The desired frequency is selected by the use of a filter on the output of the helix. A TWT mixer has the added advantage of providing gain as well as simply acting as a mixer.

TWT MODULATION

A TWT may be modulated by applying the modulating signal to a modulator grid. The modulator grid may be used to turn the electron beam on and off, as in pulsed microwave applications, or to control the density of the beam and its ability to transfer energy to the traveling wave. Thus, the grid may be used to amplitude modulate the output.

TWT OSCILLATOR

A forward wave traveling wave tube may be constructed to serve as a microwave oscillator. Physically, a TWT amplifier and oscillator differ in two major ways. The helix of the oscillator is longer than that of the amplifier and there is no input connection to the oscillator.

The operating frequency of a TWT oscillator is determined by the pitch of the tube's helix. The oscillator may be tuned, within limits, by adjusting the operating potentials of the tube.

Operation

The electron beam, passing through the helix, induces an electromagnetic field in the helix. Although initially weak, this field will, through the action previously described, cause bunching of succeeding portions of the electron beam. With the proper potentials applied, the bunches of electrons will reinforce the signal on the helix. This, in turn, increases the bunching of succeeding portions of the electron beam. The signal on the helix is sustained and amplified by this positive feedback resulting from the exchange of energy between electron beam and helix.

GUNN (GALLIUM ARSENIDE) OSCILLATOR

The Gunn oscillator is a solid-state (gallium arsenide crystal) bulk effect source of microwave energy. The discovery that microwaves could be generated by applying a steady voltage across a chip of N type gallium arsenide crystal was made in 1963 by J.B. Gunn. The operation of this device results from the excitation of electrons in the crystal to energy states higher than those they normally occupy.

In a gallium arsenide semiconductor there exist empty electron valance bands, higher than those occupied by electrons. These higher valance bands have the property that electrons occupying them are less mobile under the influence of an electric field than when they are in their normal state at a lower valance band (valance bands were discussed in chapter 3 of volume I).

To simplify the explanation of this effect, assume that electrons in the higher valance band have essentially no mobility. If an electric field is applied to the gallium arsenide semiconductor, the current that flows will increase with an increase in voltage, provided the voltage is low. However, if the voltage is made high enough, it may be possible to excite electrons from their initial band to the higher band where they become immobile. If the rate at which electrons are removed is high enough, the current will decrease even though the electric field is being increased.

If a voltage is applied across an unevenly doped N type gallium arsenide crystal, the crystal will break up into regions with different intensity electric fields across them. In particular, a small domain will form within which the field will be very strong, whereas in the rest of the crystal, outside this domain, the electric field will be weak.

It is not difficult to see that such a domain is unstable. Consider the result of momentarily disturbing the electron density in such a crystal. Suppose there is a sudden increase in electron density at some point in the crystal which tends to reduce the electric field to the left of the disturbance while increasing the electric field to the right. If the material has a positive resistance (an increase in current with an increase in voltage) the decreasing electric field to the left will result in a decreasing current flowing in to the region from the left, and the increasing field to the right will result in an increasing current flowing out of the disturbed region. The excess electrons will drain away. Thus, a disturbance caused by a temporary local increase in electron density will be dissipated as a result of the changing pattern of currents.

The situation is quite different in a negative resistance material. In a negative resistance material the decreasing field to the left of the disturbance will cause an increase in current flowing into the disturbed region, whereas the increase in the field to the right will tend to lower the current outside this region. This current pattern will have the effect of building up the charge disturbance even more; hence, the situation will become unstable and will result in a redistribution of the electric field within the crystal.

This concept may be more familiar in a somewhat different connection. It is possible to obtain negative resistance by accelerating electrons to such a velocity that when they collide with atoms in the system they produce more free electrons. Once this happens, the voltage necessary to produce a given current declines. If voltage is applied across the material, different regions of the material may conduct different quantities of current. In fact, filaments form across the material, each containing a different current. The extreme example of this situation is an electric spark in a gas, which consists of narrow filaments of high current, while the rest of the gas in the region is transporting much smaller currents. The spark is in some sense, a current domain, whereas this discussion concerns electric field domains.

The domains formed in the gallium arsenide crystal will not be stationary, since the electric field acting on the electron energy will cause the domain to move across the crystal. This is illustrated in figure 8-35. The domain will travel across the crystal from one electrode to the other, and as it disappears at the anode a new domain will form near the cathode.

The Gunn oscillator will have a frequency inversely proportional to the time required for a domain to cross the crystal. This time is proportional to the length of the crystal, and to

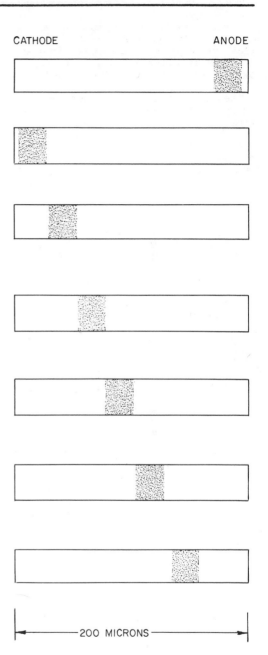

179.731

Figure 8-35. — Gallium arsenide crystal domain.

163

some degree to the potential applied. Each domain results in a pulse of current at the output, hence, the output of the Gunn oscillator is a microwave frequency which is determined, for the most part, by the physical length of the chip.

The Gunn oscillator has delivered power outputs of 65 milliwatts at 2 GHz (continuous operation) and up to 200 watts in pulsed operation. The power output capability of this device is limited by the difficulty of removing heat from the small chip. It is conceivable that much higher power outputs may be achieved by using many wafers of gallium arsenide as a single source.

The advantages of the Gunn oscillator are its small size, ruggedness, low cost of manufacture, lack of vacuum or filaments, and relatively good efficiency. These advantages open a wide range of application for this device in all phases of microwave transmissions. It and other solid-state bulk effect microwave devices are still new, and much research is being carried on in this area. It is conceivable that these devices will rival the conventional electron tube microwave devices in future applications.

CHAPTER 9

MICROWAVE RECEIVERS

The RF echo pulses reflected by a distant object are similar to the transmitted pulses discussed in chapter 8, but are considerably diminished in amplitude. These minute signals are amplified and converted into video pulses by the receiver. A voltage amplification in the order of 10^{10} is required to produce a video pulse of sufficient amplitude to intensify the beam of a crt. The receiver must accomplish this amplification with minimum introduction of noise voltages.

In addition to having high gain and a low noise figure, the receiver must provide a sufficient bandwidth to pass the many harmonics contained in the video pulses. This is to minimize distortion of the pulses. The receiver must also accurately track, or follow, the transmitter in frequency since any amount of drift will diminish the reception of the echo signal. The receiver tuning range need only be equal to that of the transmitter.

BASIC SYSTEM

The superheterodyne receiver is exclusively used in microwave systems. A simple block diagram of a microwave receiver based on the superheterodyne principle is shown in figure 9-1. The echo signal enters the system via the antenna shown in the upper left-hand corner of figure 9-1. It passes through the duplexer (discussed in chapter 10) and is amplified by the low-noise RF amplifier. (TWT's, parametric amplifiers, and masers are representative devices which are used as low noise, high gain RF amplifiers.) When external noise is negligible, the noise generated by the input stage of the receiver largely determines the receiver sensitivity. In early radar receivers and many present day receivers, an RF amplifier is not used and the mixer is the first stage (as indicated by the dashed path in fig. 9-1). The function of the mixer stage, or first detector, is to translate the RF to a lower intermediate frequency, usually 30 or 60 MHz. This is accomplished by heterodyning

the returning RF signal echo with a local oscillator signal in a nonlinear device (mixer) and extracting the signal component at the difference frequency (IF). The purpose of this process is explained in an earlier chapter. It is thus sufficient here merely to state that by using the intermediate frequency, the necessary gain is easier to obtain than by using the higher RF. It is also easier to develop the response function (or bandpass characteristic) of the receiver IF stages. The second detector, which is sometimes an electron tube but more often a crystal rectifier, extracts the video modulation from the carrier. The modulation is amplified in the video stages to a level high enough to operate the indicator or display devices.

One of the requirements of the radar receiver is that its internal noise be kept to a minimum. It is important, therefore, that the input stages of receivers be designed with low noise figures. If the mixer is the first stage, its crystal characteristics will include low conversion loss and a low noise to temperature change ratio. Any noise generated by the local oscillator must be kept out of the mixer stage, either by the insertion of a narrowband filter between the local oscillator and the crystal or by the use of a balanced mixer.

As the bandwidth of the RF portion of the receiver is relatively wide, the frequency-response characteristic of the IF amplifier determines the overall response characteristic of the receiver. It is in the design of the IF portion of the receiver that the response characteristics are accomplished as is also the signal-to-noise ratio.

AUTOMATIC FREQUENCY CONTROL

The AFC system (fig. 9-1) normally employed to keep the receiver in tune with the transmitter is known as a DIFFERENCE-FREQUENCY SYSTEM. In this system a portion of the transmitter

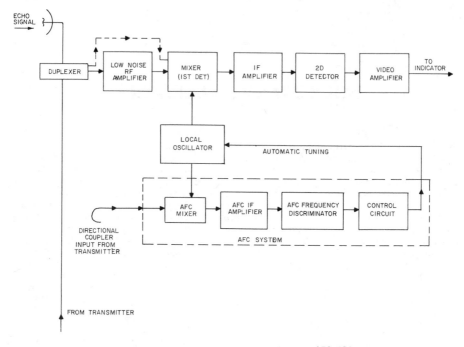

179.732
Figure 9-1.—Microwave receiver based on the
superheterodyne principle.

signal is coupled into the AFC mixer and is heterodyned with the local oscillator signal. If the transmitter and receiver are correctly in tune, the
resultant difference frequency will be at the
correct IF. If the receiver is not in tune with
the transmitter, the difference frequency will not
be correct. Any deviation from the correct IF
signal is detected by the AFC frequency dis-
criminator, which, in turn, generates an error
voltage which magnitude is proportional to the
deviation from the correct IF, and which polarity
determines the direction of the error. The error
voltage corrects the frequency of the local oscillator common to both the receiver mixer and AFC
mixer.

LOCAL OSCILLATOR

In radar, most receivers use 30- or 60-MHz
intermediate frequencies. A highly important factor in receiver operation is the tracking stability
of the local oscillator which generates the fre-

quency that beats with the incoming signal to
produce the IF. For example, if the local oscillator frequency is 3000 MHz, a frequency shift
of as much as 0.1% would be a 3-MHz frequency
shift. This is equal to the bandwidth of most
receivers and would cause a considerable loss
in gain.

In receivers which use crystal mixers, the
power required of the local oscillator is small,
being only 20 to 50 milliwatts in the 4000-MHz
region. Due to the very loose coupling, only about
one milliwatt actually reaches the crystal.

Another requirement of a local oscillator is
that it must be tunable over a range of several
megahertz. This is to compensate for changes in
the transmitted frequencies and in its own frequency. It is desirable that the local oscillator
have the capability of being tuned by varying the
voltage applied to it.

Operation

Because the reflex klystron meets the above requirements, it is used as a local oscillator in microwave receivers. The following deals with its operation. The cavity resonators are treated as parallel resonant circuits. The schematic diagram of a reflex klystron is illustrated in figure 9-2.

Electrons accelerated by the accelerator grid will be velocity modulated as they pass through the cavity grids (grids 2 and 3). The electrons, after passing through the cavity grids, will move at different velocities. Since the repeller plate is made highly negative, the electrons progressing toward it will stop and reverse their direction. The high velocity electrons will come physically closer to the repeller plate than either the medium or low velocity electrons. After repulsion, they will be directed back toward the cavity grids. In the reflex klystron, bunching action occurs on the return trip of the electrons. In fact, bunching occurs immediately before the electrons come under the influence of the RF field about the cavity grids. The distance that the electrons move before they are repelled by the negative repeller plate is a function of the voltage values of the accelerating grid, the d.c. value of the voltage applied to the cavity grids, the d.c. voltage applied to the repeller plate, and the magnitude of the RF voltage coupled to the cavity grids by the cavity resonator. The voltages applied and the physical construction of the klystron should be of such values that the electrons will return to the cavity grids in bunches.

To explain the bunching process a diagram is illustrated in figure 9-3A. This diagram is sometimes referred to as an APPLEGATE DIAGRAM. Figure 9-3B shows a velocity-time diagram of the electrons during transit. The electron at time 3 (center electron) passes the cavity as the RF field (bunching voltage) is zero, and its velocity is unaffected. Therefore, an electron bunch will form about it. The time 1 and 2 electrons pass the cavity with higher velocity, because they pass through the cavity at a time when the RF field (bunching voltage) across the cavity grids accelerates them. Therefore, they penetrate farther into the retarding field, and return to the cavity at essentially the same time as the center electron (fig. 9-3A). Similarly, the electrons at times 4 and 5 leave with lower velocity, penetrate a shorter distance and return at essentially the same instant as the three previous electrons.

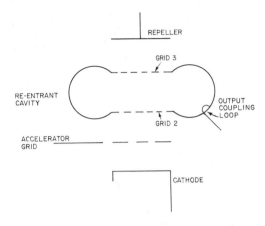

179.733
Figure 9-2.—Reflex klystron schematic.

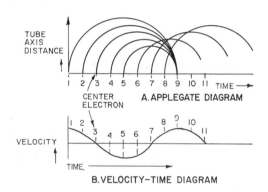

179.734
Figure 9-3.—Bunching process in reflex klystron.

The potential of the cavity grids when the repeller electrons return is important. The bunched electrons should be returned when the potential applied to the cavity grids is such that the energy of the returning bunches will be absorbed. The maximum absorption of energy will occur when the bunched electrons reach the midpoint between the cavity grids in coincidence with the maximum positive peak of RF voltage between these grids. As the electron bunch reaches the midpoint, the grid nearest the repeller plate must be positive in relation to the other buncher grid for correct alignment of the electrostatic field. The electron bunch will be decelerated in this field, thus expending some of it's energy in sustaining RF oscillations within the grid cavity. For example, in figure 9-3B, the center electron (time 3) remains in the repelling field three-quarters of a cycle and the bunch (electrons at time 1, 2, 3, 4, and 5) return at time 9, when the cavity field has a maximum value in the direction to decelerate the returning bunch.

Under these conditions, electrons leaving the cathode will receive maximum acceleration from the cavity field, while returning electron bunches will receive maximum deceleration. If the grids are separated by approximately one-half a wavelength, the electron bunch would pass through the first grid (one nearest the repeller plate) as its RF potential is zero and changing from negative to positive. The electron bunch would pass through the second grid when its potential is zero and is changing from negative to positive. After the returning electron bunches have given their energy to the cavity, they are absorbed by the cavity grid nearest the cathode and are returned to the power supply.

The cavity grids perform a dual function—velocity modulation and that of a catcher grid. The output from the tube is taken by use of the coupling loop shown in the diagram.

By proper adjustment of the negative voltage applied to the repeller plate, the electrons which have passed through the bunching field may be made to pass through the resonator again at the proper time to deliver energy to this circuit. Thus the feedback needed to produce oscillations is obtained and the tube construction is greatly simplified. Spent electrons are removed from the tube by the positive accelerator grid or by the grids of the resonator. The operating frequency of the tube can be varied over a small range by changing the voltage on the repeller plate. This potential determines the transit time of the electrons between their first and second passages through the resonator. However, the output power of the oscillator is affected considerably more than the frequency by changes in the magnitude of the repeller voltage. This is because the output power depends upon the fact that the electrons are bunched at exactly the decelerating half-cycle of oscillating grid voltage. The volume of the resonant cavity is changed to change the oscillator frequency. The repeller voltage may be varied over a narrow range to provide minor adjustments in frequency.

It was mentioned that the electron bunches should arrive at the grids midpoint when the RF swing is at its maximum positive value on the grid closest to the repeller plate. It is not necessary for the electron bunches to return on the first positive half-cycle. They may be returned on the second, third, or fourth positive half-cycles. The positive half-cycle in which the electrons are returned and bunching occurs determines the MODE OF OPERATION.

The mode of operation is determined by the transit time of the electrons. Transit time here means the time between which electrons leave the bunching grids and the time when the bunches deliver their energy to the cavity grids. Figure 9-4 shows the electrons being returned for the different operational modes. For the first mode, the bunching should occur three-fourths of a cycle after the average velocity electrons leave the bunching grids, the second mode of operation occurs one and three-fourths cycles after the average velocity electrons leave the bunching grids, the third after two and three-fourths cycles, and the fourth after three and three-fourths cycles. In practical operation, either the second, third, or fourth modes are used.

Since the mode of operation is determined by the transit time of the electrons, and the transit time is a function of both the accelerator voltage and the repeller plate voltage; the mode of

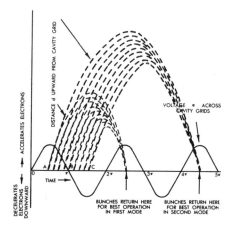

25.193
Figure 9-4.—Modes of operation.

operation is controlled by the repeller plate voltage because the accelerating voltage is a fixed quantity.

The variations of power output and frequency that occur as the repeller voltage of a reflex klystron is changed are particularly important characteristics of the tube. Figure 9-5 illustrates curves that may be obtained if the power output and frequency of a reflex klystron are measured as its repeller voltage is varied from zero to a large negative value, accelerating voltage and load remaining fixed. Oscillations occur only for certain ranges of the reflector voltage (corresponding to the voltage modes) for which the electron bunches return to the cavity in the proper phase to deliver energy to the cavity.

The center points of the modes, labeled A, B. and C, in figure 9-5, correspond to reflector voltages for which the time spent by electrons in the retarding field is correct. That is, an integral number of cycles plus three-quarters of a cycle. At these points the oscillation frequency is the resonant frequency of the cavity, and the power output is the maximum power of that mode. Note that the power outputs for the various modes at the resonant frequency are not the same and that the output is least in the highest mode. This can be explained by examining the factors which limit the amplitude of oscillations and which, in turn, limit the power output.

Power and amplitude limitations are due to overbunching as well as the usual losses in the oscillatory circuit. Overbunching occurs as oscillations build up and the bunching voltage becomes greater, thus increasing the amount of acceleration and deceleration. This causes bunching to occur in a shorter period of time (before the electrons reach the grids on the return trip) which tends to reduce the magnitude of oscillations. In the higher modes of oscillations where the bunches are formed more slowly, the electrons are more susceptible to overbunching.

As shown in figure 9-5, the frequency of oscillations in a reflex velocity-modulated tube is variable to a limited degree in any of the modes of operation by varying the repeller voltage. When the repeller voltage is varied, it causes a bunch to return either a little sooner or a little later than normal. Off resonance, the amplitude of oscillations decreases by an amount depending on the Q of the cavity. In this tube the tuning range is small in comparison with the frequency of oscillations and varies somewhat from one mode to another. It is greatest in the highest mode, because bunching and debunching take place at a slower rate and because greater variation from the ideal time of return is possible without debunching, which would cause the amplitude of oscillations to drop below the usable output level.

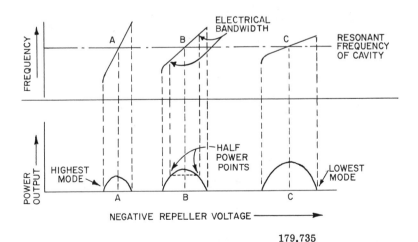

179.735

Figure 9-5.—Power and frequency characteristics
of reflex klystron.

169

Another way to look at this is to consider that in the highest mode, the time required by an electron leaving the grids to return is greater, and the change in period and its accompanying change in frequency occur in a shorter period. To illustrate, in the third mode the interval before return must be about two and three-fourths cycles. A small change in the period of the bunching voltage would therefore by only 3/11 as great a portion of the interval as it would if operation were in the first mode where the ideal time interval is three-fourths of a cycle.

The band of frequencies which can be obtained by varying the repeller voltage lies between the half power points shown in figure 9-5. This range of frequencies is known as the ELECTRICAL BANDWIDTH. The power output curve of the bandwidth is unsymmetrical for the lower order voltage mode. This results from the fact that if the repeller voltage is increased, not only does the bunching voltage decrease and cause bunches to form at a later time, but the repeller voltage causes a quicker return. The effects of the two actions combine to cause poor bunching at the the electrons return, resulting in a rapid drop in output on the high side of the hump. At lower voltages, however, even though the bunching voltage decreases and causes slower bunching, the decreased repeller voltage causes a later return to the grids. In this way, the two effects are counteracting and a greater change in repeller voltage is possible before the output drops below the usable level. The asymmetry is not noticeable in the higher order modes because the percentage change in bunching that can occur in a higher order mode is negligible.

As the local oscillator in a microwave receiver, a reflex klystron need not supply large amounts of power but should oscillate at a frequency that is relatively stable and easily controlled.

The need for a wide electronic tuning range suggests the use of a voltage mode of a high order. However, if a mode of excessively high order is selected, the power available is too small for local oscillator applications, and a compromise between wide range and power is necessary. Use of a very high order mode is undesirable also because the noise output of a reflex klystron is essentially the same for all voltage modes. Thus, the closer coupling to the mixer required with high order, low power modes increases the receiver noise figure. Usually the 1-3/4 or 2-3/4 voltage mode is found suitable. Since the modes are nonsymmetrical, the point of operation is usually a little below the

resonant frequency of the cavity. This makes possible tuning above the operating frequency to a greater degree than if the precise resonant frequency was used.

In practice, the reflex klystron is used in conjunction with an automatic frequency control circuit. Since, the repeller voltage is effective in making small changes in frequency, the AFC circuit is used to control the repeller voltage to maintain the correct intermediate frequency. It should be noted that the coarse frequency of oscillation is determined by the dimensions of the cavity and there is, on most reflex klystrons, a coarse frequency adjustment which varies the cavity size.

Tubes

Table 9-1 gives some of the operating characteristics of some representative reflex velocity-modulated tubes. The data in this table will give you an idea of the order of magnitude of tube quantities. As can be seen, there is a wide variation between different tubes and different conditions of operation.

The K417 reflex klystron is one of the earlier types that was used for 10-cm. operation. One feature of this tube was that in its early application it did not have provision for controlling the frequency through a change in the repeller voltage since both coarse and fine frequency controls changed the cavity grid spacing.

Another 10-cm. tube is the 707A (McNally) tube shown in figure 9-6. In it the cavities are external to the tube and are not evacuated. This makes them susceptible to changes in temperature which results in changes in frequency. To get good frequency stability it is necessary to control the cavity temperature. The coarse frequency control consists of plugs which, when screwed into or out of the cavity, change its size. Fine frequency is controlled by a variable repeller plate voltage.

The Shepherd-Pierce tube (fig. 9-7) is an all metal tube which is available for both 10-cm. (726A) and 3-cm. (723A) operation. In it the cavities are located inside the tube. Mechanical coarse tuning is accomplished by varying the size of the cavities. The cavity size is varied by screw adjusting the tuning struts. The repeller voltage control also serves as the fine frequency control. The 10-cm. and 3-cm. type Shepherd-Pierce tubes differ in the shape of the cavity and in the method of coupling the output. The one illustrated employs a coaxial output lead.

Table 9-1.—Representative Reflex Klystrons Operating Characteristics

Type No.	Frequency (MHz)	Accelerating voltage, volts	Beam current, ma	Repeller voltage, volts	Power output, mw	Electronic tuning range, (MHz)
K417	3000	300-600	5-30	+50 to −500	150	5
707A	3000	250-325	25-35	0 to −250	75	30
726A	3000	300	22	−20 to −300	100	20
723A	9400	300	18-25	−20 to −300	20	45

179.736

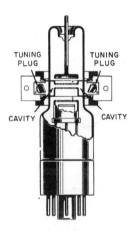

179.737
Figure 9-6.—707A (McNally) external cavity reflex klystron.

MICROWAVE RECEIVER MIXERS

Many microwave receivers do not employ RF amplifiers. They simply use a crystal mixer stage as the receiver front end. A crystal is used rather than an electron tube diode because at microwave frequencies the tube would generate excessive noise. Most electron tubes would also be limited by the effects of transit time.

A type of crystal commonly used is the POINT CONTACT SILICON CRYSTAL DIODE. Unlike the

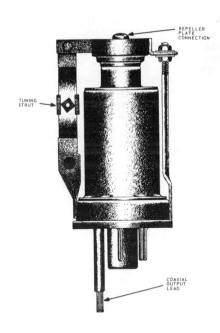

179.738
Figure 9-7.—Shepherd-Pierce internal cavity reflex klystron.

junction diode, the point contact diode depends on the pressure of contact between a point and a semiconductor crystal for its operation. Figure 9-8A and B illustrates a cutaway view of a point contact diode. One section of the diode consists of a small rectangular crystal on N type silicon. A fine berylium-copper, bronze-phosphor, or tungsten wire called the CATWHISKER presses against the crystal and forms the other part of the diode. During the manufacture of the point contact diode, a relatively large current is passed from the catwhisker to the silicon crystal. The result of this large current is the formation of a small region of P material around the crystal in the vicinity of the point contact. Thus, there is a PN junction formed which behaves in the same way as the PN junctions previously described in chapter 5 of volume I.

The reason for using the pointed wire instead of a flat metal plate is to produce a high intensity electric field at the point contact without using a large external source voltage. It is not possible to apply large voltages across the average semiconductor because of excessive heating.

The end of the catwhisker is one of the terminals of the diode. It has a low resistance contact to the external circuit. A flat metal plate on which the crystal is mounted forms the lower contact of the diode with the external circuit. Both contacts with the external circuit are low resistance contacts.

The characteristics of the point contact diode under forward and reverse bias are somewhat different from those of the junction diode. With forward bias, the resistance of the point contact diode is higher than that of the junction diode. With reverse bias, the current flow through a point contact diode is not as independent of the voltage applied to the crystal as it is in the junction diode. The point contact diode has an advantage over the junction diode in that the capacitance between the catwhisker and the crystal is less than the capacitance between the two sides of the junction diode. As such, the capacitive reactance existing across the point contact diode is higher and the capacitive current that will flow in the circuit at high frequencies is smaller. A cutaway view of the entire point contact diode is shown in figure 9-8C. The schematic symbol of a point contact diode is shown in figure 9-8D.

The simplest type of radar mixer is the SINGLE ENDED or UNBALANCED CRYSTAL MIXER. This type of mixer is illustrated in figure 9-9. The mixer illustrated uses a tuned section of coaxial transmission line one-half

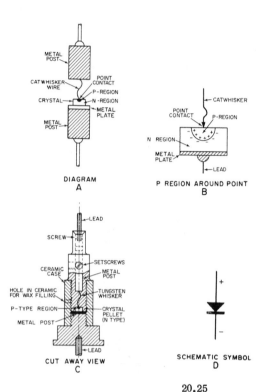

DIAGRAM A

P REGION AROUND POINT B

CUT AWAY VIEW C

SCHEMATIC SYMBOL D

20.25

Figure 9-8.—Point contact diode.

wavelength long which matches the crystal to the signal echo and the local oscillator (LO) inputs. Local oscillator injector is accomplished by means of a probe, while the signal is injected by means of a slot in the coaxial assembly. This slot would normally be inserted in the duplexer waveguide assembly (to be discussed in chapter 10) and properly oriented to provide coupling of the returned signal. In this application, the unwanted signals at the output of the mixer (the carrier, local oscillator, and the sum of these two signals) are effectively eliminated by a resonant circuit tuned to the intermediate, or difference, frequency. One advantage of the unbalanced crystal mixer is its simplicity. It has, however,

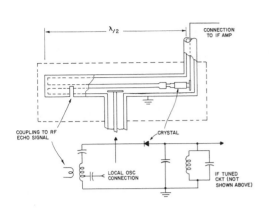

179.739
Figure 9-9.—Single ended crystal mixer.

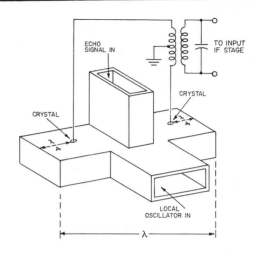

67.31
Figure 9-10.—Balanced hybrid crystal mixer.

one major disadvantage. That is its inability to cancel local oscillator noise. Recall that a klystron generates a high degree of noise. Difficulty in detecting weak signals will exist should noise be allowed to pass through the mixer along with the signal.

One type of mixer which cancels local oscillator noise is the BALANCED, or HYBRID, MIXER (sometimes called the MAGIC TEE). Figure 9-10 illustrates this type of mixer. In hybrid mixers, crystals are inserted directly into the waveguide. The crystals are located one-quarter wavelength from their respective short-circuited waveguide ends. This is a point of maximum voltage along a tuned line. The crystals are also connected to a balanced transformer, the secondary of which is tuned to the desired IF. The local oscillator signal is introduced into the waveguide local oscillator arm and distributes itself as shown in figure 9-11A. Observe that the LO signal is in phase across the crystals. The echo signal is introduced into the echo signal arm of the waveguide and is out of phase across the crystals, as illustrated in figure 9-11B. The resulting fields are illustrated in figure 9-11C.

Since there is a difference in phase between echo signals applied across the two crystals, and because the signal applied to the crystals from the LO is in phase, there will be a condition when both signals applied to crystal # 1 will be in phase, and the signals applied to crystal # 2 will be out of phase. This means that an IF signal of one polarity will be produced across crystal # 1 and an IF signal of the opposite polarity will be produced

across crystal # 2. When these two signals are applied to the balanced output transformer (fig. 9-10) they will add. Outputs of the same polarity will cancel across the balanced transformer.

It is this action which eliminates the LO noise. Noise components which are introduced from the LO are in phase across the crystals and are therefore cancelled in the balanced transformer. It is necessary that the RF admittances of the crystals be nearly equal or the LO noise will not completely cancel. It should be noted that only the noise produced by the LO is canceled. Noise arriving with the echo signal is not affected.

IF AMPLIFIERS

The IF section of a microwave receiver determines the receivers gain, signal-to-noise ratio, and effective bandwidth. The IF amplifier stages must have sufficient gain and dynamic range to accommodate the expected variation of echo signal power. They must have a low noise figure and a bandpass wide enough to accommodate the range of frequencies associated with the echo pulse.

Input Stage

The most critical stage of a microwave receiver's IF section is the input or first stage. Upon the excellence of this stage, depends the noise figure of the receiver and the performance

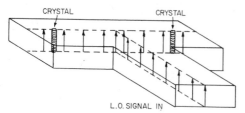

A. WAVEFORM AND LOCAL OSCILLATOR ARM

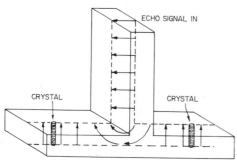

B. WAVEGUIDE AND ECHO SIGNAL ARM

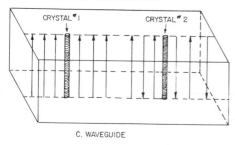

C. WAVEGUIDE

179.740

Figure 9-11.—Balanced mixer fields.

of the entire receiving system with respect to detection of small objects at long ranges. Not only must gain and bandwidth be considered in the design of the first IF stage, but also, and perhaps of more importance, noise generation in this stage must be low. Noise generated in the input IF stage will be amplified by succeeding stages and may exceed the echo signal in strength.

A circuit configuration used to satisfy most requirements of the IF input stage, is a grounded-cathode triode followed by a grounded-grid triode.

This configuration is illustrated in figure 9-12 and is called a CASCODE AMPLIFIER.

Both triode stages form one cascode amplifier stage. Triodes are used instead of pentodes since triodes generate less tube noise. The noise figure of the cascode amplifier is approximately equal to the noise figure of the first triode stage, V1, since V2 is a grounded grid amplifier and contributes very little noise. V2 loads V1 to the point where the first-stage gain is essentially unity. In addition, this enhances the stability of the circuit since, with such low gain the first stage will not oscillate and need not be neutralized. V2, a grounded grid amplifier, does not require neutralization. The gain of the stage is primarily that of V2.

L1, L2, L3 and L4 are tuned so that, with their associate stray capacitances, they resonate at the desired midband frequency. The resonant circuit composed of L3 and its associated stray capacitance provides a d.c. path for V2 while representing a high impedance to the RF. A cascode amplifier provides the stability, gain, and bandwidth of a pentode and the low noise figure of a triode.

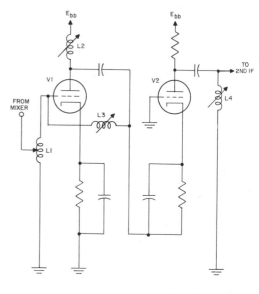

179.741

Figure 9-12.—IF input stage.

Second and Final
IF Stages

IF stages succeeding the first stage may use pentodes to achieve higher gain. This is possible because the signal level has been sufficiently increased by the low noise input stage so as to preclude problems caused by noise generation of pentodes. A commonly used IF circuit is the single tuned amplifier illustrated in figure 9-13. Note that each stage includes only one tuning adjustment. Inductance L is varied until resonance between it and the total shunt capacitance of the stage occurs at the desired IF.

As was previously stated, the IF stages require a wide bandwidth to accommodate the many frequencies which form the echo pulse. Insufficient bandwidth results in what is known as TRANSIENT DISTORTION. Transient distortion is the inability of the stages to amplify transients linearly. One type of transient distortion of a pulse is illustrated in figure 9-14. Transient distortion may result

in ambiguities in the range of the target, due to the nonlinear rise of the leading edge of the reproduced echo pulse.

The cascading of amplifier stages to achieve the high gain required in microwave If amplifiers results in an overall bandwidth reduction. To compensate for this effect, the bandwidth of separate stages must be increased. This may be accomplished by several methods. One method is stagger tuning. Stagger tuning was discussed in detail in chapter 17 of volume I. Briefly reviewing, the resonant frequencies of the various stages combine so that together they pass the frequency band to be amplified. The product of each stage's amplitude response curve forms the overall response curve. A response curve for stagger tuning is illustrated in figure 9-15.

Another method of increasing individual stage bandwidth is by using low Q tank circuits. This is possible since the stage bandwidth depends primarily on the tank circuit Q. A resistor called a SWAMPING RESISTOR is placed across the tank

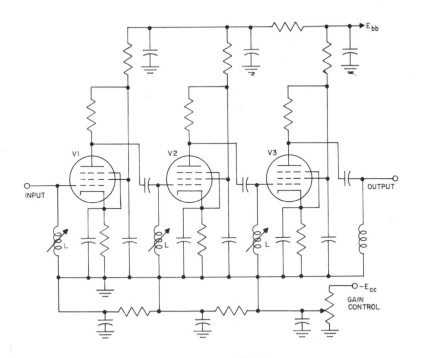

179.742
Figure 9-13.—IF strip.

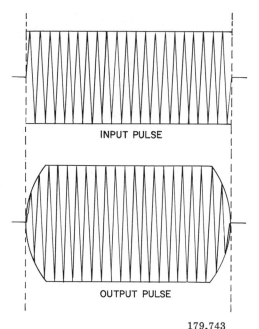

INPUT PULSE

OUTPUT PULSE

179.743
Figure 9-14.—Transient distortion.

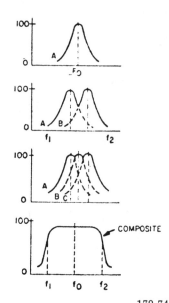

COMPOSITE

179.744
Figure 9-15.—Stagger tuning response curve.

and thereby reduces the circuit Q. The reduction in Q also results in a reduction in gain and additional stages may be required to achieve the necessary amplification. This method was also covered in chapter 17 of volume I.

A third method of increasing bandwidth, discussed in chapter 17 of volume I, is double-tuned, stagger-tuned coupling. The response characteristic of such a double-tuned stage is dependent upon the degree of coupling. Amplitude response curves for varying degrees of coupling are illustrated in figure 9-16. The degree of coupling usually employed is a form of over-coupling. Overcoupling occasionally produces oscillations in the vicinity of the pulse edges. Thus, if all stages of the IF amplifier section are double stagger-tuned, coupling greater than transition (optimum) coupling should be avoided. An amplifier comprising only double stagger-tuned stages is difficult to adjust. Therefore, such stages are ordinarily used in combination with a greater number of single tuned stages. In such an amplifier, the double tuned stages may be overcoupled. The combination of double tuned and single tuned response curves yields a flat overall characteristic response.

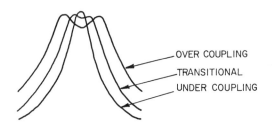

OVER COUPLING

TRANSITIONAL

UNDER COUPLING

20.73.2(179)
Figure 9-16.—Double-tuned stagger-tuned response curves.

The gain of an IF amplifier is high. Because of this high gain, caution must be used to prevent the least amount of regenerative feedback. A small value of voltage in the proper phase fed back to a previous IF stage will cause the amplifier to break into oscillations. Because of this, great care is taken to minimize the possibility of positive feedback. Extensive decoupling in the plate and heater circuits of microwave IF amplifiers is necessary and generally provided. IF amplifiers in microwave receivers are usually isolated and well shielded to prevent any undesired feedback or coupling.

The tuning of an IF amplifier section is generally critical. Whenever required, it should be done in accordance with the instructions contained in the manual for the particular equipment concerned, and care should be exercised.

DETECTOR

The detector in a microwave receiver serves to convert the IF pulses into video pulses. After amplification, these are applied to the indicator. The simplest form of detector, and that one most commonly used in microwave receivers is the diode detector.

A diode detector circuit is illustrated in figure 9-17A. The secondary of T1 and C1 form a tuned circuit which is resonant at the intermediate frequency. Should an echo pulse of sufficient amplitude be received, the voltage, e_i, developed across the tuned circuit is an IF pulse. The amplitude of this pulse is in the order of several volts. Its shape is indicated by the dashed line in figure 9-17B. Positive excursions of e_i cause no current to flow through the diode. However, negative excursions result in a flow of diode current and a subsequent negative voltage, e_o, to be developed across R1 and C2. Between peak negative voltage excursions of the e_i wave, capacitor C2 discharges through R1. Thus, the e_o waveform is a negative video pulse with sloping edges and superimposed IF ripple, as indicated by the solid line in figure 9-17B. Negative polarity of the output pulse is ordinarily preferred, but a positive pulse may be obtained by reversing the connections of the diode. Inductance L1 in figure 9-17A in combination with wiring capacitance and C2 forms a low-pass filter. This filter attenuates the IF components in the e_o waveform but results in a minimum loss of video high frequency components.

VIDEO AMPLIFIERS

The video amplifier receives the pulses from the detector and amplifies these pulses for application to the indicating device. A video amplifier is fundamentally an RC coupled amplifier which uses high gain transistors or pentodes. However, a video amplifier must be capable of a relatively wide frequency response. Stray and interelectrode capacitances reduce the high frequency response of an amplifier, and the reactance of the coupling

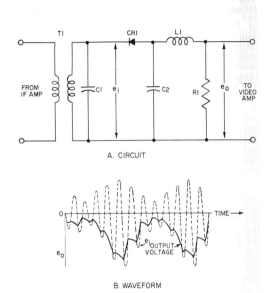

A. CIRCUIT

B. WAVEFORM

179.745
Figure 9-17.—Typical diode detector.

capacitor diminishes the low frequency response. To overcome these problems, a video amplifier utilizes frequency compensation networks as discussed in chapters 6 and 7 of volume I for transistors and electron tubes respectively. All or any one of the types of compensation may be used in a video amplifier stage.

INDICATOR

After the video pulse has been amplified, it must be supplied to the indicator. The indicator may be some distance from the radar receiver. The output of video amplifiers must be transmitted, by coaxial cable, to the indicator. To accomplish this with a maximum transfer of power requires that the output impedance of the video amplifier be matched to the impedance of the coaxial cable. As the characteristic impedance of coaxial cable is relatively low, a cathode follower is used to accomplish the impedance match.

If the video signal must be transmitted a considerable distance, there may be a significant cable attenuation loss. To overcome such line losses, indicators contain additional video amplifiers.

177

MOVING TARGET INDICATOR
(MTI) SYSTEM

The moving target indicator (MTI) system effectively cancels clutter caused by fixed target signals, and thus reveals moving target signals. The unwanted echoes can consist of ground clutter (echoes from surrounding objects on the ground), sea clutter (echoes from the irregular surface of the sea), or even echoes from storm clouds. The problem is to find the desired echo in the midst of the clutter. To do this, the MTI system must be able to distinguish between fixed and moving targets and then eliminate only the fixed targets. This is accomplished by phase detection and pulse-to-pulse comparison.

Target echo signals from fixed objects have a definite fixed phase relationship from one receiving period to the next. Moving objects produce echo signals that have a different phase relationship from one receiving period to the next. It is on this principle that the MTI system operates to discriminate between fixed and moving targets.

Elimination of fixed targets is obtained by delaying the signals received from each transmitted pulse for a period of time exactly equal to the pulse recurrence time, and then combining the delayed signals with the signals received from the next transmitted pulse in such a manner that the amplitudes subtract from each other. Since the fixed targets give the same amplitude on each successive pulse, they will be eliminated. But the moving target signals are, in general, of different amplitude on each successive pulse so that the signals do not completely cancel but leave a small signal that can be amplified and presented on the indicators.

In figure 9-18, 30-megahertz signals from the signal mixer are applied to the 30-megahertz amplifier, where they are amplified, limited, and fed to the phase detector. Another 30-megahertz signal, obtained from the coherent oscillator (coho) mixer, is applied as a lock pulse to the coho. The coho lock pulse is originated by the transmitted pulse and is used to synchronize the coho to a fixed phase relationship with the transmitted frequency at each transmitted pulse. The 30-megahertz CW reference signal output of the coho is applied, together with the 30-megahertz echo signal, to the phase detector.

The phase detector produces a video signal whose amplitude is determined by the phase difference between the coho reference signal and the IF echo signals. This phase difference is the same as that between the actual transmitted pulse

and its echo. The resultant video signal may be either positive or negative. This video output, called coherent video, is applied to the 14-megahertz carrier oscillator.

The 14-megahertz CW carrier frequency is amplitude modulated by the phase-detected coherent video. The modulated signal is amplified and applied to two channels. One channel delays the 14-megahertz signal for a period equal to the time between transmitted pulses. The signal is then amplified and detected. The delay required (the period between transmitted pulses) is obtained by using a mercury delay line or a fused quartz delay line, which operates ultrasonically at 14 megahertz.

The signal to the other channel is amplified and detected with no delay introduced. This channel includes an attenuating network that introduces the same amount of attenuation as does the delay line in the delayed video channel. The resulting nondelayed video signal is combined in opposite polarity with the video signal resulting from the preceding echo signal, that is, the delayed signal. The amplitude difference at the comparison point, if any, between the two video signals is amplified, and since the signal is bipolar, it is made unipolar. The resultant video signal, which represents only moving targets, is sent to the indicator system for display.

An analysis of the MTI system operation just described shows that signals from fixed targets, which have an unchanging phase relationship to their respective transmitted pulses, produce in the phase detector recurring video signals of the same amplitude and polarity. Thus, when one video pulse is combined with the preceding pulse of opposite polarity, the video signals cancel, and no information is passed on to the indicator system.

Signals from moving targets, however, will have a varying phase relationship with the transmitted pulse. As a result, the signals from adjacent receiving periods produce signals of different amplitudes in the phase detector. When such signals are combined, the difference in signal amplitude provides a video signal that is sent to the indicator system for display.

The timing circuits, shown in figure 9-18, are used to accurately control the transmitter pulse recurrence frequency to insure that the pulse repetition period remains constant from pulse to pulse. This is necessary, of course, for the pulses arriving at the comparison point to coincide in time and thus achieve cancellation of fixed targets.

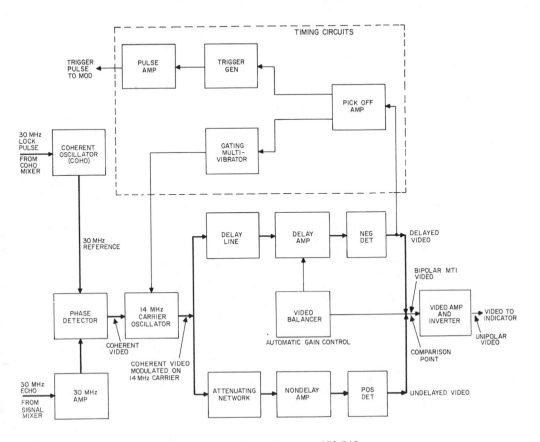

179.746

Figure 9-18.—MTI block diagram.

As shown in figure 9-18, a feedback loop is used from the output of the delay channel, through the pickoff amplifier, to the trigger generator and gating multivibrator circuits. The leading edge of the square wave produced by the detected carrier wave in the delayed video channel is differentiated at the pickoff amplifier and used to activate the trigger generator and gating multivibrator. The trigger generator sends an amplified trigger pulse to the modulator, causing the set to transmit.

The gating multivibrator is also triggered by the negative spike from the differentiated square wave. This stage applies a 2400-microsecond negative gate to the 14-MHz carrier oscillator. Thus, the 14-megahertz oscillator operates for 2400 microseconds and then is cut off. Since the delay line time is 2500 microseconds, the 14-megahertz oscillations stop before the initial waves reach the end of the delay line. This wave train, when detected and differentiated, turns the gating multivibrator on, producing another 2400-microsecond wave train. The 100 microseconds of the delay period when no signals are applied to the delay line is necessary in order that the mechanical waves within the line have time to damp out before the next pulse repetition period. In this manner the pulse repetition period of the set is controlled by the delay of the mercury or quartz delay line. Since this delay line is also common to the video pulses going to the comparison point, the delayed and the undelayed video pulses will arrive at exactly the same time.

179

CHAPTER 10

DUPLEXERS AND ANTENNAS

Whenever a single antenna is used for both transmitting and receiving, as in a radar system, there arises the problem of ensuring that maximum use is made of the available energy. The simplest solution is to use a mechanical switch to transfer the antenna connection from the receiver to the transmitter during the transmitted pulse and back to the receiver during the return (echo) pulse. However, the high pulse repetition rate of radar systems makes the use of a mechanical switch impossible. Therefore, electronic switches, commonly called TR (transmit-receive) SWITCHES, are used. The device which performs the switching is known as a DUPLEXER. The duplexer is the subject of the first part of this chapter.

DUPLEXER

When selecting a switch for the task of switching the antenna from transmitter to receiver, it must be remembered that protection of the receiver is as important as the power-efficiency consideration. At frequencies where receiver RF amplifier tubes are used, such tubes are chosen to withstand relatively large input powers without damage. However, in microwave receivers, the crystal mixer at the input circuit is easily damaged by large signals and must be carefully protected.

In general, if the receiver input circuit is properly protected, the remaining receiver circuits can be prevented from blocking or overloading as the result of strong signals. However, a very strong main pulse signal will appear in the receiver's output unless additional precautions are taken to eliminate it. This can be done by a receiver gate signal that turns on the receiver during the desired time.

The requirements of a radar duplexing switch are:

1. During the period of transmission the switch must connect the antenna to the transmitter and disconnect it from the receiver.

2. The receiver must be thoroughly isolated from the transmitter during the transmission of the high power pulse to avoid damage to the sensitive converter elements.

3. After transmission, the switch must rapidly disconnect the transmitter and connect the receiver to the antenna. If targets close to the radar are to be seen, the action of the switch must be extremely rapid.

4. The switch should absorb an absolute minimum of power, both during transmission and reception.

Therefore, a radar duplexer is the microwave equivalent of a fast, low loss, single-pole double-throw switch. The devices which have been developed for this purpose are similar to spark gaps where high-current microwave discharges furnish low-impedance paths. A duplexer usually contains two switching tubes (spark gaps) connected in a microwave circuit with three terminal transmission lines; one each for the transmitter, the receiver, and the antenna (fig. 10-1). These circuits may be connected in parallel (A) or series (B). Both systems will be discussed in detail in this chapter. One tube is called the TRANSMIT-RECEIVE TUBE or TR TUBE; The other is called the ANTI-TRANSMIT-RECEIVE TUBE or ATR TUBE. The TR tube has a primary function of disconnecting the receiver, the ATR tube of disconnecting the transmitter.

TR TUBE

The spark gap used in a given TR system may vary from a simple one formed by two electrodes placed across the transmission line, to one enclosed in an evacuated glass envelope with special features to improve operation. The requirements of the spark gap are high impedance prior to the arc and very low impedance during arc time. At the end of the transmitted pulse the arc should be extinguished as rapidly as possible to remove the loss caused by the arc, and to permit signals from nearby targets to reach the receiver.

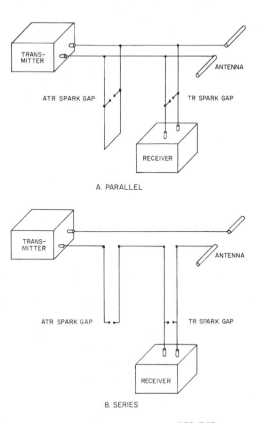

A. PARALLEL

B. SERIES

179.747

Figure 10-1.—Duplexer systems.

conducted through the ionized gas or vapor. The magnitude of voltage necessary to break down a gap may be lowered by reducing the pressure of the gas which surrounds the electrodes. There is an optimum pressure which achieves the most efficient TR operation. The recovery time, or DEIONIZATION TIME, of the gap can be reduced by introducing water vapor into the TR tube. A TR tube containing water vapor at a pressure of 1 mm. of mercury will recover in 0.5 usec. It is important for a TR tube to have a short recovery time in order to reduce the range at which targets near the radar can be detected. If, for example, echo signals reflected from nearby objects return to the radar before the TR tube has recovered, those signals will be unable to enter the receiver.

TR tubes for use at microwave frequencies are built to fit into, and become a part of, a resonant cavity. The speed with which the gap breaks down after the transmitter fires may be increased by placing a voltage across the gap electrodes. This potential is known as KEEP-ALIVE VOLTAGE and ranges from 100 volts to 1000 volts. A glow discharge is maintained between the electrodes. (The term "glow discharge" refers to the discharge of electricity through a gas filled electron tube. This is distinguished by a cathode glow and a voltage drop much higher then the gas ionization voltage in the cathode vicinity.) This action provides for rapid ionization when the transmitter pulse arrives.

Failure of the TR tube is primary determined by two factors. The first and most common cause of failure is the gradual buildup of metal particles which have been dislodged from the electrodes. Such metal bits become spattered on

The simple gap formed in air has a resistance during conduction of from 30 to 50 ohms. This is usually too high for use with any but an open-wire transmission line. The time required for the air surrounding the gap to completely deionize after the pulse voltage has been removed is about 10 usec. During this time the gap acts as an increasing resistance across the transmission line to which it is connected. However, in a TR system using an air gap, the echo signals reaching the receiver beyond the gap will be permitted to increase to half their proper magnitude 3 usec. after the pulse voltage has been removed. This interval is known as RECOVERY TIME.

TR tubes are conventional spark gaps enclosed in partially evacuated, sealed glass envelopes (fig. 10-2). The arc is formed as electrons are

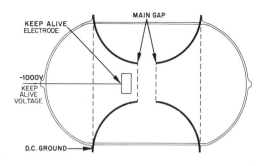

55.106.1

Figure 10-2.—TR tube with keep-alive electrode.

the inside of the glass envelope. These particles act as small, conducting areas and tend to lower the Q of the resonant cavity and to dissipate power. If the tube is continued in use for too long a period, the particles will form a de-tuning wall within the cavity and will eventually prevent the tube from functioning. A second cause of failure is the absorption of gas within the enclosure by the metal electrodes. This results in a gradual reduction of pressure within the tube to the point where gap breakdown becomes very difficult. The final result is that extremely strong signals (from the transmitter) are coupled to the receiver. Because both types of failures develop gradually, the TR tube must be checked carefully and periodically to determine its level of performance. TR tubes contain radioactive material. The precautions to be taken when handling these tubes are listed in chapter 2 of Volume I.

ATR TUBE

The ATR tube is usually a simpler device than a TR tube. An ATR tube might use a pure inert gas such as argon since recovery time generally is not a vital factor. Furthermore, a priming agent such as keep-alive voltage is not needed. The absence of either a chemically active gas or a keep-alive voltage results in ATR tubes having a longer duty life than TR tubes. ATR tubes also contain radioactive materials and the same precautions as are applicable to TR tubes apply to ATR tubes.

PARALLEL CONNECTED
DUPLEXER OPERATION

In order to analyze the overall action of the TR-ATR circuits, recall that a quarter-wave-length section of transmission line (or odd multiple thereof) inverts the impedance from end to end. (Wavelength impedances are discussed in chapter 28 of Volume I.) The TR spark gap in figure 10-3 is located in the receiver coupling line one-quarter wavelength from the tee (T) junction. A quarter-wavelength from the T junction (in the direction of the transmitter) a half-wavelength, closed-end section of transmission line, called a STUB, is shunted across the main transmission line. An ATR spark gap is located in this line one-quarter wavelength from the main transmission line, and one-quarter wavelength from the closed end of the stub. As shown in figure 10-3, the antenna impedance, the line impedance, and the transmitter output impedance when transmitting are all equal.

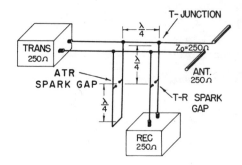

67.20(179)

Figure 10-3.—Parallel connected duplexer system showing distance and impedance.

During the transmitting pulse, an arc appears across both spark gaps and causes the TR and ATR circuits to act as a shorted (closed-end) quarter-wave stub and reflect an open circuit to the TR and ATR circuit connections to the main transmission line (fig. 10-4).

None of the transmitted energy can pass through these reflected opens into the ATR stub or into the receiver. Therefore, all of the transmitted pulses is directed to the antenna.

During reception, the amplitude of the received echo is not sufficient to cause an arc across either spark gap (fig. 10-5). Under this condition, the ATR circuit now acts as a half-wave transmission line terminated in a short-circuit. This is reflected as an open circuit at the receiver T junction, three-quarter wavelengths away. The received echo sees an open circuit in the direction of the transmitter. However, the receiver input impedance is matched to the transmission line impedance so that the entire received signal will go to the receiver with a minimum amount of loss.

SERIES CONNECTED
DUPLEXER OPERATION

In the series connected duplexer system (fig. 10-6), the TR spark gap is located one-half wavelength from the receiver T junction. The ATR spark gap is located one-half wavelength from the transmission line and three-quarters wavelength from the receiver T junction.

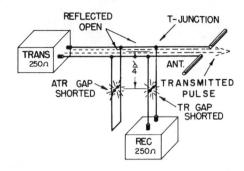

179.749

Figure 10-4.—Parallel connected duplexer system during transmission.

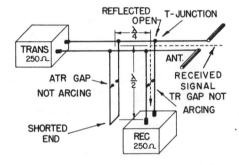

179.750

Figure 10-5.—Parallel connected duplexer system during reception.

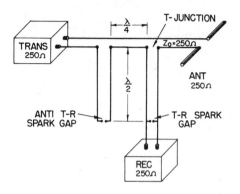

179.751

Figure 10-6.—Series connected duplexer system showing distance and impedance.

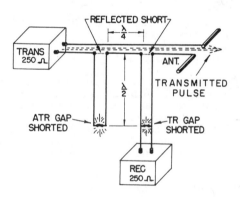

179.752

Figure 10-7.—Series connected duplexer system during transmission.

During transmission, the TR and ATR gaps fire in the series connected duplexer system (fig. 10-7). This causes a short-circuit to be reflected at the series connection to the main transmission line one-half wavelength away. The transmitted pulse sees a low impedance path in the direction of the antenna, and does not go into the ATR stub or the receiver.

During reception, neither spark gap is fired (fig. 10-8). The ATR acts as a half-wave stub terminated in an open. This open is reflected as a short-circuit at the T junction three-quarters of a wavelength away. Consequently, the received signal sees a low impedance path to the receiver, and none of the received signal is lost in the transmitting circuit.

HANDLING AND DISPOSAL OF TR AND ATR TUBES

Many TR and ATR tubes contain radioactive material and should be handled with care to avoid breakage. The level of radioactivity of an unbroken tube is small and presents no danger to personnel during normal handling. All tubes, damaged or not, should be disposed of in accordance with BUSHIPS instruction 5100.5 and NAVSHIPS Technical Manual, chapter 9670.

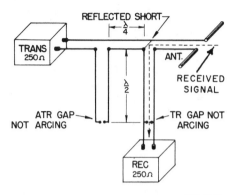

179.753

Figure 10-8.—Series connected duplexer system during reception.

The greatest danger to personnel in the field is physical contact with radioactive material from accidentally broken tubes. The danger can be minimized by becoming familiar with and following the procedures set forth in Radiation, the Health Protection Manual, NAVMED-P-5055, and observing the following safety practices:

1. Remove radioactive tubes from the cartons immediately prior to installation.
2. Place the tubes in an appropriate container upon removal from equipment.
3. Carry the tube in such a way as to minimize possible breakage; never in your pocket.
4. If a tube should break, avoid breathing the vapor or dust released by the tube; notify the cognizant authority and qualified radiological personnel, and isolate the exposure area.

5. During clean-up and decontamination use rubber or plastic gloves, forceps, a vacuum cleaner with an approved disposal collection bag and any other approved means to keep the contaminated material fron contacting your body.
6. Seal all debris, cloths, and collection bags in a container such as a plastic bag or glass jar and place this in a steel can for disposal by a naval shipyard or supply activity.
7. Do not eat any food, drink any liquid, or smoke any tobacco product which has been near the contaminated area, and when leaving the area remove any contaminated clothing and wash your hands and arms with soap and water followed by a clean water rinse.

8. If injured by a sharp radioactive object notify the medical officer. While waiting for his arrival stimulate mild bleeding by pressure about the wound and by suction bulbs. Do not use the mouth. If the wound is small or a puncture type make a cut to allow free bleeding and to facilitage cleaning and flushing.
9. Decontaminate all tools used for radioactive material removal with soap and water. They should emit less than 0.1 mr./hr. at the surface when monitored with a radiac set.

RADAR ANTENNAS

Antenna's fall into two general classes, OMNIDIRECTIONAL and DIRECTIONAL. Omnidirectional antennas radiate RF energy in all directions simultaneously. They are seldom used with modern radars, but are commonly used in radio equipment, in IFF (Identification Friend or Foe) equipment, and in countermeasures receivers for the detection of enemy radar signals. Directional antennas radiate RF energy in LOBES or BEAMS, that extend outward from the antenna in one direction for a given antenna position. The radiation pattern contains small minor lobes, but these lobes are weak and normally have little effect on the main radiation pattern. The main lobe may range in angular width from 1° or 2° in some radars to several degrees in other radars, depending on the system's purpose or the degree of accuracy required. Main lobe energy is confined to a few degrees in the horizontal and/or vertical plane. The minor lobes are made as small as possible in order to concentrate maximum energy into the main lobes.

Directional antennas have two important characteristics. One is DIRECTIVITY. The directivity of an antenna refers to the degree of sharpness of its beam. If the beam is narrow in either the horizontal or vertical plane, the antenna is said to have high directivity in that plane. Conversely, if the beam is broad in either plane, the directivity of the antenna in that plane is low. Thus, if an antenna has a narrow horizontal beam and a wide vertical beam, the horizontal directivity is high and the vertical directivity is low.

When the directivity of an antenna is increased, that is, when the beam is narrowed, less power is required to cover the same range because the power is concentrated. Thus, another characteristic of an antenna is brought to light. This characteristic is called POWER GAIN, and is directly related to directivity.

Power gain of an antenna is the ratio of its radiated power to that of a reference (basic) dipole. Both antennas must have been excited or fed in the same manner and each must have radiated from the same position. A single point of measurement for the power gain ratio must lie within the radiation field of each antenna. An antenna with high directivity has a high power gain, and vice versa. The power gain of a single dipole with no reflector is unity. An array of several dipoles in the same position as the single dipole and fed with the same line would have a power gain of more than one, the exact figure depending on the directivity of the array.

The measurement of the bearing of a target as seen by the radar is usually given as an angular position. The angle may be measured either from true north (true bearing), or with respect to the heading of a vessel or aircraft containing the radar set (relative bearing). The angle at which the echo signal returns is measured by utilizing the directional characteristics of the radar antenna system. Radar antennas consist of radiating elements, reflectors, and directors to produce a narrow unidirectional beam of energy. The pattern produced in this manner permits the beaming of maximum energy in a desired direction. The transmitting pattern of an antenna system is also its receiving pattern. An antenna can therefore be used to transmit energy, receive energy, or accomplish both.

The simplest form of antenna for measuring azimuth or bearing is a rotating antenna which produces a single lobe pattern. The remaining plane necessary to locate absolutely an object in space may be expressed either as elevation angle or as altitude. If one is known, the other can be calculated from basic trigonometric functions. A method of determining the angle of elevation or the altitude is shown in figure 10-9. The slant range is obtained from the radar scope as the distance to the target. The angle of elevation is the angle between the axis of the radar beam and the earths surface. The altitude in feet is equal to the slant range in feet multiplied by the sine of the angle of elevation. For example if the slant range in figure 10-9 is 2000 feet and the angle of elevation is 45°, the altitude is 1414.2 feet (2000 x .7071). In some radar equipments using antennas that may be varied in angle of elevation, altitude determination is automatically computed electronically.

PARABOLIC REFLECTORS

A spherical wavefront spreads out as it travels as described in chapter 29 of Volume I. This produces a pattern that is not too sharp or directive.

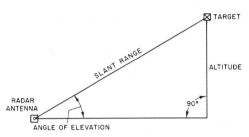

20.285

Figure 10-9.—Radar determination of altitude.

On the other hand, a plane wavefront does not spread out because all of the wavefront moves forward in the same direction. For a sharply defined radar beam, the need exists to change the spherical wavefront from the antenna into a plane wavefront. A parabolic reflector is one means of accomplishing this.

Radio waves behave similarly to light waves. Microwaves travel in straight lines as do light rays. They may be focused and/or reflected just as light rays can. In figure 10-10, a point radiation source is placed at the focal point F. The field leaves this antenna with a spherical wavefront. As each part of the wavefront reaches the reflecting surface, it is shifted 180° in phase and sent outward at angles that cause all parts of the field to travel in parallel paths. Because of the shape of a parabolic surface, all paths from F to the reflector and back to line XY are the same length. Therefore, all parts of the field arrive at line XY the same time after reflection.

If a dipole is used as the source of radiation, there will be radiation from the antenna into space as well as toward the reflector. Energy which is not directed toward the paraboloid has a wide beam characteristic which would destroy the narrow pattern from the parabolic reflector. To prevent this occurrence, a hemispherical shield (not shown) is used to direct most radiation toward the parabolic surface. By this means, direct radiation is eliminated, the beam is made sharper, and power is concentrated in the beam. Without the shifled, some of the radiated field would leave the radiator directly. Since it would not be reflected, it would not become a part of the main beam and thus could serve no useful purpose. Another method of accomplishing the same end is through the use of a parasitic array which directs the radiated field back to the reflector.

The radiation of a parabola contains a major lobe, which is directed along the axis of revolution, and several minor lobes, as shown in figure

185

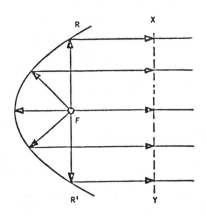

13.45(179)
Figure 10-10.—Parabolic reflector radiation.

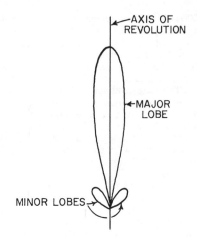

59.26(179)
Figure 10-11.—Parabolic radiation pattern.

10-11. Very narrow beams are possible with this type of reflector. Figure 10-12A illustrates the paraboloid reflector.

Truncated Paraboloid

Figure 10-12B shows a horizontally truncated paraboloid. Since the reflector is parabolic in the horizontal plane, the energy is focused into a narrow beam. With the reflector truncated, or cut, so that it is shortened vertically, the beam spreads out vertically instead of being focused. Such a fanshaped beam is used to determine azimuth accurately. Since the beam is wide vertically, it will detect aircraft at different altitudes without changing the tilt of the antenna. It also works well for surface search to overcome the pitch and roll of the ship.

The truncated paraboloid reflector may be used in height-finding systems if the reflector is rotated 90 degrees (fig. 10-12C). Since the reflector is now parabolic in the vertical plane, the energy is focused into a narrow beam vertically. With the reflector truncated, or cut, so that it is shortened horizontally, the beam spreads out horizontally instead of being focused. Such a fan shaped beam is used to determine elevation very accurately.

Orange-Peel
Paraboloid

A section of a complete circular paraboloid, often called an ORANGE-PEEL REFLECTOR because of its shape, is shown in figure 10-12D.

Since the reflector is narrow in the horizontal plane and wide in the vertical, it produces a beam that is wide in the horizontal plane and narrow in the vertical. In shape, the beam resembles a huge beaver tail. The RF energy is sent into the parabolic reflector by a horn radiator fed by a waveguide. The horn nearly covers the shape of the reflector, so almost all of the RF energy illuminates the reflector, very little escaping at the sides. This type of antenna system is generally used in height-finding equipment.

Cylindrical
Paraboloid

When a beam of radiated energy noticeably wider in one cross-sectional dimension than in the other is desired, a cylindrical paraboloidal section approximating a rectangle can be used. Figure 10-12E illustrates this antenna. A parabolic cylinder has a parabolic cross section in one dimension only; therefore, the reflector is directive in one plane only. The cylindrical paraboloid reflector is either fed by a linear array of dipoles, a slit in the side of a waveguide, or by a thin waveguide radiator. Rather than a single focal point, this type of reflector has a series of focal points forming a straight line. Placing the radiator, or radiators, along this focal line produces a directed beam of energy. As the width of the parabolic section is changed, different

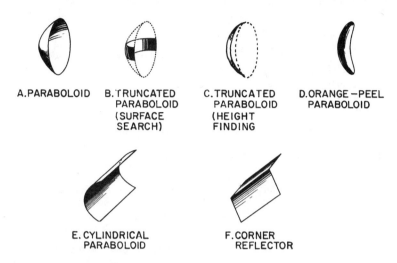

A. PARABOLOID

B. TRUNCATED
PARABOLOID
(SURFACE
SEARCH)

C. TRUNCATED
PARABOLOID
(HEIGHT
FINDING

D. ORANGE – PEEL
PARABOLOID

E. CYLINDRICAL
PARABOLOID

F. CORNER
REFLECTOR

25.228
Figure 10-12.—Reflector shapes.

beam shapes are obtained. This type of antenna systems is used in search and in Ground Control Approach (GCA) systems.

Corner Reflector

The corner-reflector antenna consists of two flat conducting sheets that meet at an angle to form a corner, as shown in figure 10-12F. This type reflector is normally driven by a half-wave radiator located on a line which bisects the angle formed by the sheet reflectors.

BROADSIDE ARRAY

The desired beam widths are provided for some VHF radars by a broadside array (fig. 10-13) which consists of two or more half-wave dipole elements and a flat reflector. The elements are placed one-half wavelength apart and parallel to each other. Because they are excited in phase, most of the radiation is broadside to the plane of the elements. The flat reflector is located approximately one-eighth wavelength behind the dipole elements and makes possible the unidirectional characteristics of the antenna system.

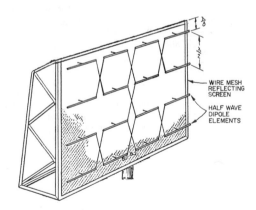

WIRE MESH
REFLECTING
SCREEN

HALF WAVE
DIPOLE
ELEMENTS

1.257(179)
Figure 10-13.—Broadside antenna array.

HORN RADIATORS

Horn radiators, like parabolic reflectors, may be used to obtain directive radiation at microwave frequencies. Because they do not involve resonant elements, horns have the advantage of being usable over a wide frequency band.

The operation of a horn as an electromagnetic directing device is analogous to that of acoustic horns. However, the throat of an acoustic horn usually has dimensions much smaller than the sound wavelengths for which is is used, while the throat of the electromagnetic horn has dimensions that are comparable to the wavelength being used.

Horn radiators are readily adaptable for use with waveguides because they serve both as an impedance-matching device and as a directional radiator. Horn radiators may be fed by coaxial or other types of lines.

Horns are constructed in a variety of shapes as illustrated in figure 10-14. The shape of the horn, along with the dimensions of the length and mouth, largely determines the field-pattern shape. The ratio of the horn length to mouth opening size determines the beam angle and thus the directivity. In general, the larger the opening of the horn, the more directive is the resulting field pattern.

AIRBORNE RADAR ANTENNAS

Airborne radar equipment is used for several specific purposes. Some of these are bombing, navigation, and search. Radar antennas for this equipment are invariably housed inside nonconducting radomes, not only for protection but also to preserve aerodynamic design. Some of these radomes are carried outside the fuselage, while others are flush with the skin of the fuselage. In the latter case, the radar antenna itself is carried inside the fuselage, and a section of the metallic skin is replaced by the nonconducting radome. The radar antenna and its radome must operate under a wide variety of temperature, humidity, and pressure conditions. As a result, mechanical construction and design must minimize any possibility of failure. Transmission lines are usually hermetically sealed to prevent moisture accumulation inside them as this would introduce losses. Since the low air pressures encountered at high elevations are very conducive to arcing, pressurization of equipment is widely used, with the pressure being maintained by a small air pump. In some radar equipments, practically all of the equipment is sealed in an airtight housing, along with the antenna and transmission line. The antenna radome forms a portion of the housing.

Airborne radar antennas are constructed to withstand large amounts of vibration and shock; the radar antenna are rigidly attached to the

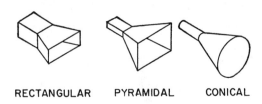

RECTANGULAR   PYRAMIDAL   CONICAL

13.41

Figure 10-14.—Horn radiators.

airframe, and the weight of the radar antenna, including the rotating mechanism required for scanning, is kept to a minimum. In addition, the shape of the radome is constructed so as not to impair the operation of the aircraft.

The airborne radar antenna must have an unobstructed view for most useful operation. Frequently, the antenna must be able to scan the ground directly under the aircraft and well out toward the horizon. To meet this requirement, the antenna must be mounted below the fuselage. If scanning toward the rear is not required, the antenna is mounted behind and below the nose of the aircraft, where the radome may be completely faired in. If only forward scanning is needed, the antenna is mounted in the nose. When an external site is required, a location at the wing tip is common. Fire-control radar antennas are frequently located near the turret guns or in a special nacelle, where it can scan toward the rear or sides of the aircraft.

RF SAFETY PRECAUTIONS

Radio frequency electromagnetic radiation from transmission lines and antennas, although usually insufficient to electrocute personnel may lead to other accidents and compound injuries. RF voltages may be induced in ungrounded metal objects such as wire guys, wire cable (hawser), hand rails or ladders. Personnel who come in contact these objects could receive a shock or RF burn. This shock can cause personnel to jump or fall into nearby mechanical equipment, or when working aloft, to fall from an elevated work area. Care should be taken to ensure that all transmission lines or antennas are deenergized before working near or on them.

Guys, cables, rails or ladders should be checked for RF shock dangers. Working aloft "chits" and safety harnesses should be used for further safety. Since signing a working aloft chit signifies that all equipment is a nonradiating status, the personnel who signs the chit should prevent RF danger in areas where men are working. Nearby ships or parked aircraft are another source of RF energy and must be considered when checking a work area for safety.

Combustible material can be ignited and cause severe fires from arcs or heat generated by RF energy. Also, RF radiation can detonate ordnance devices by inducing currents in the internal wiring of the device or into the external test equipment or leads connected to the device.

All personnel must obey RF hazards warning signs and keep a safe distance from radiating antennas for any type of work involved. NAVSHIPS Technical Manual, chapter 9670 and chapter 2 of Volume I in this training manual contain additional material which you should know concerning minimum distances to be maintained by personnel to ensure safety.

RF Burns

Close or direct contact with RF transmission lines or antennas may result in RF burns. There are usually deep penetrating third degree burns which, to heal properly, must heal from the in-side, or bottom of the burn, to the skin surface. To prevent infection, proper medical attention must be given to all RF burns including the small "pinhole" burns. Petrolatum gauze can be used to cover burns temporarily before reporting to medical facilities for further treatment.

Dielectric Heating

Dielectric heating is the heating of a nominally insulating material by placing it in a high-frequency electric field. The heat results from internal losses during the rapid reversal of polarization of molecules in the dielectric material.

In the case of a human in an RF field, the body acts as a dielectric and if the power exceeds 10 milliwatts per centimeter, there will be a noticeable rise in body temperature. The eyes are highly susceptible to dielectric heating. For this reason, it is not advisable to look directly into devices radiating RF energy. The vital organs of the body are also susceptible to dielectric heating. In the interest of health, you must not stand directly in the path of RF radiating devices.

CHAPTER 11

PLAN POSITION INDICATOR DISPLAY

Cathode ray tube (CRT) presentations, in electronic ranging and navigational systems, are used to present visual indications of data relating such measurable quantities as range, bearing, height, depth, speed, and time. Figure 11-1 shows the most common scans.

The type-A presentation (fig. 11-1A) is used to determine range. The screen of this scope has a short persistence. The echo causes a vertical displacement of the electron beam, the amplitude of which depends on the strength of the returned signal pulse. The point on the horizontal base line at which the vertical displacement occurs indicates the range.

The type-B presentation (fig. 11-1B) indicates both range and azimuth angle (bearing) within 90° on either side of the antenna 0° bearing. The vertical displacement of the echo signal indicates range, and the horizontal displacement of the echo signal indicates azimuth angle. This scope has long persistence.

The E scan (fig. 11-1C) presentation is another type of scan for presenting range and height information. The E scan is also known as the RHI (Range Height Indicator) scan. The type-E scan is a modification of the type-B scan on which an echo appears as a bright spot with the range indicated by the horizontal coordinate and the elevation (height) as the vertical coordinate. This type is in directing planes in blind landing, for ground-controlled approach, for carrier-controlled approach, and in determining altitude. This scan is also used in weapons control systems.

The PPI (Plan Position Indicator) presentation (fig. 11-1D) is the most common type of scan. A discussion of this scan follows.

BASIC PPI

Figure 11-2 illustrates the basic block diagram of a plan position indicator. Synchronization of events is particularly important in the presentation system. At the instant a radar or similar transmitter fires (or some predetermined time thereafter), circuits which control the presentation on the indicator must be activated. To ensure accurate range determination, these events must be performed to a high degree of accuracy.

The gate circuit develops pulses which synchronize the indicator with the transmitter. The gate circuit itself, is synchronized by trigger pulses from the modulator. The sweep control circuit converts mechanical bearing information from the antenna into voltages which control sweep circuit azimuth.

The sweep generator circuit produces currents which deflect an electron beam across the CRT.

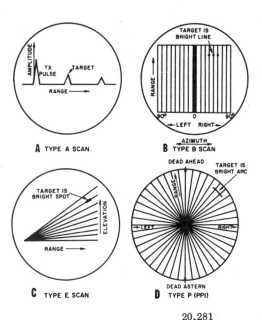

20.281
Figure 11-1.—Types of scans.

190

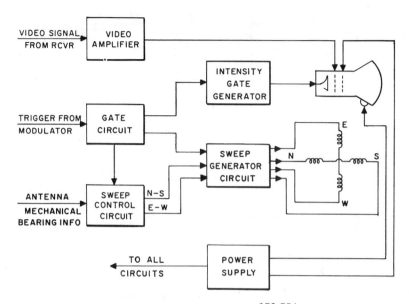

179.754
Figure 11-2.—Basic PPI indicator.

Varying voltages from the sweep control circuit are applied to deflection elements. Gate voltages determine sweep rate, and therefore, the effective distance (range) covered by each sweep. Sweep potentials consist of separate north-south and east-west voltages whose amplitudes determine sweep azimuth. The sweep generator is synchronized by an input from the gate circuit.

The intensity gate generator provides a gate which unblanks the CRT during sweep periods. The intensity of the trace appearing on the CRT is determined by the d.c. level of this gate. This circuit is also synchronized by the gate circuit.

The video amplifier circuit amplifies the video signal from the receiver and applies it to the CRT intensity modulating element. The power supply produces all voltages needed to operate the indicator. It also includes protective devices and metering circuits.

Although not shown in the basic block diagram, many indicators contain circuits which aid in range and bearing determination. These circuits are also synchronized by the gate circuit.

ELECTROMAGNETIC DEFLECTION

In modern presentation systems, electromagnetic deflection of the CRT electron beam is preferred to electrostatic deflection. Reasons for this choice are increased control of the beam, improved deflection sensitivity, better beam position accuracy, and simpler construction of CRT.

The primary difference between electromagnetic and electrostatic cathode ray tubes is the method of controlling deflection and focusing of an electron beam. Both types employ an electron gun and use an electrostatic field to accelerate and control the flow of electrons. Physical construction of a CRT employing electromagnetic deflection is similar to an electrostatic type. The construction of a CRT employing electromagnetic deflection is shown in figure 11-3.

The electron gun is comprised of a heater, cathode, control grid, second or screen grid, focus coil, and anode (aquadag coating). Focusing the electron beam on the face of the screen is accomplished by use of a focus coil. A direct current through the winding sets up a strong magnetic field at the center of the coil. Electrons moving

191

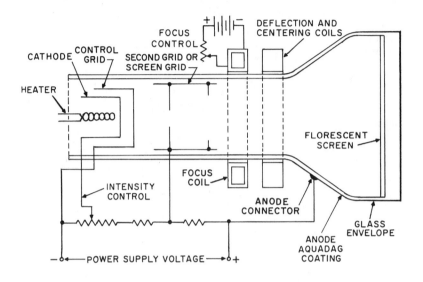

179.755
Figure 11-3.—Electromagnetic CRT construction.

exactly along the axis of the tube pass through the focusing field with no deflection, since they move parallel to the magnetic field at all times.

An electron which enters the focusing field at an angle to the axis of the tube has a force exerted on it perpendicular to its direction of motion. A second force on this electron is perpendicular to the magnetic lines and is, therefore, constantly changing in direction. These forces cause the electron to move in a helical or corkscrew path (fig. 11-4). With the proper velocity and strength of magnetic field, the electron will move at an angle which causes it to converge with other electrons at some point on the CRT screen. Focusing is accomplished by adjusting the intensity of current flow through the focusing coils.

The focused electron beam is deflected by a magnetic field. This field is formed by current flow through a set of deflection coils. These coils are mounted around the outside surface of the neck of the tube as illustrated in figure 11-3. Normally, four deflection coils are used (fig. 11-2). Two coils in series are positioned in a manner which causes the magnetic field produced to be in a vertical plane. The other two coils, also connected in series, are positioned so that their magnetic field is in a horizontal plane. The coils which produce a horizontal field are

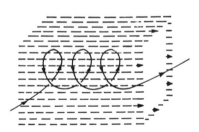

179.756
Figure 11-4.—Helical motion of electron passing through a uniform magnetic field.

called the VERTICAL DEFLECTION COILS and the coils which produce a vertical field are called the HORIZONTAL DEFLECTION COILS. This may be more clearly understood if it is recalled that an electron beam will be deflected at right angles to the deflecting field. The entire coil assembly is called a DEFLECTION YOKE. The physical and schematic representation of the deflection coils is illustrated in figure 11-5.

Electron deflection in the electromagnetic type

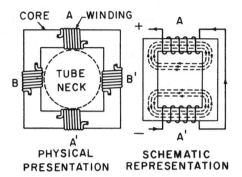

Figure 11-5.—Deflection yoke.

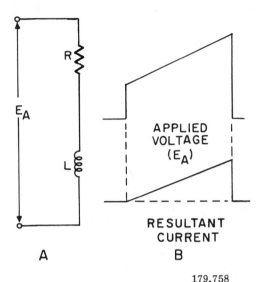

RESULTANT
CURRENT

A   B

179.758
Figure 11-6.—Deflection coil equivalent
circuit and waveforms.

CRT is proportional to the strength of the magnetic fields. Therefore, the sweep circuits associated with electromagnetically deflected cathode ray tubes must provide currents, rather than voltages, to produce the desired beam deflection.

To produce a linear trace, a sawtooth of current is required. A deflection coil may be considered equivalent to the circuit shown in figure 11-6A. Due to the inductance of the coil, a voltage of trapezoidal form must be applied across the coil in order to produce a sawtooth of current through it. This is illustrated in figure 11-6B.

TRAPEZOIDAL VOLTAGE
SWEEP GENERATOR

A simple trapezoidal voltage sweep generator is illustrated in figure 11-7. A negative rectangular pulse, from the gate circuit, is applied to the grid of V1. When the tube is cut off by the driving pulse, plate voltage rises toward the supply voltage, E_{bb}, at the rate of charge of C1. However, at the instant V1 is cut off, the output voltage (taken across C1 and R2) jumps to some value, determined by the ratio of R1 to R2. As capacitor C1 begins to charge, the output voltage rises further, exponentially, toward the supply voltage. This action produces the required trapezoidal sweep voltage.

ROTATING SWEEP

PPI azimuth indication requires that the range trace rotate about the center of the screen. A very simple means of achieving sweep rotation

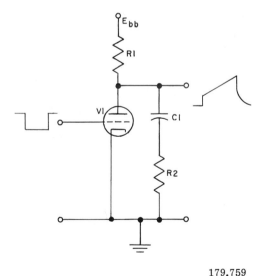

179.759
Figure 11-7.—Trapezoidal sweep generator.

is to cause the deflection coil to rotate about the neck of the CRT in synchronization with the antenna motion.

Figure 11-8 shows the tube and rotating coil. The yoke carrying the deflection coil is mounted

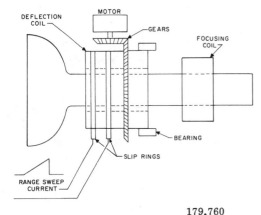

179.760
Figure 11-8.—Rotating deflection coil.

in bearings and driven by a motor. Slip rings are provided so that range sweep currents may be sent through the coil. Figure 11-9 illustrates the way in which azimuth is provided. The magnetic field, established by range sweep currents, rotates with the deflection coil, and the sweep trace is always in a direction perpendicular to this field.

Synchronizing the yoke with antenna rotation may be accomplished by using synchronous motors connected to a common power supply or through the use of electromechanical repeaters or servomechanisms.

Most modern PPI systems employ fixed deflection coils and use special circuits to rotate the magnetic field. Figure 11-10 illustrates a method of electronically producing a rotating sweep. In figure 11-10A, a range sweep current, i, is applied to the horizontal deflection coils only, and the resulting magnetic field, Φ lies along the axis of these coils. The resulting range trace is horizontal because the electron beam is deflected perpendicular to the magnetic field. In figure 11-10B, range sweep currents are applied to both sets of coils, and the resultant magnetic field takes a position between the axis of the two sets of coils. Because of this shift of the magnetic field, the range trace is rotated clockwise from its original position. The two current waveforms

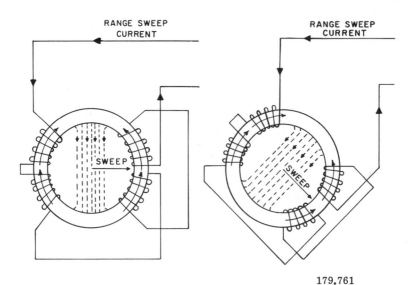

179.761
Figure 11-9.—Relation of sweep direction to deflection coil position.

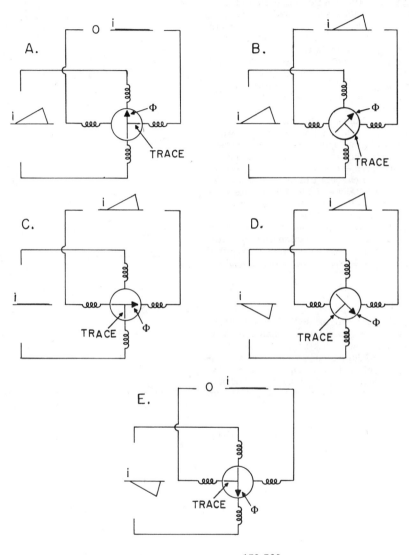

179.762
Figure 11-10.—Trace rotation.

195

must be exactly alike and must be applied simultaneously; otherwise, the trace would not be a straight line. In figure 11-10C, sweep current is applied to the vertical deflection coils only, and the range trace lies 90-degrees clockwise from its original position. Further rotation is obtained if the deflection coil currents are reversed in proper sequence, as illustrated in figure 11-10D and E.

In order to synchronize sweep rotation with antenna rotation, mechanical bearing information from the antenna is converted into electrical signals, which control the amplitude and polarity of the sweep currents applied to the deflection coils.

Figure 11-11 illustrates the waveforms of current required to produce a rotating range sweep. Note that the two envelopes (which are the result of antenna bearing information) differ in phase by 90 degrees and have equal peak values. Individual sawtooth waves of current are applied to the deflection coils simultaneously; the phase difference of 90-degrees applies only to the variation of amplitude from one sawtooth to the next.

In the preceeding discussion, currents of sawtooth form were utilized for explanation. It should be recalled that the voltage waveform required to produce these sawtooth currents is trapezoidal.

CRT SCREEN
PERSISTANCE

A PPI system requires a CRT whose screen is coated with a long persistence phosphor. This is necessary because each target is reflected for only a short time during each rotation of the antenna. The target indication on the face of the CRT must continue to glow during the portion of antenna rotation when it is not directly reflected.

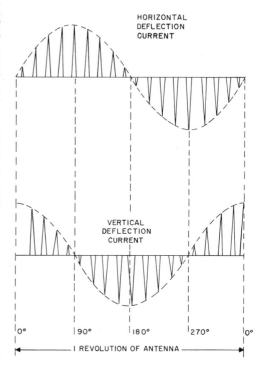

179.763
Figure 11-11.—Deflection coil currents.

196

CHAPTER 12

SYNCHRO AND SERVOSYSTEMS

Synchros play a very important part in the operation of U.S. Navy equipment. Almost every radar, sonar, or fire control equipment contains synchro devices which are vital to the operation of the equipment.

The synchro is a small a.c. electromechanical device used for the electrical transmission of angular position data. Synchros are also known by various trade names such as Selsyn, Autosyn, and Synchrotie; but in the U.S. Navy the name synchro has become universal.

The purpose of the synchro is to transmit angular position data between remote locations, such as from a radar antenna to an indicator or from the master gyro compass to repeaters found in various locations aboard ship. The synchro also has an important application as an error sensing or error detecting device in the input controller of a control system. Synchros are also utilized as computers, for example, modifying the angular position input information to a fire control system with continually changing data such as windage and the roll or pitch of ship. Speed and accuracy of data transmission are most important. Synchros provide this speed and accuracy along with good reliability, adaptability, and compactness.

SYNCHRO CLASSIFICATIONS

Synchros work in teams. Two or more synchros are interconnected electrically to form a synchro system. There are two general classifications of synchro systems: TORQUE SYSTEMS and CONTROL SYSTEMS. Torque systems are used for light loads such as the positioning of dials, pointers, or similar indicators. The positioning of these devices requires a relatively low torque.

The control systems are used where it is desired to move large loads. The output of the control synchro system is an electrical error signal that indicates the direction and amount of error in the load to be positioned.

In addition to the two general classifications, synchros are grouped into seven basic functional classes as shown in table 12-1. Four of these are of the torque type and three are of the control type. Each of the functional classifications will be described by name abbreviation, input, output, and the other synchro units which may be connected to it. Generally, torque and control synchros may not be interchanged.

1. TORQUE TRANSMITTER (TX)—electrically transmits angular position data. In other words, it converts an angular position into an electrical signal that is sent through interconnecting wires to other synchro units. It consists of a rotor which has a single winding and a stator with three windings displaced 120 degrees. The rotor is supplied from an a.c. source and is mechanically moved to the angular position that is to be transmitted. The voltages induced in the stator windings, as a result of transformer action from the alternating field set up by current in the rotor winding, are representative of the angular position of the rotor at any instant. This electrical output from the stator may be sent to a torque receiver, torque differential receiver, or torque differential transmitter through interconnecting wires.

2. CONTROL TRANSMITTER (CX)—is functionally the same as the torque transmitter, except the electrical output is sent to a control differential transmitter or a control transformer.

3. TORQUE DIFFERENTIAL TRANSMITTER (TDX)—electrically transmits angular position data equal to the algebraic sum or difference (depending on external connections) of two inputs. It consists of a rotor with three windings displaced 120 degrees and a stator with three windings displaced 120 degrees. The rotor is positioned mechanically for one input. The other input is an electrical signal applied to the stator from a TX or another TDX. The output, representing the algebraic combination of the two inputs, is an electrical signal induced in the

Table 12-1.—Synchro Information

FUNCTIONAL CLASSIFICATION	ABBREVIATION	INPUT	OUTPUT
Torque transmitter	TX	Mechanical input to rotor (rotor energized from AC source)	Electrical output from stator representing angular position of rotor to TDX, TDR, or TR.
Control Transmitter	CX	Same as TX	Same as TX except it is supplied to CDX or CT.
Torque differential transmitter	TDX	Mechanical input to rotor, electrical input to stator from TX or another TDX	Electrical output from rotor representing algebraic sum or difference between rotor angle and angle represented by electrical input to TR, TDR, or another TDX.
Control differential transmitter	CDX	Same as TDX except electrical input is from CX or another CDX	Same as TDX except output to CT or another CDX.
Torque Receiver	TR	Electrical input to stator from TX or TDX. (Rotor energized from AC source).	Mechanical output from rotor note: rotor has mechanical inertia damper.
Torque differential receiver	TDR	Electrical input to stator from TX or TDX, another electrical input to rotor from TX or TDX.	Mechanical output from rotor representing algebraic sum or difference between angles represented by electrical inputs. Has inertia damper.
Control transformer	CT	Electrical input to stator from CX or CDX, mechanical input to rotor.	Electrical output from rotor proportional to the sine of the angle between rotor position and angle represented by electrical input to stator. Called error signal.

12.328

rotor windings by the alternating field set up by current in the stator windings. This electrical output may be sent to a torque receiver, torque differential receiver, or another TDX.

4. CONTROL DIFFERENTIAL TRANSMITTER (CDX)—is functionally the same as the torque differential transmitter, except that it is used in control rather than torque systems. The electrical input to the stator comes from a CX or another CDX, the electrical output from the rotor is sent to a control transformer or another CDX.

5. TORQUE RECEIVER (TR)—is electrically similar to the TX. Its function is to convert the electrical data applied to its stator back to mechanical angular position through movement of

the rotor which is free to turn. The rotor is connected to the same a.c. source as the TX and assumes a position determined by the alternating magnetic field of the rotor and the magnetic field produced by currents in the stator. The stator is connected electrically to the output of a TX or TDX. The only important difference between the TX and TR is that the TR has an inertia damper which prevents mechanical oscillation, often referred to as HUNTING, of the rotor when there are changes in its position.

6. TORQUE DIFFERENTIAL RECEIVER (TDR)—is electrically similar to the TDX. One electrical input from a TX or TDX is applied to the stator; a second electrical input from

another TX or TDX is applied to the rotor. The rotor then turns to a position which is the algebraic sum or difference of the angles represented by the two electrical inputs. The principal difference between the TDX and TDR is that the TDR has an inertia damper to prevent mechanical rotor oscillations.

7. CONTROL TRANSFORMER (CT)—consists of a stator with three windings displaced 120 degrees, which is supplied with an electrical signal from a CX, or CDX, and a single winding rotor, which can be positioned manually but will not turn by itself. The output is an electrical signal induced in the rotor by the alternating field set up in the stator with current from the electrical input. The output is proportional to the sine of the angle between the rotor and the angle represented by the electrical input to the stator. This output is called the ERROR SIGNAL in control applications.

SYNCHRO CONSTRUCTION

All synchro units use similar stators. The stator consists of slotted laminations located inside of a housing as shown in figure 12-1. Three coils are wound in the slots with their axes 120-degrees apart. The three stator windings will be connected either delta or wye and the leads from the stator case are labeled S1, S2, and S3. Figure 12-2 shows the schematic representation and lead identifications of delta and wye connected stators. The two types work basically the same but wye connected is the most common, so only wye connected will be utilized in the explanation of synchros.

Control transformer windings differ from the others mainly in that CT windings consists of more turns of finer wire, as explained later. The lamination slots are skewed (fig. 12-1) to eliminate resultant flux concentrations between rotor and stator that would cause the rotor to slot-lock in certain positions. Either, but not both, rotor or stator laminations may be skewed.

The housing with the laminations and windings is completed with end bells that contain rotor bearing supports, brush holders, and brushes for the rotor electrical connections. Leads from the stator windings and the rotor brushes are connected to terminals on one end bell. The stators are not connected directly to an a.c. source, they receive their excitation from the alternating fields of TX, CX, or TR rotors.

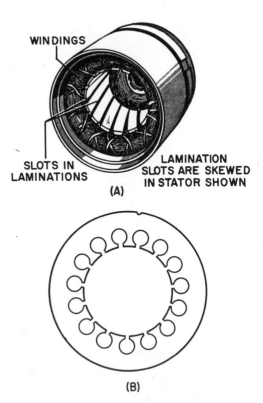

72.42.3:73.147
Figure 12-1.—Synchro stator; (A) Stator; (B) Laminations.

There are two basic types of rotors as shown in figure 12-3. The SALIENT POLE ROTOR consists of a single winding on a bobbin or dumbbell shaped laminated core. This type of rotor is used in the TX, CX, and TR synchro units.

The DRUM, or WOUND, ROTOR is used in the TDX, CDX, and TDR. The rotor windings on these differential units consist of three coils, wye connected. The CT also uses a drum rotor but has a single winding rather than three. Electrical connection to the rotor windings is made by a brush riding on a slip ring mounted at one end of the rotor shaft.

The TR and TDR rotors have an inertia damper. One of the most common types consists of a heavy brass flywheel which is free to rotate around a bushing which is attached to the rotor

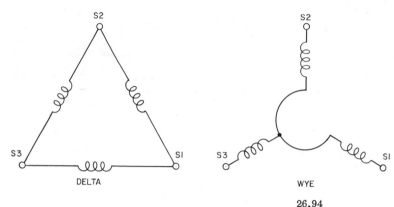

26.94
Figure 12-2.—Delta and wye connected stators.

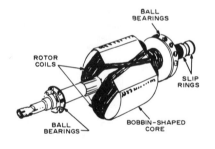

A. SALIENT-POLE ROTOR

B. DRUM OR WOUND ROTOR

72.42.2:51.289
Figure 12-3.—Salient pole and drum
type rotors.

shaft. A tension spring on the bushing rubs against the flywheel so that they turn together during normal operation. When the rotor shaft tends to change its speed or direction of rotation suddenly, the inertia of the damper opposes the changing condition.

SYNCHRO OPERATION

Synchro systems receive their excitation from external a.c. sources connected to the rotors of the TX, CX, and TR units. The a.c. source is generally 115 volts, of either 60- or 400-hertz frequency. Synchro units are designed to work with a particular frequency. The 400-hertz synchros are generally small and lighter than the 60-hertz synchros and find many applications in military equipment.

Synchro units may be compared with transformers, differing from conventional transformers by having one winding that is movable. Before considering the induced voltages in an actual synchro unit, we will analyze the behavior of a simple transformer when one of its coils is rotated through 360 degrees. Assume the transformer has a one to one turns ratio and no losses. The secondary voltage will then equal the primary voltage when the two windings are parallel as shown in figure 12-4A. At this 0-degree position the flux linkage is maximum and P1 and S1 are opposite in phase. As the secondary is rotated, in respect to the primary, the induced voltage is proportional to the cosine of the angular displacement between the windings, or:

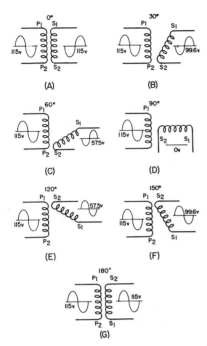

51.295

Figure 12-4.—Amplitude and phase relationships.

$$e_{sec} = e_{max} \cos \theta$$

where e_{sec} is the induced voltage, e_{max} is the maximum induced voltage, and $\cos \theta$ is the cosine of the angle between the two windings. Figure 12-4B through D shows the induced voltage decreasing to zero when the windings are at right angles, and E through G shows the voltage increasing to maximum when the windings are again parallel. Note that the P1 and S1 phase is opposite from 0 degrees to 90 degrees and the same phase from 90 degrees to 180 degrees. If the secondary is rotated further, P1 and S1 will remain in phase until the 270-degree point is reached and then will be opposite again from 270 degrees to 360 degrees, the starting point.

In a TX, CX, or TR the rotor is the primary connected to the 115-v.a.c. line. The three stator windings act as three secondaries. The standard turns ratio between the rotor and a single stator is 2.2 to 1 step-down. Therefore, there will be a maximum of 52 v.a.c. developed across any one stator coil occuring when the rotor is aligned with

that coil. Voltages at various angular displacements are shown in figure 12-5.

Because the common connection between the stator coils is not accessible, it is possible to measure only the terminal-to-terminal voltages. When the maximum terminal-to-terminal effective voltage is known, the terminal-to-terminal effective voltage for any rotor displacement can be determined. Figure 12-6 shows how these voltages vary as the rotor is turned. Values are above the line when the terminal-to-terminal voltage is in phase with the R1 to R2 voltage and below the line when the voltage is 180° out of phase with the R1 to R2 voltage; thus negative values indicate a phase reversal. As an example, when the rotor is turned 50 degrees from the reference (zero degree) position, the S3 to S1 voltage will be about 70 volts and in phase with the R1 to R2 voltage, the S2 to S3 voltage will be about 16 volts and also in phase with the R1 to R2 voltage, and the S1 to S2 voltage will be about 85 volts; 180° out of phase with the R1 to R2 voltage. Although the curves of figure 12-6 resemble timegraphs of a.c. voltages, they show only the variations in effective voltage amplitude and phase as a function of the mechanical rotor position.

The schematic in figure 12-7 represents either a TX, CX, or TR. The rotor is shown in the zero-degree position, aligned with the S2 stator coil, and rotor lead R1 on top. This position is called electrical zero and is important as a reference point, as discussed later.

In figure 12-7, voltages of the amplitude and instantaneous polarities shown will be induced in the stator windings. No current will flow in the stators, however, since the windings constitute an open circuit.

BASIC TX-TR SYSTEM

Figure 12-8 shows the TX-TR system with the TR rotor removed. On the half-cycle when R1 is positive, instantaneous voltages and magnetic field polarities are as shown. On the next half-cycle all polarities are reversed. (The direction of stator current may be determined, if the direction, CW or CCW, of the coil around the core is known, by applying the left-hand rule as explained in Basic Electricity, NAVPERS 10086-B.) One important relationship is apparent, the resultant TR stator field is established in the same direction as the TX rotor field.

If a bar magnet is now inserted into the TR, it will align with the stator magnetic field as shown in figure 12-9. Of course, this simple system is not practical as the resultant TR stator

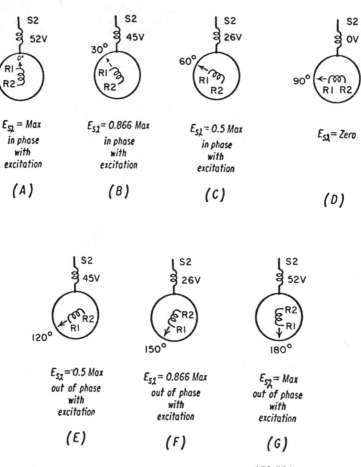

179.764
Figure 12-5.—Relationship between induced
voltage in S2 and position of rotor.

magnetic field will be alternating at the line frequency. However, if the bar magnet was replaced with a coil connected to the a.c. line, its magnetic polarity would change right in step with the stator field polarity change and the system would function properly. When we connect the rotors of the TX and TR to the same 115-v.a.c. line, we have formed a simple TX-TR synchro system as shown in figure 12-10. An outstanding characteristic of the system is that as soon as both rotors are connected in parallel to the same source, the TR rotor

immediately turns to the same position as the TX rotor and the induced voltages in both sets of stators are exactly equal and in opposition to each other. Therefore, no stator current will flow and the rotors are said to be in correspondence. Unless the rotors of both units are connected to the same a.c. source, however, the system will not function properly due to phase differences in the two units.

If we now turn the TX rotor while holding the TR rotor in place, we will be able to illustrate the torque that is developed that causes the TR

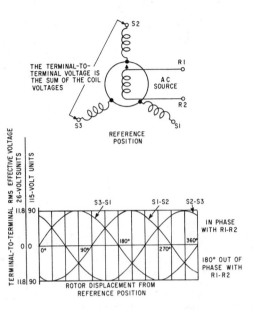

THE TERMINAL-TO-TERMINAL VOLTAGE IS THE SUM OF THE COIL VOLTAGES

72.44.1:83.71.2
Figure 12-6.—Terminal-to-terminal voltage versus rotor position.

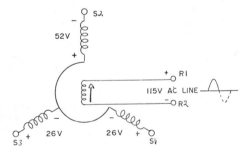

1.119-.123
Figure 12-7.—TX, CX, or TR schematic diagram.

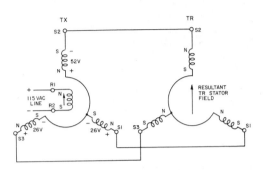

179.765
Figure 12-8.—TX-TR system with TR rotor removed.

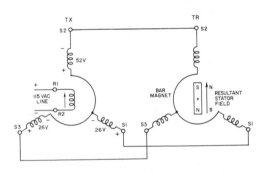

179.766
Figure 12-9.—TX-TR with bar magnet in place of TR rotor.

the current flow is determined by the algebraic sum of the stator voltages. The direction of current flow and the resultant magnetic fields are determined by the polarities of the induced voltages. Without getting concerned with the rather complex vectorial combinations of the various individual magnetic fields, just apply the relationship that was pointed out earlier, that is, the resultant TR stator field will be established in the same direction as the TX rotor field. When the TR rotor is released, due to the attraction of unlike poles, it will immediately turn 30-degrees clockwise to the new position of correspondence. The new induced voltages in the TR stator will exactly equal and oppose those in the TX stator, and stator current will cease. In

rotor to turn to the TX rotor position when it is released. Assume the TX rotor is turned 30-degrees clockwise as shown in figure 12-11. The system is now unbalanced and currents will flow in the stator windings. Remember that both sets of stators act as sources and the amplitude of

203

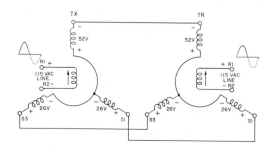

53.11

Figure 12-10.—TX-TR system with rotors
in correspondence.

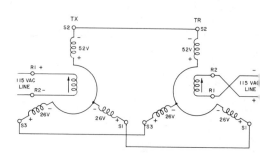

179.767

Figure 12-12.—TX-TR system with rotor
leads reversed.

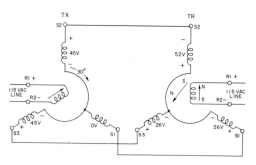

53.12

Figure 12-11.—TX-TR system with rotors
not in correspondence.

actual practice, the TR rotor would not be held
and would move right in synchronization with the
TX rotor, never lagging it by more than a fraction of a degree.

LEAD REVERSALS IN A TX-TR SYSTEM

The operation of a TX-TR system will be affected by any wiring reversals or changes taking
place in the five interconnecting leads between the
units. The effect of reversing the rotor connections
of either the TX or TR is a 180-degree displacement between both shafts, with both shafts still
rotating in the same direction. Figure 12-12 shows
the TR rotor leads reversed. This causes the
R1 leads of the TX and TR to have opposite
instantaneous polarities and the TR must turn 180
degrees to bring the system back to a condition of
equilibrium. The direction of rotation is not affected because that is determined by the stator
connections where no changes have been made.

When the S1 and S3 leads are reversed, the
field component set up in S3 of the TX will now be
transferred to S1 of the TR and S1 of the TX to
S3 of the TR. Both rotors will be in correspondence
at 0 and 180 degrees. However, the TR rotor will
turn in the opposite direction from the TX rotor.
For example, figure 12-13 shows the TX was
turned 60-degrees counterclockwise. This induced
maximum voltage in the S1 stator. The system is
brought back to equilibrium when an equal counter-
voltage is induced in S3 of the TR. For this to
occur, the TR rotor turns 60-degrees CW to the
position where it is in line with the resultant
stator field.

When the S2 leads are reversed with the S1
or S3 leads the TR rotor turns opposite to the
TX rotor and the TR rotor will be 120 degrees
out of correspondence when the TX rotor is at the
0-degree position as shown in figures 12-14 and
15. When there are two reversals in the stator
connections as shown in figure 12-16, the TR rotor
will still be 120 degrees out of correspondence with
the TX rotor but both rotors will now turn in the
same direction.

In summary, reversed rotor leads introduce
a 180-degree error into the system but do not
affect the direction of rotation. One pair of stator
leads reversed will cause the TR rotor to turn in a
direction opposite to the TX rotor. Two reversals
in the stator leads will not change the direction of
rotation. The position of the TR rotor, when the
TX is at 0 degrees, can be determined by tracing
the TX-S2 stator lead over to the TR. The TR rotor
will align with the TR stator that is connected
to the TX-S2 stator lead. The direction it turns
from that position is determined by the number of
reversals between the two stators.

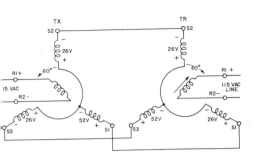

53.14(179)

Figure 12-13.—TX-TR system with S1 and S3 leads reversed.

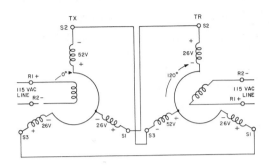

179.769

Figure 12-16.—TX-TR system with S1, S2, and S3 leads reversed.

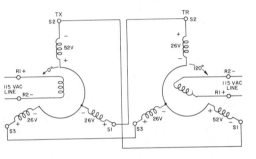

179.768

Figure 12-14.—TX-TR system with S2 and S1 leads reversed.

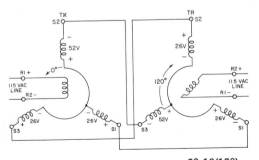

53.12(179)

Figure 12-15.—TX-TR system with S2 and S3 leads reversed.

TX-TDX-TR SYSTEM

When the TR position is to be determined by a single input of angular data, the TX-TR system is sufficient. When the TR position is to be controlled by two or more inputs of angular data, a torque differential transmitter will be employed. The TDX is not connected directly to the 115-v.a.c line but receives its excitation from a TX or another TDX. The stator windings of the TDX act as primaries and the rotor windings act as secondaries with a 1:1 voltage ratio between them. The amplitude and polarity of the voltages induced in the rotor leads are determined by two factors; the position of the resultant stator field and the position of the rotor relative to the stator field.

The resultant stator field of the TDX is in the same direction as the rotor field of the TX, just as it was in the TX-TR system. If the TDX is at zero degrees, as shown in figure 12-17, it merely passes the TX signal on to the TR and the TR turns in synchronization with the TX, just as if the circuit was a basic TX-TR system.

If on the other hand, the TX is at zero degrees and the TDX rotor is turned, the TR rotor will turn the same number of degrees but in the opposite direction. The reason the TR rotor turns in a direction opposite to that of the TDX rotor is that when the TDX rotor is turned it produces the same effect relative to the stator field as turning the stator field in the opposite direction (fig. 12-18).

When both the TX and TDX are rotated, the TR will turn to the difference of the two inputs. For this reason, the circuit connected in this manner is called SUBTRACTIVE. The formula TR° = TX° - TDX° is applied to determine the position

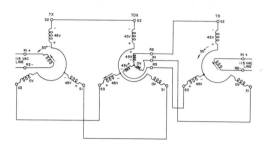

179.770

Figure 12-17.—TX-TDX-TR system with TDX
at zero degrees.

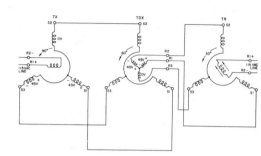

179.772

Figure 12-19.—TX-TDX-TR system (TR° =
90° – 60° = 30°).

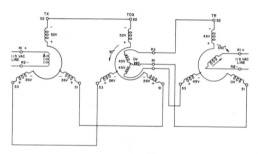

179.771

Figure 12-18.—TX-TDX-TR system with TX
at zero degrees.

of the TR. CCW rotation is assigned a positive
value and CW rotation is negative as shown in
figure 12-19.

As with the TX-TR system, if R1-S1 and R3-
S3 leads between the TDX and TR are reversed,
there will be a change in the TR direction of
rotation and it will now turn in the same direction
as the TDX rotor. Since this reversal is also
between the TX and TR, it will change the TR
direction of rotation with respect to the TX also.
If the TR and TX are to continue to turn in the
same direction, the S1 and S3 leads must be reversed between the TX and TDX. The TX now
has two reversals between it and the TR, so the
original direction of rotation is preserved. The
TX-TDX-TR system is now ADDITIVE and the
formula TR° = TX° + TDX° can be applied to

determine the TR rotor position when the TX and
TDX positions are known (fig. 12-20).

TX-TDR-TX SYSTEM

As previously explained, the differential receiver differs from the differential transmitter
mainly in its application. The TDR receives two
electrical inputs and provides the mechanical
output. Both rotor and stator receive energizing
signals from TX's or TDX's and the mechanical
output is the sum or difference of the angles
represented by the inputs. The TDR is identical
to the TDX electrically. Mechanically the only
difference is the inertia damper required by the
TDR.

In considering the operation of the TDR, (fig.
12-21), it is important to remember that its rotor
currents do not flow as a direct result of rotor
voltages induced by the changing stator field, but
as the result of an unbalance between these induced voltages and the induced stator voltages of
TX2 to which the TDR rotor is connected. When
the rotor of TX2 is turned, its stator voltages are
changed and current flows in the TX2 stator and
the TDR rotor coils. The TDR rotor field established by these currents rotates in the same
direction as the TX2 rotor. The TDR stator and
rotor fields are displaced with respect to each
other and a strong magnetic torque brings the two
fields back into alignment. Since the TDR rotor
is free to move, it rotates accordingly, restores
the voltage balance in the TDR rotor circuits and
reduces current flow to a low value. As shown in
figure 12-21, the signal from TX1 connected to
the TDR stator rotates the resultant stator field
75-degrees CCW. In a similar manner, the signal

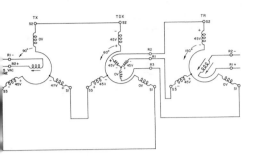

179.773

Figure 12-20.—TX-TDX-TR system (TR° = 90° + 60° = 150°).

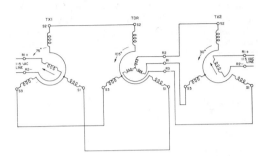

179.775

Figure 12-22.—TX-TDR-TX system (TDR° = TX1° + TX2° = 75° + 30° = 105°).

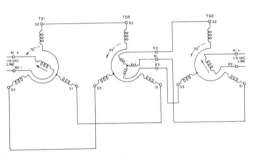

179.774

Figure 12-21.—TX-TDR-TX system (TDR° = TX1° - TX2° = 75° - 30° = 45°).

from TX2 rotates the resultant rotor field 30-degrees CCW. Since the two resultant fields are not rotated equal amounts, torque is developed to bring them into alignment. The rotor therefore turns to 45 degrees, at which point the two fields are aligned. To bring its resultant field into alignment, the TDR rotor need only be turned through an angle equal to the difference of the signals supplied by the two TX's as expressed by the formula TDR° = TX1° - TX2°.

To set the TDR system for addition, it is only necessary to reverse the R1-S1 and R3-S3 leads between the TDR rotor and the TX2 stator as shown in figure 12-22. The TDR stator field still rotates 75-degrees CCW with the TX1 input, but because of the reversed connections between the TDR rotor and TX2 stator, the rotor field turns

30-degrees CCW with the TX1 input. The angular placement of the two fields, with respect to each other, is the sum of the signals sent by the two TX's, and the magnetic force, pulling the TDR rotor field into alignment with the stator field, turns the rotor to the 105-degree position.

CONTROL TRANSFORMER

The distinguishing unit of any synchro control system is the control transformer. The CT is a synchro designed to supply, from its rotor terminals, an a.c. voltage whose magnitude and phase is dependent on the rotor position, and on the signal applied to the three stator windings. The behavior of the CT in a system differs from that of the synchro units previously considered in several important respects.

Since the rotor winding is never connected to the a.c. supply, it induces no voltage in the stator coils. As a result, the CT stator currents are determined only by the voltages applied to them. The rotor itself is wound so that its position has very little reflected effect on the stator currents. Also, there is never any appreciable current flowing in the rotor, because its output voltage is always applied to a high-impedance load, 10,000 ohms or more. Therefore, the rotor does not turn to any particular position when voltages are applied to the stators.

The rotor shaft of a CT is always turned by an external force, and produces varying output voltages from its rotor winding. Like synchro transmitters, the CT requires no inertia damper, but unlike either transmitter or receivers, rotor coupling to S2 is minimum when the CT is at electrical zero.

The electrical zero position is located where the rotor coil is perpendicular to the S2 stator winding. Since at electrical zero the rotor is at a 90-degree angle to the S2 stator and resultant magnetic field, there will be no voltage induced into the rotor by S2. Actually S1 and S3 induce voltages, but they are equal and opposite in phase and cancel.

When the rotor is turned 90 degrees from electrical zero, it is lined up with S2, and the resultant field and the voltage output is maximum, or 55 volts, as determined by the turns ratio of the CT. There are two positions for zero voltage and two positions for maximum voltage. The error voltage varies not only in magnitude but also in phase with respect to the a.c. line voltage. The output will either have the same phase as the line or the opposite. In effect, the amplitude of the error voltage is proportional to the amount of error, while the phase with respect to the line voltage shows the direction of error, clockwise or counterclockwise.

The rotor of a CT is turned by a shaft usually connected through a gear arrangement to the load as shown in figure 12-23. This is actually a closed loop servosystem. When the rotor of the CX is turned, the CT provides the phase sensitive detector/amplifier with an error signal. The phase sensitive detector/amplifier produces a d.c. power output of the correct polarity to cause the d.c. motor to turn the load in a direction determined by the error signal's instantaneous polarity with respect to a.c. line voltage. As the load turns, the mechanical gearing between the load and CT rotor provides a feedback that reduces the error to zero as the load moves to a position designated by the angular position of the CX rotor. Servosystems are discussed in more detail later in this chapter.

SYNCHRO CAPACITORS

In a circuit where a transmitter is supplying a signal to a differential unit or control transformer, the transmitter is supplying a lagging current to the stator of the differential unit or control transformer. To minimize the amount of current lag and, thereby, improve the accuracy of the system, a capacitive network is placed in the circuit. This network is generally termed a SYNCHRO CAPACITOR, even though the network contains three individual capacitors and the total capacitance of all three capacitors is the rated value of the network. The three capacitors are delta connected as shown in figure 12-24 and are connected in parallel with the stator windings of the differential unit and CT (fig. 12-24). The connections are made as short as possible, as high currents in long leads increase the transmitter load and reduce the system accuracy.

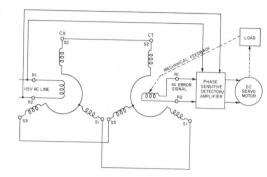

179.776
Figure 12-23.—CX-CT closed loop servosystem.

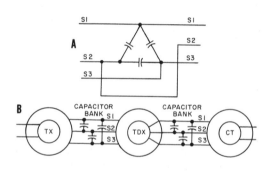

12.279
Figure 12-24.—Synchro capacitors.

Synchro capacitors decrease the line current drawn by synchro systems and, in effect, increase the torque of the synchro receivers in the system. This effective increase in torque near the point of synchronization increases the accuracy of the overall system.

Currents present in the stator circuits of a TX-TR synchro system are a result of a voltage difference between the stators. If the synchros are in correspondence there is no voltage difference and hence no current in the stator circuits. Therefore, capacitors are not required in a TX-TR synchro system.

SYSTEM SPEEDS

A single-speed synchro system must turn one complete revolution to transmit a full range of alues. One revolution of the input shaft produces one revolution of the synchro transmitter rotor which will in turn produce one revolution of the synchro receiver rotor. The single-speed synchro system has the advantage of one electrical zero position which makes the system self synchronous. This means that when the system is energized, all of the units are immediately in correspondence. A disadvantage of the single-speed system, however, is inaccuracy.

The accuracy of a synchro system can be greatly improved by using a 36-speed system. In this system, one rotation of the input shaft produces 36 rotations of the rotors in the synchros and one rotation of the output shaft. The ratio between input and output shafts, and the synchro rotors is obtained through gearbox arrangements. The system error is reduced by a factor of 36 but the system now has 36 correspondence points and therefore cannot be self-synchronous. When the system is energized the output shaft could be in any of 36 positions with respect to the input shaft.

It is very common to combine the single-speed and 36-speed systems into a 1:36 or dual-speed system. This system is shown in figure 12-25. Only one of the two synchro systems is energized at a given instant. The single-speed or coarse system is energized when the error is greater than a few degrees. The 36-speed or fine system is energized when the error is small. The dual-speed system has the accuracy of a 36-speed system while retaining the self-synchronous feature of the single-speed system. A mechanical

or electrical sensing device responds to the amount of system error and determines which system is energized by applying line voltage to the synchro rotors.

ZEROING SYNCHROS

If synchros are to work together properly in a system, it is essential that they be correctly connected and aligned in respect to each other and to the other devices with which they are used. Electrical zero is the reference point for alignment of all synchro units. The mechanical reference point for the units connected to the synchros depends upon the particular application of the system. When a synchro system is used to repeat ship's course data, the reference point would be true north. For radar and sonar equipment, the reference point would be the ship's bow or zero degrees relative. In a range or azimuth transmission system, a specific distance or angle would be used for the reference point. Whatever the application, the electrical and mechanical reference points must be aligned. The mechanical position is set first and then the synchro device must be aligned to electrical zero.

The two methods used to electrically zero synchro devices are the voltmeter method and the electrical lock method. Generally the TX, CX, TDX, CDX, and CT are zeroed by the voltmeter method, while the devices with a free rotor, the TR and TDR, are zeroed with the electrical lock method. With either method, the voltage applied to the synchro device should be 78 volts for 115 volt synchros if a source of this value is available.

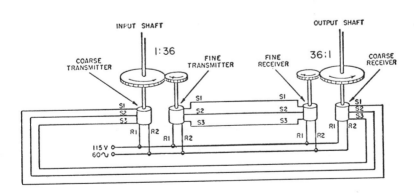

12.57(179)

Figure 12-25.—1 and 36 speed system.

209

This prevents overheating of the unit. If a 78 volt source is not available, 115 volts may be used providing that it is not applied for more than 1 or 2 minutes.

When using the voltmeter method, the most accurate results can be obtained by using an electronic or precision voltmeter having 0 to 250 and 0 to 5 volt ranges. On the 0 to 5 volt range, the meter should be able to measure voltages as low as 0.1 volt. The procedure is divided into two parts. The first is a coarse setting to ensure the device is zeroed on the zero-degree position

rather than the 180-degree position. The second part is a fine setting where the unit is accurately zeroed. The lead and voltmeter connections for coarse and fine setting are shown in figure 12-26 through 12-28 for units that are normally zeroed with the voltmeter method. The voltmeter range is indicated in the diagram. The procedure, which applies to all units, is as follows:

1. Connect the leads and voltmeter as shown for coarse setting the unit.

2. Adjust the rotor or stator, depending on

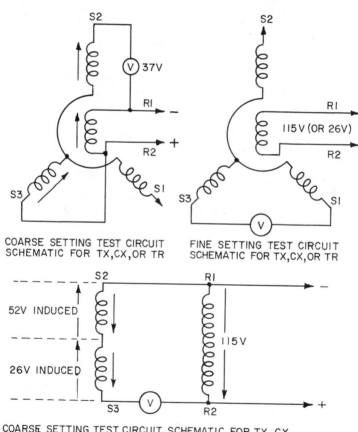

COARSE SETTING TEST CIRCUIT
SCHEMATIC FOR TX,CX,OR TR

FINE SETTING TEST CIRCUIT
SCHEMATIC FOR TX,CX,OR TR

COARSE SETTING TEST CIRCUIT SCHEMATIC FOR TX, CX,
OR TR (SIMPLIFIED) 115V – 78V (INDUCED) = 37V
(APPROXIMATE ZERO)

1.280(179)

Figure 12-26.—Zeroing a synchro by the voltmeter method.

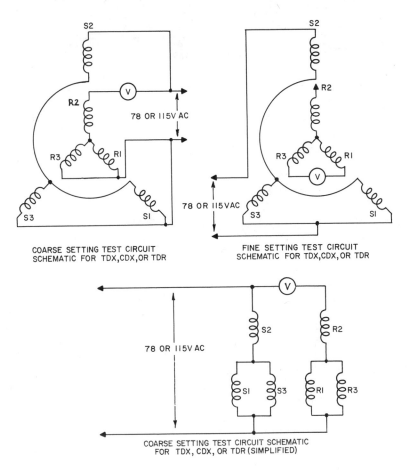

COARSE SETTING TEST CIRCUIT
SCHEMATIC FOR TDX,CDX,OR TDR

FINE SETTING TEST CIRCUIT
SCHEMATIC FOR TDX,CDX, OR TDR

COARSE SETTING TEST CIRCUIT SCHEMATIC
FOR TDX, CDX, OR TDR (SIMPLIFIED)

12.277(179)

Figure 12-27.—Zeroing a differential by the voltmeter method.

particular installation, for a minimum voltmeter reading.

3. Reconnect the leads and voltmeter as shown for fine setting the unit.

4. Adjust the rotor or stator for a minimum voltmeter reading.

5. Remove the lead and voltmeter connections.

The synchro device is now properly zeroed and should function accurately when connected to other properly zeroed units in the synchro system.

The TR may be aligned using the electrical lock method if the rotor is free to move. When the unit is connected as shown in figure 12-29, the rotor will be locked at the zero-degree position by the magnetic fields set up by resultant currents in the stator and rotor windings. The dial or pointer may then be loosened and turned to the zero or reference position. The dial or pointer is then tightened and the connections are removed.

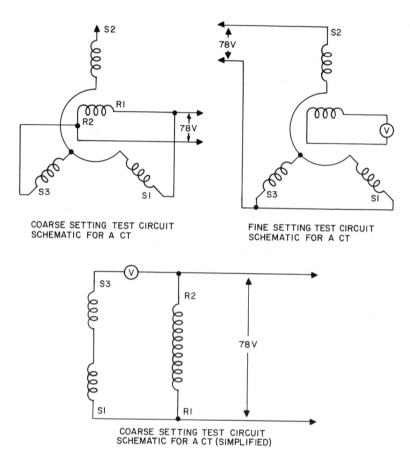

COARSE SETTING TEST CIRCUIT
SCHEMATIC FOR A CT

FINE SETTING TEST CIRCUIT
SCHEMATIC FOR A CT

COARSE SETTING TEST CIRCUIT
SCHEMATIC FOR A CT (SIMPLIFIED)

179.777
Figure 12-28.—Zeroing a CT by the voltmeter method.

RESOLVERS

In appearance and construction the resolver is similar to the synchro. The use of the term resolver for this unit comes from the fact that the unit is used to resolve a vector quantity, which has been represented electrically, into sine and cosine components. The resolver contains a rotor and a stator. Uusually the stator is composed of two coils whose axes are oriented at right angles to each other. The rotor is wound with one or two coils, depending on the application. The resolver used for an example here uses a single rotor coil.

The operation of the resolver is somewhat analogous to that of a synchro control transformer in that it is used to produce voltage outputs rather than rotation. Figure 12-30 is a schematic diagram of a synchro resolver with two stator windings, S1 and S2, used as secondaries and a single rotor used as a primary. Assume that with the rotor in the position shown the angular position represented is zero degrees. In this position there is zero voltage induced in S1 and maximum voltage induced in S2. As the antenna to which the rotor is connected turns, the voltage in S1 increases while the voltage in S2 decreases. These two voltages

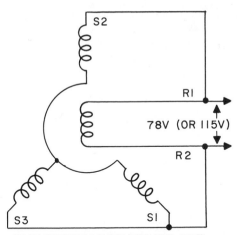

TEST CIRCUIT SCHEMATIC FOR THE
ELECTRICAL LOCK METHOD

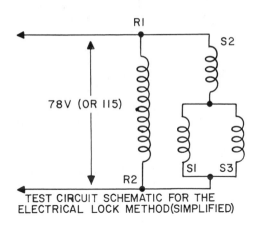

TEST CIRCUIT SCHEMATIC FOR THE
ELECTRICAL LOCK METHOD(SIMPLIFIED)

1.119A(179)
Figure 12-29.—Zeroing a synchro by the
electrical lock method.

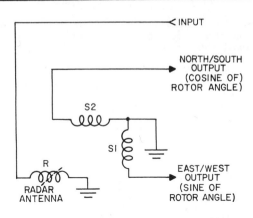

179.778
Figure 12-30.—Resolver schematic.

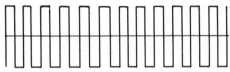

I. INPUT TO SWEEP RESOLVER PRIMARY (500 Hz)

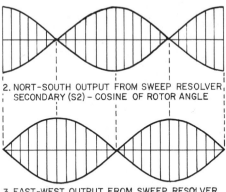

2. NORT-SOUTH OUTPUT FROM SWEEP RESOLVER,
SECONDARY (S2) - COSINE OF ROTOR ANGLE

3. EAST-WEST OUTPUT FROM SWEEP RESOLVER
SECONDARY-(SI)-SINE OF ROTOR ANGLE

4. ANGULAR POSITION OF PPI SWEEP

179.779
Figure 12-31.—Sweep resolver voltages.

represent the sine and cosine of the antenna rotor
angle and can be used to determine the position
of a PPI sweep.

The rotating sweep application of a synchro
resolver is shown in figure 12-31. The 5000-Hz
square wave is applied to the rotor input. The
output of the resolver is sine wave modulated as
the rotor is turned at a constant speed producing
the sine and cosine outputs shown. The outputs

213

are detected and the two sine waves are used to control the amplitude of the outputs from trapezoidal voltage sweep generators. The output from the sweep generators will be 90 degrees out of phase and vary in amplitude at a sinusoidal rate. This output applied to the CRT deflection coils produces the rotating trace synchronized with the rotating antenna for accurate azimuth indication.

SERVOMECHANISMS

A synchro system used alone is ideal for the electrical transmission of angular position data. The synchro system develops sufficient torque to move pointers, dials, and other small loads. When it is necessary to position large loads much more power must be provided. The device used in this situation is called a SERVOMECHANISM or SERVOSYSTEM. A servomechanism is an electro-mechanical system which positions a load in accordance with a variable signal. There are many applications for servomechanisms in the U. S. Navy. A few of these applications are listed below:

1. radar antenna position
2. ship's rudder control
3. gun or missile launcher position
4. aircraft and submarine controls
5. elevator control on carriers

The servomechanism must detect and correct error. The error detection device provides a variable signal which is proportional to the difference between the actual position of the load and the desired position. This variable signal is called the error signal. A commonly used error detection device is the synchro control system consisting of a control transmitter and a control transformer. The error signal is the electrical output of the CT. This error signal is limited in power and must be amplified before it is applied to the error correction device.

The error corrector of a servomechanism must provide the mechanical force necessary to position the load. Amplified error signals, in many cases, must be modified to meet the input requirements of the error corrector which may be a d.c., single phase a.c., or polyphase a.c. electric motor. In summary, the overall purpose of the servomechanism is to detect an error, amplify the weak error signal, and convert it into a form which will meet the input requirements of the error correcting device, which then positions the load, reducing the error to zero.

The essential components of a servomechanism are the input and output controllers. The input controller may be a synchro control system which detects the error and produces an electrical error signal output. The input controller has already been discussed as the error detection device.

The output controller consists of the servo-amplifier and the error correction device, usually an electric servomotor. The output controller's function is to amplify and convert the error signal into one suitable for the servomotor which then moves the load to the desired position.

Both a.c. and d.c. motors are used in servo-systems depending upon the requirements of the system. Systems required to position heavy loads with a wide speed range use d.c. motors, whereas the light loaded and fairly constant speed systems use a.c. motors.

The most common type of a.c. servomotor is the 2-phase induction motor. This motor has a reference (fixed) field, and a control (variable) field spaced 90 electrical degrees apart as shown in figure 12-32. The rotor is usually of the squirrel-cage type, however, other types are sometimes used.

If two a.c. supply voltages 90° out of phase with each other are applied to the reference and control fields (fig. 12-32), the currents in the fields flowing as a result of the applied voltages will also be 90° out of phase with each other; and since the magnetic fields produced by the currents will be in phase with currents producing them, the magnetic fields will be 90° displaced from each other. These magnetic fields will add vectorially to produce a resultant rotating field

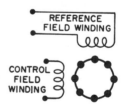

SHORT-CIRCUITED
ROTOR WINDING

55.39(175)
Figure 12-32.—2-phase a.c. servomotor.

as shown in figure 12-33. The rotating magnetic field induces voltages in the rotor causing the rotor to rotate in the direction of the magnetic field as a conventional induction motor.

The direction of rotation of the 2-phase servomotor depends upon the phase of the control field voltage (which either leads or lags the reference voltage by 90°), as shown by figure 12-33. Varying the magnitude of the current in either the control or reference field will vary the motor torque. In addition, if either field is deenergized, there will no longer be a rotating magnetic field established and the motor will stop. In actual practice in a servosystem, the reference field is supplied from a constant voltage source and the control field is supplied from the output of a servoamplifier. Thus, the motor speed and direction is controlled by the servoamplifier output.

Split-phase a.c. motors are used as servomotors in some applications. The capacitor-run type discussed in Basic Electricity, NAVPERS 10086 (revised) is the type generally used. Reversal and control are accomplished in the same manner as for the 2-phase induction motor. Other types of a.c. motors used as servomotors in certain applications, which are also discussed in Basic Electricity, NAVPERS 10086 (revised), are the shaded pole, salient pole, and universal motors.

Another type of a.c. servomotor which is used in very low power applications is the dragcup servomotor. This motor has a 2-phase (reference and control) stator winding. The rotor, however, consists of a thin aluminum or copper cup, with the rotor flux being carried by a stationary magnetic core as shown in figure 12-34.

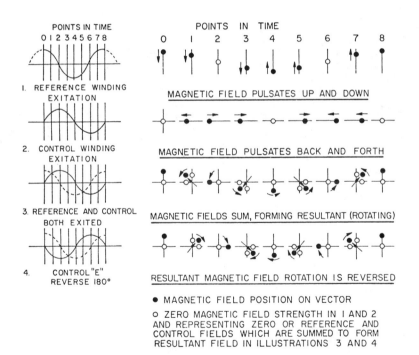

175.1

Figure 12-33.—Reference and control field winding excitation and magnetic field vectors.

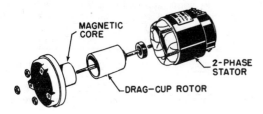

175.2

Figure 12-34.—Drag-cup servomotor.

OPEN AND CLOSED
LOOP SYSTEMS

There are two basic types of servomechanisms; OPEN LOOP and CLOSED LOOP. In the open loop system, the input controller receives no feedback from the load and therefore does not sense the load's position. The open loop system is only used to control the load's speed and direction.

In the closed loop system, the input controller does receive feedback from the load and senses the load's instantaneous position. The closed loop system may be used to stop the load at any specified position as determined by the operator.

An essential difference between open loop and closed loop systems is feedback from load to input controller in closed loop systems. This feedback is usually mechanical. The block diagram of a closed loop system is shown in figure 12-35.

The remainder of this chapter discusses servoamplifiers.

AMPLIDYNE

The amplidyne is an electro-mechanical servoamplifier which is used to amplify a low power d.c. error signal sufficiently to drive a d.c. servomotor. The amplidyne is basically a modified d.c. generator. However, much less control field power is dissipated. Where the conventional d.c. generator has a power gain of 25 to 100, the amplidyne has a power gain of 3000 to 10,000.

The amplidyne is fundamentally a two-stage generator combined in a single machine using a single armature. Development of an amplidyne from a conventional d.c. generator may be used to illustrate the high gain characteristics of the amplidyne by contrast. Figure 12-36 shows the magnetic fields of a conventional d.c. generator.

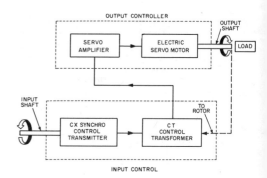

72.46(179)

Figure 12-35.—Closed loop
servomechanism.

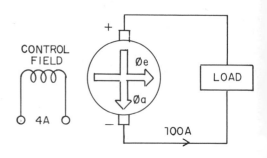

179.780

Figure 12-36.—Conventional d.c.
generator.

The induced voltage in the armature produces a load current of 100 amperes. This armature current produces an armature reaction field, $\emptyset a$, that in this machine has been made equal to the control field, $\emptyset e$.

Figure 12-37 illustrates that, by removing the load and short-circuiting the brushes, a much smaller control field current is required to cause 100 amperes of current flow through the low resistance armature and produce the same armature reaction field strength obtained in figure 12-36.

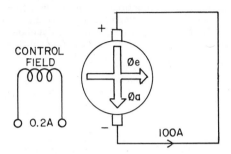

179.781
Figure 12-37.—D.c. generator with
short-circuited brushes.

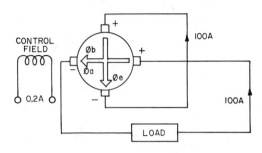

179.782
Figure 12-38.—Basic amplidyne.

Armature reaction field $\emptyset$a will induce a voltage in the armature which will be maximum at right angles to the short-circuited brushes. Since $\emptyset$a is equal in strength to $\emptyset$e, the voltage induced by a $\emptyset$a would cause 100 amperes to flow through the load as in figure 12-38.

This in turn induces a second armature reaction field, $\emptyset$b, which would cancel the control field and make the device useless. However, this problem is eliminated by using a compensating winding through which load current flows. This compensating winding induces a field equal and opposite to $\emptyset$b. Figure 12-39 shows the basic amplidyne with compensating field.

Figure 12-40 illustrates the basic amplidyne system performing as an output controller section of a servomechanism. The a.c. error signal (fig. 12-40) is amplified and rectified by the control amplifier whose d.c. output polarity is determined by the phase of the error signal with respect to the reference voltage. The amplidyne armature is driven at a constant speed by a three-phase a.c. motor. The error signal phase determines the direction of the control field which in turn determines the amplidyne output polarity and the direction of rotation of the d.c. servomotor.

When the error detector produces an alternating error signal and the error corrector requires such an input, the servoamplifier senses the direction of error and amplifies the error signal to the level required. A representative transistor circuit of this type is shown in figure 12-41 and the comparative electron tube circuit is shown in figure 12-42.

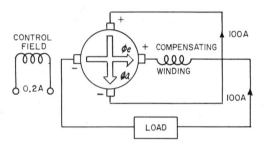

179.783
Figure 12-39.—Amplidyne with compensating winding.

Transistor
Servoamplifier

Transistors Q1, Q2, and Q3 and their associated circuit components make up the preamplifier section. Q4 and Q5 comprise the driver stage, and Q6 and Q7 are used in the output stage.

The circuits of Q1 and Q2 are common-emitter amplifiers with the output of Q1 directly coupled to Q2. Emitter resistors R2 and R5 provide bias stabilization. The degenerative feedback path from the emitter of Q2 to the base of Q1 through R3 reduces gain changes due to circuit parameter variations. The output from Q2 is RC coupled to the common-emitter amplifier Q3, whose output is directly coupled to Q4. A degenerative signal is fed back through R10 from the emitter of Q4 to the base of Q3.

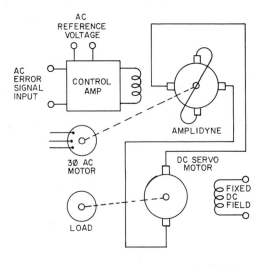

179.784
Figure 12-40.—Basic amplidyne drive.

Q4 and Q5 provide push-pull action across the primary of T1. Q4 functions as an emitter follower whose output is coupled through R_F and C7 to the emitter of Q5 which performs as a common-base amplifier. In this combination (an emitter follower feeding into a common-base amplifier), the collector currents of the two transistors vary in opposite directions; when the collector current of Q4 increases, the collector current of Q5 decreases and vice-versa. For example, when a positive going signal is applied to the P type base of Q4, the transistor conducts more, increasing the collector current through the upper primary winding of T1. The increasing positive voltage on the Q4 emitter is coupled to the N type emitter of Q5, decreasing the forward bias and the collector current through the lower primary of T1. The overall polarity of the T1 primary would then be negative on top, positive on the bottom. With a negative signal applied to the Q4 base, the action would be just the opposite. Q4 would conduct less, Q5 would conduct more, and the polarity across the T1 primary would reverse. The primary of T1 and capacitor C6 form a resonate circuit.

The circuit of Q6 and Q7 is a push-pull common-base amplifier. The emitters receive

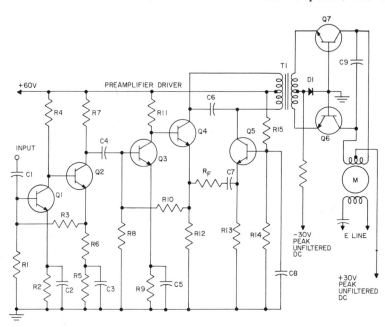

179.785
Figure 12-41.—Transistor servoamplifier.

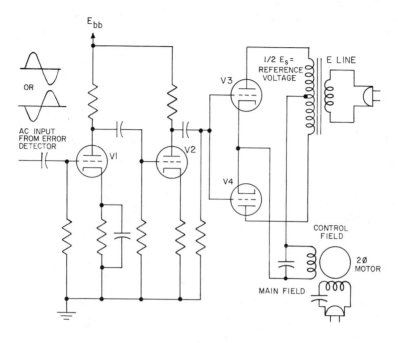

179.786
Figure 12-42.— Electron tube servoamplifier.

their inputs from opposite ends of the T1 secondary, and the collectors are connected to opposite ends of the tapped control field for the servomotor. With no input error signal, Q4 and Q5 conduct equally and in opposite directions through the T1 primary. No signal is developed across the secondary, so Q6 and Q7 are conducting equally and in opposite directions through the tapped control field winding resulting in zero control field.

When an error signal is applied and the top of the T1 secondary is swinging positive, Q7 will conduct less and Q6 will conduct more. The higher current path is from the -30 volt peak supply to the T1 secondary center tap, down through the lower T1 secondary winding, through the emitter-collector junction of Q6 and up through the lower control winding to the +30-volt peak supply. On the negative half-cycle Q7 conducts more and Q6 conducts less, so an alternating field is established in the motor control windings by the Q6 and Q7 collector currents.

The alternating current in the motor control windings will either be in phase with the line voltage or 180° out of phase. Line voltage applied to the motor main field is shifted in phase approximately 90° by the capacitor in series with the field, so the control field will lead or lag the main field by 90° and determine the direction of rotation.

Diode D1 protects Q6 or Q7 when the amplitude of the input signal is too large. For example, if the top of the T1 secondary is positive, Q6 will conduct more. If the input continues to increase it will eventually overcome the reverse bias on D1, provided by the -30-volt peak supply, and forward bias D1. D1 will then shunt the -30-volt peak supply and reduce the conduction level of Q6.

Capacitor C9 is in parallel with the motor control field and is used to develop high circulating current for maximum motor torque.

The collector and emitter supply voltages of Q6 and Q7 are obtained from full-wave unfiltered rectifier circuits operating from the same 400 Hz supply as the servomotor and synchro control system. The use of an unfiltered supply voltage reduces the heat dissipation of Q6 and Q7.

Electron Tube
Servoamplifier

When analyzing the operation of the electron tube circuit assume that the voltage reference is the top of the transformer secondary and the plate of V3. When the error detector produces an output, the error signal applied to the grid of V1 will either be in phase with the reference voltage or 180° out of phase. V1 and V2 are class A amplifiers, so the signal applied to the grids of V3 and V4 is a sine wave in phase with the error signal on the V1 grid. Since the plates of V3 and V4 are connected to opposite sides of the transformer secondary, the instantaneous plate potentials are opposite in polarity. Only one of the two plates will be positive during the positive alternation of the grid signal therefore only that tube will conduct at this time.

Figure 12-43 shows the error signal in phase with the reference voltage. During the first alternation the grids are going positive, the plate of V4 is going negative. Only V3 can conduct and current flows down through the control field winding.

On the next alternation the grids are going negative so neither tube will conduct. The result is half-wave pulses of plate current. The capacitor in parallel with the control field winding (fig. 12-42) produces a flywheel effect which restores the missing half-cycles and keeps the control field current sinusoidal rather than pulsating d.c.

Figure 12-44 illustrates the error signal 180° out of phase with the reference voltage. During the first half-cycle neither tube will conduct since both grids are going negative. During the second alternation both grids are going positive, the plate of V3 is negative, and V4's plate is positive. V4 will conduct and current again flows down through the control field winding.

The capacitor in series with the main field winding (fig. 12-42), produces an approximate 90° phase shift with respect to the reference source. Since the control field is energized by the plate current of either V3 or V4 and the plate voltages of these two tubes are 180° out of phase from each other, the control winding is excited either 90° leading or 90° lagging the main field winding. Direction of rotation of the servomotor therefore depends on the phase of the error signal which in turn determines which of the two output tubes will conduct.

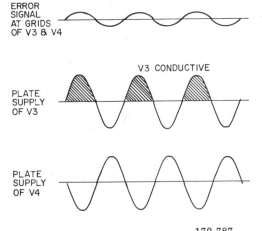

179.787
Figure 12-43.—Error signal in phase with reference voltage.

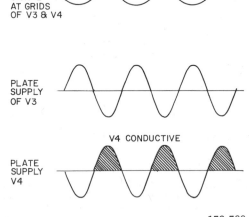

179.788
Figure 12-44.—Error signal 180° out of phase with reference voltage.

HUNTING AND
ANTI HUNT CIRCUITS

A servomechanism is susceptible to instability in the form of sustained mechanical oscillations called HUNTING. When this type of instability is present, the output shaft will swing back and forth through the desired rest position, even though the input shaft is held stationary. Such oscillations are produced by too large a time lag in the system. Since the servomechanism cannot respond in zero time to an input change, the corrective action lags behind the change of the input shaft when the error has been reduced to zero. The servomechanism overcorrects, producing an error signal in the opposite direction, overcorrects again, and so on. The output shaft oscillates around the desired angular position.

A simple anti-hunt circuit is shown in figure 12-45. When an error exists in the circuit, the potential difference between the command and followup wiper arms is applied to the RC network. While the error is increasing capacitor C charges, which increases the voltage across R2 resulting in a greater input to the servoamplifier. The corrective action is exaggerated permitting the output shaft to accelerate rapidly. When the output shaft begins to catch up, the error begins to reduce and the capacitor begins to discharge. This reduces the voltage across R2 causing the amplifiers error input to appear smaller than it actually is. Corrective action is therefore decreased as the error approaches zero, preventing overshoot and hunting.

An anti-hunt circuit used in many Navy applications with a synchro control system error detector is the tachometer error rate control shown in figure 12-46. The d.c. output of the tachometer, a d.c. generator, depends on the speed of the servomotor. The faster the motor turns the greater the output from the generator. A change in the output shaft speed produces a change in the generator output voltage. This change is coupled through a differentiator to a modulator which converts the d.c. rate signal to an a.c. signal at the frequency of the synchro control system. When the error signal and resultant output shaft speed are increasing, the feedback from the tachometer through the differentiator and modulator to the servoamplifier is in phase with the error signal from the synchro transformer making the error appear to be larger than it actually is. This causes the output shaft to be accelerated rapidly. When the amount of error is decreasing the feedback from the tachometer error rate control is out of phase with the error signal causing the error at the servoamplifier input to appear smaller than it actually is, thus reducing overshoot and the resultant hunting.

Rate generators (usually referred to as tachometer generators) used in servosystems are small a.c. or d.c. generators which develop an output voltage (proportional to the generator r.p.m.) whose phase or polarity is dependent upon the direction of rotation. Direct current rate

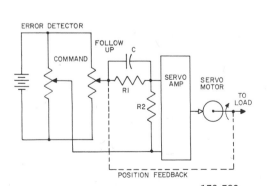

179.789

Figure 12-45.—Simplified anti-hunt circuit.

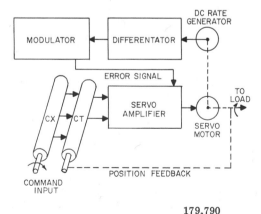

179.790

Figure 12-46.—Tachometer error rate control.

generators usually have permanent magnetic field excitation, whereas the a.c. units are excited by a constant a.c. supply.

The most common type of a.c. rate generator is the drag-cup type constructed similar to figure 12-34.

The generator has two stator windings 90° apart, and an aluminum or copper cup rotor. The rotor rotates around a stationary soft-iron magnetic core. One stator winding is energized by a reference a.c. source. The other stator winding is the generator output or secondary winding.

The voltage applied to the primary winding creates a magnetic field at right angles to the secondary winding when the rotor is stationary, as shown in figure 12-47A. When the rotor is turned, it distorts the magnetic field so that it is no longer 90 electrical degrees from the secondary winding. Flux linkage is created with the secondary winding and a voltage is induced (fig. 12-47B and C). The amount of magnetic field that will be distroted is determined by the angular velocity of the rotor. Therefore, the magnitude of the voltage induced in the secondary winding is proportional to the rotor's velocity.

The direction of the magnetic field's distortion is determined by the direction of the rotor's motion. If the rotor is turned in one direction, the lines of flux will cut the secondary winding in one direction. If the motion of the rotor is reversed, the lines of flux will cut the secondary winding in the opposite direction. Therefore, the phase of the voltage induced in the secondary winding, measured with respect to the phase of the supply voltage, is determined by the direction of the rotor's motion.

The frequency of the generator output voltage is the same as the frequency of the reference voltage. This is true because the magnetic field produced by the primary winding fluctuates at the supply's frequency. The output voltage is generated by the alternating flux field cutting the secondary winding; therefore, the output voltage must have the same frequency as the supply voltage.

Other types of a.c. rate generators have a squirrel-cage rotor. Otherwise their construction and principles of operation are identical to the drag-cup type.

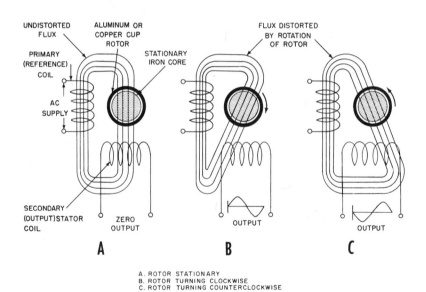

A. ROTOR STATIONARY
B. ROTOR TURNING CLOCKWISE
C. ROTOR TURNING COUNTERCLOCKWISE

55.40
Figure 12-47.—A.c. drag-cup rate generator.

The d.c. rate generator employs the same principles of magnetic coupling between the reference winding and the output winding as the a.c. generator. The d.c. rate generator, however, has a stationary primary magnetic field. This magnetic field is usually supplied by permanent magnets as stated previously. The amount of voltage induced in the rotor winding is proportional to the magnetic flux lines the winding cuts.

The polarity of the output voltage is determined by the direction in which the rotor cuts the lines of magnetic flux.

Rate generators are used in servosystems to supply velocity or damping signals and are sometimes mounted on the same shaft with, and enclosed within, the same housing as the servomotor.

CHAPTER 13

NUMBER SYSTEMS AND LOGIC

This chapter will serve as a basis for the understanding of machines designed to perform logical functions (computers and associated devices). A knowledge of number systems other than the decimal system is a necessity as is a knowledge of Boolean algebra.

NUMBER SYSTEMS

A number system is any set of symbols or characters used for the purpose of enumerating objects and performing mathematical computations such as addition, subtraction, multiplication, and division. All number systems are related to each other by symbols or characters commonly referred to as digits.

NOTE: All modern number systems will have certain digits in common; however, they do not all use the same number of digits as illustrated in table 13-1.

Our most commonly used system is the HINDU-ARABIC SYSTEM, which uses the digits 0, 1, 2, 3, 4, 5, 6, 7, 8, and 9. Since most measurements are made using this system, it will be used as the basis for a discussion of other number systems.

In ancient times number systems were used primarily for the purpose of making measurements and keeping records since mathematical computations using early number systems were extremely difficult. This lack of an adequate number system probably was a major factor in hampering scientific development in those early civilizations.

The acceptance of two basic concepts has greatly simplified mathematical computations and led to the development of modern number systems. These two concepts are (1) the use of zero to signify the absence of an object or unit and (2) the principle of positional value.

The principle of positional value consists of assigning a digit a value which depends on the digit's own value and a weighting value which is determined by the digits position within a given number. For example, in each of the decimal numbers 456, 654, and 564 the digit 6 will have a different value; in the first number the digit 6 has its basic value (6); in the second number, it has a value of 600 (6 x 10 x 10) and in the third it has a value of 60 (6 x 10). Sometimes a position within a given number will not have a value. However, if this position is left out (omitted) then there is no way to distinguish between two different quantities such as 505 and 55. Thus, the zero is used to signify that a particular position within a given number has no value assigned. As may be seen, the use of these two concepts has greatly simplified counting and mathematical computations. Thus, they are used in all modern number systems.

Before continuing, consideration should be given to the following definitions which are applicable to all numbering systems:

1. UNIT...A single object or thing.
2. NUMBER...An arbitrary symbol or group of symbols
3. NUMBER SYSTEM...A method of indicating the number of units counted.

    a. All modern number systems include the zero.

    b. The RADIX, or BASE, of a number system is the number of characters or symbols it possesses, including the zero.

4. QUANTITY...A number of units (implies both a unit and a number).

5. MODULUS...The total number of different numbers or stable conditions that a counting device can indicate. (For example, the odometer on most automobiles has a modulus of 100,000 since it indicates all numbers from 00,000 to 99,999. The modulus of the hour hand on most watches is 12, and that of the minute hand is 60.)

Table 13-1.— A Comparison Of Four Of the More Commonly Used Number Systems

BINARY	OCTAL	DECIMAL	DUODECIMAL
0	0	0	0
1	1	1	1
10	2	2	2
11	3	3	3
100	4	4	4
101	5	5	5
110	6	6	6
111	7	7	7
1000	10	8	8
1001	11	9	9
1010	12	10	t
1011	13	11	e
1100	14	12	10
1101	15	13	11
1110	16	14	12
1111	17	15	13

179.801

POSITIONAL NOTATION

The standard shorthand form of writing numbers is known as POSITIONAL NOTATION. As mentioned before concerning this subject, the value of a particular digit depends not only on the digit value, but also on the position of the digit within the number. Consequently, the decimal number 9751.68 is the standard shorthand form of the quantity nine thousand seven hundred fifty-one and sixty-eight hundredths. What the shorthand form really states is best illustrated by an example as follows:

$$9751.68 = (9 \times 10^3) + (7 \times 10^2) + (5 \times 10^1) +$$
$$(1 \times 10^0) + (6 \times 10^{-1}) + (8 \times 10^{-2}).$$

A quantity may be expressed in positional notation (standard shorthand form) in any number system. This is true since the general form for expressing a quantity regardless of base (radix) is as follows:

$$Q = (d_n \times r^n) + \ldots + (d_2 \times r^2) + (d_1 \times r^1) +$$
$$(d_0 \times r^0) + (d_{-1} \times r^{-1}) + (d_{-2} \times r^{-2}) +$$
$$\ldots + (d_{-n} \times r^{-n})$$

where: Q is the quantity expressed in positional notation form; r is the base or radix of the number system raised to a power; and d_2, d_1, d_0, d_{-1}, d_{-2}, etc...... are the characters of the radix.

Note that the radix point in the general expression (known as the decimal point in the decimal system) is not required because the exponent goes negative. In the shorthand form the radix point is placed between the $d_0 \times r^0$ and $d_{-1} \times r^{-1}$ values.

THE RADIX

Every number system has a radix, or base. When the radix (r) is ten, the decimal system is indicated; when r is eight, the octal system is indicated; and when r is two, the binary system is indicated. The division between integers and fractions is recognized by the position of the radix point. Additional characteristics of the radix are as follows:

1. The radix of a numbering system is equal to the number of the different characters which are necessary to indicate all the various magnitudes a digit may represent. For example, the decimal system, with a radix ten, has ten digits of magnitudes, 0 through 9.

2. The value of the radix is always one unit greater than the largest basic character being used. This is because the radix is equal to the number of characters, whereas the characters themselves start from zero. Thus, the octal system (discussed later) has a radix of eight and uses digits 0 through 7.

3. The positional notation does not, in itself, indicate the radix. The symbol "312" could represent a number written in the quartic (base four), octal, or decimal system, or in any system having a radix of four or greater. Binary numbers are usually recognizable from their string of ones and zeros. To avoid confusion, numbers written in systems other than the decimal system should have the radix noted as a subscript, i.e.,

$$315.72_8$$

The radix subscript is always written as a decimal (base ten) number.

4. Any number can easily be multiplied or divided by the radix of its number system. In decimal notation, to multiply a number by ten, move the decimal (radix) point one digit to the right of its former position, as follows:

$$\begin{array}{r} 34.564 \\ \underline{\times \text{ ten}} \\ 345.64 \end{array}$$

A fact often overlooked is that the radix point could remain stationary while the digits are moved and accomplish the same thing.

To divide a number by ten, move the decimal (radix) point one digit to the left of its former position, or move the digits one digit space to the right relative to the radix point as follows:

$$\frac{34.564}{(\text{ten})} = 3.4564$$

In the same fashion, a binary number is multiplied by two when the binary (radix) point is moved to the right one position value or the number is shifted to the left as follows:

$$10101.01 \times (\text{two}) = 101010.1$$

COUNTING

The rules for counting numbers written in a system of positional notation are the same for every radix. (See table 13-1.) The octal system is used in the following example to illustrate these rules.

1. Starting from zero, add "one" to the least significant digit until all basic characters have been used:

$$0, 1, 2, 3, 4, 5, 6, 7 \ldots \ldots$$

Note: A series of all the characters in sequence, in a given number system is called a CYCLE.

2. Since seven is the largest character in this system, a larger number requires two digits. Start the series of two-digit numbers with zero as the least significant digit and a "1" to left of the zero:

$$\ldots \ldots 6, 7, 10, 11, 12, 13, 14, 15, 16, 17, \ldots \ldots$$

3. Whenever any digit reaches its maximum value (seven, in this case), replace it with zero and add "1" to its next more significant digit:

$$\ldots 16, 17, 20, 21, \ldots 26, 27, 30,$$
$$31, \ldots 66, 67, 70, 71$$

4. When two or more consecutive digits reach the maximum value, replace them with zeros and add "1" to the next more significant digit:

$$\ldots 76, 77, 100, 101, \ldots 176, 177,$$
$$200, \ldots 776, 777, 1000$$

NOTE: The symbol "10" always represents the radix in its own system. This is true because the radix is one unit larger than the largest character, and by the rules of counting, this value is written as "10."

For example:

Binary "10" = two (the radix of the binary system)
Octal "10" = eight (the radix of the octal system)
Decimal "10" = ten (the radix of the decimal system)

DECIMAL SYSTEM

Since the decimal system uses ten symbols, or digits, (tab. 13-1), it has a radix, or base, of 10. This system is thought to have evolved and found common usage as a result of our having ten fingers (digits).

Because this system is used almost universally throughout the world, basic mathematical computations done by a person in one country are easily understood by a person from another country. In other words the decimal system serves as sort of a universal language.

BINARY SYSTEM

The simplest possible number system is based on powers of two and is known as the binary system. This system, which is keyed to the decimal and octal systems, is used in the majority of modern computers and in all digital devices.

By a convenient coincidence, the two binary conditions (1 and 0) can be easily represented by many electrical/electronic components if the 1 binary state is indicated when the component is conductive and the 0 state is indicated when the component is nonconductive. The reverse of this will work equally as well, i.e., the nonconducting state of a component can be used to represent a 1 binary condition and a conducting

state the 0 condition. Both procedures are used in digital computer applications and frequently within a single computer. Numerous devices are used to provide representation of binary conditions. These include switches, transistors, relays, diodes and magnetic devices.

The quantity represented using binary characters (or the characters in any numbering system) cannot be determined without knowing the positional weighting value of each character (digit). The positional values of binary characters from 2^0 (1) to 2^9 (512_{10}, or 512 base ten) are illustrated in figure 13-1.

Consider the following: a number of flip-flops (multivibrators) may be interconnected in such a fashion as to form a chain, or register. The incoming pulses to this chain, or register, are then gated in such a fashion as to cause certain of the flip-flops to be driven to a SET, or 1, OUTPUT while the rest are driven to a CLEAR, or 0, OUTPUT. (Note: The characteristics and operation of registers will be discussed in more detail later in this chapter.) If it is assumed that there are ten flip-flops in

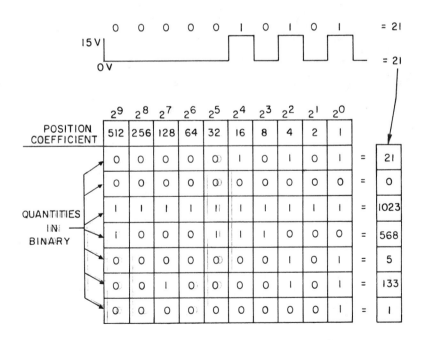

164.64

Figure 13-1.— Positional weighting value.

the chain, the binary condition (number) represented could be 0000010101. The value of this number can be determined by simply summing the positional values as indicated by 1's in the table of figure 13-1. By this procedure, the top number in the table yields:

$$(1 \times 2^4) + (1 \times 2^2) + (1 \times 2^0) = 21$$

Thus,

$$00000101012_2 = 21_{10},$$

and it is now known that the flip-flop chain is storing the decimal equivalent of 21.

The values of the other binary numbers in figure 13-1 are determined in the same manner.

Regardless of the number system in use, weighting values are by convention always arranged in the same order when counting. That is, the MOST SIGNIFICANT DIGIT (MSD) will be on the left while the LEAST SIGNIFICANT DIGIT (LSD) will be on the right, and the weighting value will progressively increase from the LSD to the MSD.

In a binary number the leftmost 1 (fig. 13-2) is referred to as the MSD as it will be multiplied by the highest positioned coefficient. Once the MSD is determined, all positions to the left of the MSD have no significance even though they may be occupied by zeros; this is true because in a given number any position which is unoccupied or is occupied by a zero does not have a value assigned.

Whether or not a value is assigned all digit, or bit, positions which are to the right of the MSD in a given number must be occupied either by a significant digit, or if no value has been assigned by a zero, if we are to distinguish one number from the other. Even though occupied by zeros all digit, or bit, positions to the right of the MSD in a given number are considered to be occupied by significant digits. Thus, the digit position at the extreme right of a given number is always considered to be occupied by the LSD even though it in fact contains a zero indicating no value has been assigned for this position. The digit, or bit, position immediately to the left of the one occupied by the LSD is occupied by the 2nd LSD, the one to the left of that by the 3rd LSD and etc.....until the MSD is reached. If we reverse the order then the digit, or bit, position immediately to the right of the MSD is considered to hold the 2nd MSD, the one to the right of that the 3rd MSD and etc...until the LSD is reached.

The terms most significant digit and least significant digit have the same meaning in any numbering system. The MSD of the decimal number 43,096 is 4, while the LSD is 6. When 43.096 is multiplied, divided, subtracted, or added to another number, the 4, being in the 10,000 place, will produce the greatest change in the answer. Thus, it is the MSD. The 6, being in the units place will produce the least change in the answer. It is the least significant digit. In most practical computation, an error in the LSD has little significant effect. An MSD error, however, can result in an incorrect answer of more consequence.

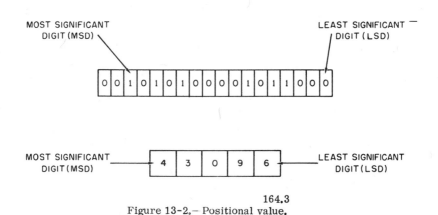

164.3
Figure 13-2.— Positional value.

OCTAL SYSTEM

The octal system has eight distinct characters (table 13-1), hence its radix is eight. The octal system is quite useful as an accessory to the binary system, because eight is an integral power of two ($8_{10} = 2_{10}^3$). One octal digit is always equal to three binary and vice versa.

Octal to Binary	Binary to Octal
2 2 5	010 010 101
010 010 101	2 2 5

This direct relationship between the two systems facilitates the programming of digital machines, since the octal system may be used for programming in place of the more cumbersome binary system which is the language of the machine. The conversion from octal to binary and vice versa is then a simple process which may be accomplished at any point in the system as desired by a relatively simple device.

DUODECIMAL SYSTEM

The duodecimal system has a radix of 12 and thus makes use of the ten symbols of the decimal system plus two additional symbols which are commonly represented by the characters t and e in order to meet system requirements. The primary use of this system at present is for error detection and correction in certain digital machines.

ARITHMETIC OPERATIONS

The operations we will discuss, for various bases, will be the basic arithmetic operations of addition, subtraction, multiplication, and division. Ease in performing these operations will facilitate the understanding of conversions from one base to another base which will be discussed later in this chapter.

Addition

In general, the rules of arithmetic apply to any number system. However, each system has a unique digit addition and digit multiplication table. These tables will be discussed with each system.

DECIMAL.—Addition facts in base ten are shown in table 13-2. The sign of operation is

Table 13-2.—Decimal addition

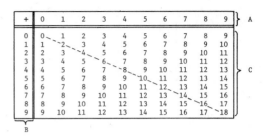

+	0	1	2	3	4	5	6	7	8	9
0	0	1	2	3	4	5	6	7	8	9
1	1	2	3	4	5	6	7	8	9	10
2	2	3	4	5	6	7	8	9	10	11
3	3	4	5	6	7	8	9	10	11	12
4	4	5	6	7	8	9	10	11	12	13
5	5	6	7	8	9	10	11	12	13	14
6	6	7	8	9	10	11	12	13	14	15
7	7	8	9	10	11	12	13	14	15	16
8	8	9	10	11	12	13	14	15	16	17
9	9	10	11	12	13	14	15	16	17	18

179.802

given in the upper left corner. The addends or augends are indicated by row A and column B. (Note: The numbers in row A may represent either the addend or the augend and the same applies to the numbers in column B. However, if the numbers in row A are to represent the addend then the number in column B must represent the augend and vice versa.) The sums are shown in the array C. To find the sum C of A + B locate the addend A and augend B. The sum C will be located where A and B intersect. The commutative principle causes the table to be symmetrical with respect to the diagonal with a negative slope. This is shown by the dotted line.

BINARY.—Binary addition facts are shown in table 13-3. Notice that a binary digit has only two possible values, 0 and 1. A carry of two is involved in binary addition. That is, when we add one and one the sum is two, but we have no two so we write 10_2 which indicates one group of $(2)^1$ and no group of one.

EXAMPLE: Add 1011_2 and 1101_2.
SOLUTION: Write

$$1011_2$$
$$+1101_2$$

Then, one and one are two, but "two" is

$$10_2$$

so we write

$$1$$
$$1011_2$$
$$+1101_2$$
$$\overline{}$$
$$0$$

229

Table 13-3.—Binary Addition

B

179.803

and carry a one. The following steps, with the carry indicated, show the completion of our addition.

$$
\begin{array}{r}
1 \\
1011_2 \\
+1101_2 \\
\hline
0_2
\end{array}
$$

$$
\begin{array}{r}
1 \\
1011_2 \\
+1101_2 \\
\hline
00_2
\end{array}
$$

$$
\begin{array}{r}
1 \\
1011_2 \\
+1101_2 \\
\hline
000_2
\end{array}
$$

$$
\begin{array}{r}
1011_2 \\
+1101_2 \\
\hline
11000_2
\end{array}
$$

Notice in the last step we added three ones which total three, and "three" is written as

$$11_2$$

To verify out answer we may proceed as follows

$$
\begin{array}{r}
1011_2 = 11_{10} \\
+1101_2 = 13_{10} \\
\hline
11000_2 = 24_{10}
\end{array}
$$

OCTAL.—The octal system has the digits 0, 1, 2, 3, 4, 5, 6, and 7. When an addition carry is made, the carry is eight. The addition facts are shown in table 13-4.

Table 13-4.—Octal Addition

+	0	1	2	3	4	5	6	7	A
0	0	1	2	3	4	5	6	7	
1	1	2	3	4	5	6	7	10	
2	2	3	4	5	6	7	10	11	
3	3	4	5	6	7	10	11	12	C
4	4	5	6	7	10	11	12	13	
5	5	6	7	10	11	12	13	14	
6	6	7	10	11	12	13	14	15	
7	7	10	11	12	13	14	15	16	

B

179.804

When we add 7_8 and 6_8 we have a sum of thirteen but thirteen in base eight is one group of eight and five groups of one. We write

$$
\begin{array}{r}
7_8 \\
+6_8 \\
\hline
15_8
\end{array}
$$

EXAMPLE: Add 765_8 and 675_8.
SOLUTION: Write

$$
\begin{array}{r}
1 \\
765_8 \\
+ 675_8 \\
\hline
2_8
\end{array}
$$

$$
\begin{array}{r}
1 \\
765_8 \\
+ 675_8 \\
\hline
62_8
\end{array}
$$

$$
\begin{array}{r}
765_8 \\
+ 675_8 \\
\hline
1662_8
\end{array}
$$

DUODECIMAL.—The addition facts for the base twelve system are shown in table 13-5. The t equals ten and the e equals eleven. When a carry is made the carry is twelve. When we add 9_{12} and e_{12} we find the sum is twenty and twenty is written as one group of twelve and eight groups of one; that is,

$$
\begin{array}{r}
9_{12} \\
+ e_{12} \\
\hline
18_{12}
\end{array}
$$

Table 13-5.—Duodecimal Addition

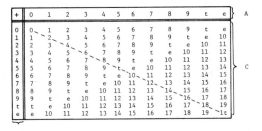

+	0	1	2	3	4	5	6	7	8	9	t	e
0	0	1	2	3	4	5	6	7	8	9	t	e
1	1	2	3	4	5	6	7	8	9	t	e	10
2	2	3	4	5	6	7	8	9	t	e	10	11
3	3	4	5	6	7	8	9	t	e	10	11	12
4	4	5	6	7	8	9	t	e	10	11	12	13
5	5	6	7	8	9	t	e	10	11	12	13	14
6	6	7	8	9	t	e	10	11	12	13	14	15
7	7	8	9	t	e	10	11	12	13	14	15	16
8	8	9	t	e	10	11	12	13	14	15	16	17
9	9	t	e	10	11	12	13	14	15	16	17	18
t	t	e	10	11	12	13	14	15	16	17	18	19
e	e	10	11	12	13	14	15	16	17	18	19	1t

A

C

B

179.805

EXAMPLE: Add $8te2_{12}$ and $9e4_{12}$.
SOLUTION: Write

$$8te2_{12}$$
$$+\ \ 9e4_{12}$$
$$\overline{\qquad 6_{12}}$$

$$1$$
$$8te2_{12}$$
$$+\ \ 9e4_{12}$$
$$\overline{\qquad t6_{12}}$$

$$1$$
$$8te2_{12}$$
$$+\ \ 9e4_{12}$$
$$\overline{\qquad 8t6_{12}}$$

$$8te4_{12}$$
$$+\ \ 9e4_{12}$$
$$\overline{98t6_{12}}$$

Subtraction

Subtraction in any number system is performed in the same manner as in the decimal system. In the process of addition we were faced with the "carry," and in subtraction we are faced with "borrowing."

Since the process of subtraction is the opposite of addition, we may use the addition tables for subtraction facts for the various bases discussed previously.

DECIMAL.—Table 13-2 is the addition table for the decimal system. Since this table indicates that

$$A + B = C$$

we may use this table for subtraction facts by writing

$$A + B = C$$

then

$$C - A = B$$

or

$$C - B = A$$

To subtract 8 from 15, find 8 in either the A row or B column. Find where this row or column intersects with a value of 15 for C, then move to the remaining row or column to find the remainder.

This problem, when written in the familiar form of

15 minuend

$$-\ \ 8$$ subtrahend

7 remainder

requires the use of the "borrow"; that is, when we try to subtract 8 from 5 to obtain a positive remainder, we cannot accomplish this. We borrow the 1 which is really one group of ten. Then, one group of ten and 5 groups of one equals 15 and 15 minus 8 leaves a remainder of 7.

BINARY.—When subtracting in base two, the addition table in table 13-3 is used. To subtract 1_2 from 10_2 the borrow of two is used. That is,

$$10_2$$
$$-\ \ 1_2$$

one group of $(2)^1$ and no group of $(2)^0$ minus one group of $(2)^0$. Thinking in base ten, this is 2 minus 1 which is 1. This may be verified by using table 13-3.

EXAMPLE: Subtract 11_2 from 101_2.
SOLUTION: Write

$$101_2$$
$$-\ \ 11_2$$

Then, 1 from 1 is 0 and write

$$101_2$$
$$-\ 11_2$$
$$\overline{\quad 0_2}$$

Now borrow the left hand 1 which has the value two when moved to the next column to the right. 1 from 2 is 1, and

$$
\begin{array}{r}
2 \\
\cancel{1}01_2 \\
-\ 11_2 \\
\hline
10_2
\end{array}
$$

OCTAL.—Table 13-4 contains the octal subtraction facts in that

$$C - B = A$$

or

$$C - A = B$$

EXAMPLE: Find the remainder when 6_8 is subtracted from 13_8.
SOLUTION: If, in table 13-4,

$$C = 13_8$$

and

$$B = 6_8$$

then

$$C - B = A$$

$$13_8 - 6_8 = 5_8$$

DUODECIMAL.—Through the use of table 13-5 we find that 13_{12} minus 9_{12} is 6_{12}. This may be explained by writing

$$
\begin{array}{r}
13_{12} \\
-\ 9_{12} \\
\hline
\end{array}
$$

We borrow one group of twelve and add it to the group of three ones to obtain fifteen. Then, nine from fifteen is six. Therefore,

$$13_{12} - 9_{12} = 6_{12}$$

Here, as before, we think in base ten and write in the base being used.
EXAMPLE: Subtract $2e9_{12}$ from $t64_{12}$.
SOLUTION: Write

$$
\begin{array}{r}
t64_{12} \\
-\ 2e9_{12} \\
\hline
\end{array}
$$

Borrow one group of twelve and add it to four to obtain sixteen. Then nine from sixteen is seven. Write

$$
\begin{array}{r}
5\ \ 12 \\
t\ \cancel{6}\ \ 4_{12} \\
-\ 2\ e\ \ 9_{12} \\
\hline
7_{12}
\end{array}
$$

Then, borrow one group from t, the $(12)^2$ column, and add it to the five groups of $(12)^1$ to obtain seventeen groups of $(12)^1$ minus e groups of $(12)^1$ for a remainder of six groups of $(12)^1$. Write

$$
\begin{array}{r}
9\ 12\ 12 \\
t\ \ 5\ \ \ 4_{12} \\
-\ 2\ \ e\ \ \ 9_{12} \\
\hline
6\ \ \ 7_{12}
\end{array}
$$

then, 2_{12} from 9_{12} is 7_{12}, therefore,

$$
\begin{array}{r}
9\ 12\ 12 \\
\cancel{t}\ \ 5\ \ \ 4_{12} \\
-\ 2\ \ e\ \ \ 9_{12} \\
\hline
7\ \ 6\ \ \ 7_{12}
\end{array}
$$

Multiplication

Multiplication in any number system is performed in the same manner as in the decimal system. Each system has a unique digit multiplication table. These tables will be discussed with each system. The rows, columns, and arrays of these tables are labeled in the same fashion as the addition tables. Only the sign of operation and array values are different.

DECIMAL.—In multiplication in the decimal system, certain rules are followed which use the decimal digit multiplication and decimal digit addition tables. These rules are well known and apply to direct multiplication in any number system. Table 13-6 shows the decimal multiplication facts.

Table 13-6.—Decimal Multiplication

x	0	1	2	3	4	5	6	7	8	9	A
0	0	0	0	0	0	0	0	0	0	0	
1	0	1	2	3	4	5	6	7	8	9	
2	0	2	4	6	8	10	12	14	16	18	
3	0	3	6	9	12	15	18	21	24	27	C
4	0	4	8	12	16	20	24	28	32	36	
5	0	5	10	15	20	25	30	35	40	45	
6	0	6	12	18	24	30	36	42	48	54	
7	0	7	14	21	28	35	42	49	56	63	
8	0	8	16	24	32	40	48	56	64	72	
9	0	9	18	27	36	45	54	63	72	81	

B

179.806

The direct method of multiplication of decimal numbers is shown in the following example.
EXAMPLE: Multiply 32 by 25
SOLUTION: Write

$$25 = 20 + 5$$

then

$$32(25)$$

$$= 32(20 + 5)$$

$$= 32(20) + 32(5)$$

$$= 640 + 160$$

$$= 800$$

The same problem written as

$$\begin{array}{r} 32 \\ \times\ 25 \\ \hline \end{array}$$

gives

$$32(5)$$
$$= 160 \quad \text{partial product}$$

then

$$32(20)$$
$$= 640 \quad \text{partial product}$$

then

$$160 + 640$$

$$= 800 \quad \text{product}$$

The technique generally used is

$$\begin{array}{r} 32 \\ \times\ 25 \\ \hline 160 \\ 64 \\ \hline 800 \end{array}$$

Notice that the 64 really represents 640 but the zero is omitted.
EXAMPLE: Multiply 306 by 762.
SOLUTION: Write

	306	factor
x	762	factor
	612	partial product
	1836	partial product
	2142	partial product
	233172	product

BINARY.—Table 13-7 shows the multiplication facts for the binary system. This is the simplest set of facts of any of the number systems and as will be seen the only difficulty in binary multiplication may be in the addition of the partial products.
EXAMPLE: Multiply 1101_2 by 1101_2.
SOLUTION: Write

$$\begin{array}{r} 1101_2 \\ \times\ \ 101_2 \\ \hline \end{array}$$

The partial products and the products are as follows:

	1101	
x	101	
	1101	partial product
	11010	partial product
	1000001	product

Table 13-7.—Binary Multiplication

x	0	1	A
0	0	0	C
1	0	1	

B

179.807

233

As in the addition section, the problem that may be encountered in the addition of the partial products is what to carry. The following example will illustrate this problem.

EXAMPLE: Multiply 1111_2 by 111_2.

SOLUTION: Write

$$
\begin{array}{r}
1111_2 \\
\times\ 111_2 \\
\hline
1111 \\
1111 \\
1111 \\
\end{array}
$$

We add the partial products by writing

$$
\begin{array}{r}
1111_2 \\
\times\ 111_2 \\
\hline
1111 \\
1111 \\
1111 \\
\hline
01 \\
\end{array}
$$

and when we add the four ones we find four is written in binary as 100_2. We write the zero, then we must carry the 10_2. The symbol 10_2 is really two, thinking in base ten; therefore, we carry two and when two is added to the next three ones we have five. Five is written as 101_2; therefore, we write 1 and carry the 10_2 or two. Two and two are four so we write zero and carry 10_2 or two. Finally, two and one are three and we write 11_2. The entire addition process is shown as follows:

$$
\begin{array}{r}
1111 \\
1111 \\
1111 \\
\hline
1101001_2 \\
\end{array}
$$

OCTAL.—Base eight multiplication facts are given in table 13-8. When multiplying 6_8 by 7_8 we find the product by thinking "six times seven is forty-two" and writing forty-two as five groups of eight and two groups of one or

$$6_8 \times 7_8 = 52_8$$

Table 13-8.—Octal Multiplication

x	0	1	2	3	4	5	6	7
0	0	0	0	0	0	0	0	0
1	0	1	2	3	4	5	6	7
2	0	2	4	6	10	12	14	16
3	0	3	6	11	14	17	22	25
4	0	4	10	14	20	24	30	34
5	0	5	12	17	24	31	36	43
6	0	6	14	22	30	36	44	52
7	0	7	16	25	34	43	52	61

179.808

EXAMPLE: Multiply 41_8 by 23_8.

SOLUTION: Write

$$
\begin{array}{r}
41_8 \\
\times\ 23_8 \\
\hline
143 \\
102 \\
\hline
1163_8 \\
\end{array}
$$

DUODECIMAL.—Multiplication facts for the base twelve system are shown in table 13-9. The process of multiplication is the same as in other bases.

EXAMPLE: Multiply 9_{12} by 5_{15}.

SOLUTION: Write

$$
\begin{array}{r}
9_{12} \\
\times\ 5_{12} \\
\hline
\end{array}
$$

Nine times five is forty-five in base ten, and forty-five is written as three groups of twelve and nine ones, in base twelve; that is,

$$
\begin{array}{r}
9_{12} \\
\times\ 5_{12} \\
\hline
39_{12} \\
\end{array}
$$

EXAMPLE: Multiply 5_{12} by 7_{12}.

SOLUTION: Write

$$
\begin{array}{r}
5_{12} \\
\times\ 7_{12} \\
\hline
\end{array}
$$

Table 13-9.—Duodecimal Multiplication

x	0	1	2	3	4	5	6	7	8	9	t	e
0	0	0	0	0	0	0	0	0	0	0	0	0
1	0	1	2	3	4	5	6	7	8	9	t	e
2	0	2	4	6	8	t	10	12	14	16	18	1t
3	0	3	6	9	10	13	16	19	20	23	26	29
4	0	4	8	10	14	18	20	24	28	30	34	38
5	0	5	t	13	18	21	26	2e	34	39	42	47
6	0	6	10	16	20	26	30	36	40	46	50	56
7	0	7	12	19	24	2e	36	41	48	53	5t	65
8	0	8	14	20	28	34	40	48	54	60	68	74
9	0	9	16	23	30	39	46	53	60	69	76	83
t	0	t	18	26	34	42	50	5t	68	76	84	92
e	0	e	1t	29	38	47	56	65	74	83	92	t1

179.809

and thirty-five in base ten is written as two groups of twelve and eleven ones, in base twelve; therefore,

$$5_{12}$$
$$\times \; 7_{12}$$
$$2e_{12}$$

Division

The process of division is the opposite of multiplication; therefore, we may use the multiplication tables for the various bases to show division facts. We will define division by writing

$$\frac{C}{B} = A \text{ if, and only if, } AB = C, \; B \neq 0$$

We show this by use of table 13-6. That is, if

$$C = 42$$

and

$$B = 7$$

then

$$A = 6$$

Notice that the value of C is the intersection of the values of A and B.

For the remainder of this section on division, examples and problems for the various number bases along with their respective multiplication tables will be used.

DECIMAL

EXAMPLE: Divide 54 by 9.

$$
\begin{array}{r}
6 \\
9 \overline{\smash{)}54} \\
\underline{54} \\
0
\end{array}
$$

EXAMPLE: Divide 252 by 6.
SOLUTION: Write

$$
\begin{array}{r}
42 \\
6 \overline{\smash{)}252} \\
\underline{24} \\
12 \\
\underline{12} \\
0
\end{array}
$$

BINARY

EXAMPLE: Divide 1111_2 by 11_2.
SOLUTION: Write

$$
\begin{array}{r}
101_2 \\
11 \overline{\smash{)}1111} \\
\underline{11} \\
011 \\
\underline{11} \\
0
\end{array}
$$

EXAMPLE: Divide 101_2 by 10_2.
SOLUTION: Write

$$
\begin{array}{r}
10_2 \\
10 \overline{\smash{)}101} \\
\underline{10} \\
1 \; \text{remainder}
\end{array}
$$

OCTAL

EXAMPLE: Divide 234_8 by 6_8.
SOLUTION: Write

$$
\begin{array}{r}
32_8 \\
6 \overline{\smash{)}234} \\
\underline{22} \\
14 \\
\underline{14} \\
0
\end{array}
$$

235

EXAMPLE: Divide 765_8 by 4_8.
SOLUTION: Write

$$175_8$$

$$
\begin{array}{r}
4\,\overline{)765} \\
4 \\ \hline
36 \\
34 \\ \hline
25 \\
24 \\ \hline
1 \quad \text{remainder}
\end{array}
$$

DUODECIMAL

EXAMPLE: Divide 446_{12} by 6_{12}.
SOLUTION: Write

$$89_{12}$$

$$
\begin{array}{r}
6\,\overline{)446} \\
40 \\ \hline
46 \\
46 \\ \hline
0
\end{array}
$$

EXAMPLE: Divide 417_{12} by 5_{12}.
SOLUTION: Write

$$9e_{12}$$

$$
\begin{array}{r}
5\,\overline{)417} \\
39 \\ \hline
47 \\
47 \\ \hline
0
\end{array}
$$

CONVERSION

If two numbers written in different numbering systems represent the same quantity, these numbers are equivalent (the represented quantities are equal although the numbers are not necessarily composed of the same characters). Any change which retains the equivalence of the original numbers results in a new set of equivalent numbers. Therefore, it is possible to convert numbers from one numbering system to another numbering system, that is, to change radices.

Nondecimal to
Decimal

To determine the decimal quantitative meaning of a number expressed in a system other than the decimal system, write the number in a power series summation, multiplying each digit of the number by its radix raised to the indicated positional power. The radix and the power must be expressed as a decimal number to obtain the decimal quantity.

Examples:

Problem: For what decimal quantity does the binary number 10101.01 stand?

Solution:

$$
\begin{aligned}
\text{Quantity} &= (1 \times 2^4) + (0 \times 2^3) + (1 \times 2^2) + \\
&\quad (0 \times 2^1) + (1 \times 2^0) + (0 \times 2^{-1}) + \\
&\quad (1 \times 2^{-2}) \\
&= (1 \times 16) + 0 + (1 \times 4) + 0 + (1 \times 1) \\
&\quad + 0 + (1 \times 0.25) = 21.25_{10}
\end{aligned}
$$

Problem: For what decimal quantity does the octal number 25.2 stand?

Solution:

$$
\begin{aligned}
\text{Quantity} &= (2 \times 8^1) + (5 \times 8^0) + (2 \times 8^{-1}) \\
&= (2 \times 8) + (5 \times 1) + (2 \times 0.125) \\
&= 16 + 5 + 0.25 \\
&= 21.25_{10}
\end{aligned}
$$

Note that the numbers 21.25_{10}, 25.2_8, and 10101.01_2 are equivalent.

Another method of nondecimal to decimal conversion is by synthetic substitution. This method is shown in the following example.

EXAMPLE: Convert 634_8 to decimal.
SOLUTION: Write

$$
8\,\big|\; 6 \quad 3 \quad 4
$$

Bring down the six

$$
8\,\big|\; 6 \quad 3 \quad 4 \\
 6
$$

Multiply the six by the base (expressed in decimal form) and carry the decimal product to the next lower place value column.

$$8 \mid \quad 6 \qquad 3 \qquad 4$$
$$\quad \quad \quad 48$$
$$\quad \quad 6$$

ADD the three and the carried product

$$8 \mid \quad 6 \qquad 3 \qquad 4$$
$$\quad \quad \quad 48$$
$$\quad \quad 6 \qquad 51$$

Multiply this sum by the base and carry to the next lower place value column.

$$8 \mid \quad 6 \qquad 3 \qquad 4$$
$$\quad \quad \quad 48 \qquad 408$$
$$\quad \quad 6 \qquad 51$$

Add the four and the carried product to find the decimal equivalent of 634_8 to be 412.

$$8 \mid \quad 6 \qquad 3 \qquad 4$$
$$\quad \quad \quad 48 \qquad 408$$
$$\quad \quad 6 \qquad 51 \qquad 412$$

A third method of converting a number from nondecimal to decimal is by use of repeated division where the remainders indicate the decimal equivalent. The denominator is ten expressed in the nondecimal number.

EXAMPLE: Convert 634_8 to decimal.

SOLUTION: Ten expressed in base eight is 12; therefore, write

$$12\overline{)634}$$

This division is carried out in base eight.

$$\begin{array}{r} 51 \\ 12\overline{)634} \\ \underline{62} \\ 14 \\ \underline{12} \\ R_1 = 2_8 = 2 \end{array}$$

The dividend is now divided by 12_8.

$$\begin{array}{r} 4 \\ 12\overline{)51} \\ \underline{50} \\ R_2 = 1_8 = 1 \end{array}$$

This process is continued until the dividend is zero.

$$\begin{array}{r} 0 \\ 12\overline{)4} \\ \underline{0} \\ R_3 = 4_8 = 4 \end{array}$$

Now, if $634_8 = X_{10}$ then

$$X_{10} = R_3 \ R_2 \ R_1$$

where

$$R_1 = 2$$
$$R_2 = 1$$
$$R_3 = 4$$

Therefore

$$X_{10} = 412$$

The reverse of this process may be used to convert a nondecimal fraction to its decimal equivalent. For example to convert an octal fraction to its decimal equivalent multiply by 12_8 (10_{10}). The integer in the produce will now be the decimal equivalent (10^{-1}). The fractional portion of the product is now used as the multiplicand in a like manner to find the decimal equivalent (10^{-2}). As can be seen this process may be repeated as many times as necessary to obtain the desired number of places (or accuracy).

Convert the octal fraction $.3137_8$ to its decimal equivalent as follows:

.31373	.76716	.65014	.22170	.66260
12	12	12	12	12
62766	175634	152030	44360	154540
31373	76716	65014	22170	66260
3.76716	11.65014	10.22170	2.66260	10.37340
3	9	8	2	8

Thus, $.31373_8 = .39828_{10}$

237

Decimal To
Nondecimal

To convert a number from decimal to non-decimal the process of repeated division is used and the remainders indicate the nondecimal number. The denominator is the nondecimal base expressed in base ten and the division process is in base ten.

EXAMPLE: Convert 319 to octal; that is, if

$$319 = X_8, \text{ then } X_8 = ?$$

SOLUTION: Base eight expressed in decimal is 8, therefore, write

$$
\begin{array}{r}
39 \\
8\overline{)319} \\
24 \\
\hline
79 \\
72 \\
\hline
R_1 = 7 = 7_8
\end{array}
$$

and

$$
\begin{array}{r}
4 \\
8\overline{)39} \\
32 \\
\hline
R_2 = 7 = 7_8
\end{array}
$$

then

$$
\begin{array}{r}
0 \\
8\overline{)4} \\
0 \\
\hline
R_3 = 4 = 4_8
\end{array}
$$

Now, if

$$319 = X_8$$

then

$$X_8 = R_3 \ \ R_2 \ \ R_1$$

In this example

$$R_1 = 7$$

$$R_2 = 7$$

$$R_3 = 4$$

therefore,

$$X_8 = 477_8$$

The format for the repeated division process may be simplified in cases where the actual division is simple. This is shown in the following example:

EXAMPLE: Convert 18 to binary.

SOLUTION: Carry out the repeated division indicating the remainder to the right of the division; that is,

$$
\begin{array}{ll}
& \text{Remainder} \\
2/18 & \\
2/9 & 0 \\
2/4 & 1 \\
2/2 & 0 \\
2/1 & 0 \\
\hline
0 & 1
\end{array}
$$

Now read the remainder from the bottom to the top to find the binary equivalent of the decimal number. In this case the binary number is 10010_2.

Nondecimal To
Nondecimal

Three approaches to the nondecimal to nondecimal conversion will be considered at this point. On method will be through base ten and the other two methods will be direct.

When going through base ten, the polynomial form is used along with repeated division.

EXAMPLE: Convert 10110_2 to base twelve.

SOLUTION: Write (polynomial form)

$$10110_2 = 1(2)^4 + 0(2)^3 + 1(2)^2 + 1(2)^1 + 0(2)^0$$

$$= 16 + 0 + 4 + 2 + 0$$

$$= 22$$

Then (repeated division),

$$
\begin{array}{ll}
& \text{Remainder} \\
12/22 & \\
12/1 & t \\
\hline
0 & 1
\end{array}
$$

therefore,

$$10110_2 = 1t_{12}$$

The second method of converting a nondecimal number to a nondecimal number is by division. The division is carried out by dividing by the base wanted, performing the calculation in the base given.

EXAMPLE: Convert $t73_{12}$ to base eight

SOLUTION: The base given is twelve and the base wanted is eight. Therefore, express eight in base twelve, obtaining 8_{12}. We carry out the division by 8_{12} in base twelve as follows:

$$
\begin{array}{r}
13t \\
8\,\overline{)\,t73} \\
8 \\
\hline
27 \\
20 \\
\hline
73 \\
68 \\
\hline
R_1 = {}^7{}_{12} = 7_8
\end{array}
$$

then

$$
\begin{array}{r}
1e \\
8\,\overline{)\,13t} \\
8 \\
\hline
7t \\
74 \\
\hline
R_2 = {}^6{}_{12} = 6_8
\end{array}
$$

and

$$
\begin{array}{r}
2 \\
8\,\overline{)\,1e} \\
14 \\
\hline
R_3 = {}^7{}_{12} = 7_8
\end{array}
$$

and

$$
\begin{array}{r}
0 \\
8\,\overline{)\,2} \\
0 \\
\hline
R_4 = {}^2{}_{12} = 2_8
\end{array}
$$

then

$$t73_{12} = R_4 \; R_3 \; R_2 \; R_1$$

$$= 2767_8$$

The last method we will discuss is called the explosion method. It consists of the following rules:

1. Perform all arithmetic operations in the desired base.

2. Express the base of the original number in terms of the base of the desired number.

3. Multiply the number obtained in step 2 by the leftmost digit and add the product to the next digit on the right of the original number.

(NOTE: It may be necessary to convert each digit of the original number to an expression conforming to the desired base.)

4. Repeat step 3 as many times as there are digits. The final sum is the answer.

EXAMPLE: Convert 347_8 to base twelve.

SOLUTION: Write

Step (2) $10_8 = 8_{12}$

Step (3)

$$
\begin{array}{ccc}
(3_{12} & 4_{12} & 7_{12})8 \\
\times\;8_{12} & +\;20_{12} & +\;168_{12} \\
\hline
20_{12} & 24_{12} & 173_{12} \\
& \times\;8_{12} & \\
\cline{2-2}
& 168_{12} &
\end{array}
$$

thus

$$347_8 = 173_{12}.$$

SPECIAL CASES OF CONVERSION

For certain applications there are special methods of conversion which may simplify the process and reduce the time and effort required for conversions.

The Table Method

The table method is an adaptation of the power series summation method for nondecimal to decimal conversions discussed previously. This method is especially applicable for binary to decimal and decimal to binary conversions since all that is necessary is to construct a table and plug in the values as shown in table 13-10. To use the table proceed as follows:

Decimal to Binary

1. Find the largest whole power of two that may be subtracted from the decimal number and place the binary character "1" underneath the decimal value of this whole power of two in the table.

2. Using the remainder, if any, from (1), again subtract the largest whole power of two possible. Place the binary character "1" underneath the decimal value of this whole power of two in the table.

3. Continue until the remainder is zero. Under every decimal value which does not have the binary character "1," place the binary character "0."

4. Rounding-off of fractions is necessary where exact conversions into binary are either not possible or necessary.

Table 13-10.—Table Method Of Conversion

POWERS OF TWO	2^9	2^8	2^7	2^6	2^5	2^4	2^3	2^2	2^1	2^0	2^{-1}	2^{-2}	2^{-3}
DECIMAL EQUIVALENT	512	256	128	64	32	16	8	4	2	1	0.5	0.25	0.125
BINARY NUMBER													

RADIX POINT

NOTE: TABLE MAY BE EXPANDED TO INCLUDE AS MANY POWERS OF TWO AS NECESSARY BOTH POSITIVE AND NEGATIVE.

179.810

EXAMPLE: Convert 75_{10} to binary.
SOLUTION:

Step #1 It is found that 64 is the largest power of two which may be subtracted from 75, therefore, the binary digit "1" is entered in the table under 64.

Step #2 The subtraction is now made leaving 11 and the largest power of two which may be subtracted from 11 is 8, therefore the binary digit "1" is entered in the table under 8.

Step #3 The subtraction is now made leaving 3 and the largest power of two which may be subtracted from 3 is 2, therefore the binary digit "1" is entered in the table under 2.

Step #4 The subtraction is now made leaving 1 and the largest power of two which may be subtracted is 1, thus the binary digit "1" is entered under 1 in the table. No further subtractions are necessary since the result of the next subtraction will be 0.

Step #5 All other spaces in the table are now filled with zeros and as can be seen 75_{10} is equal to or the equivalent of 1001011_2.

Binary to Decimal

1. Enter the binary number into the table.
2. Make a summation of the decimal equivalents under which a binary digit "1" falls.

EXAMPLE: Convert 1001011_2 to decimal.

SOLUTION:

$$
\begin{aligned}
1 \times 2^6 &= 64 \\
0 \times 2^5 &= 0 \\
0 \times 2^4 &= 0 \\
1 \times 2^3 &= 8 \\
0 \times 2^2 &= 0 \\
1 \times 2^1 &= 2 \\
1 \times 2^0 &= 1 \\
\hline
&\quad 75_{10}
\end{aligned}
$$

Thus $1001011_2 = 75_{10}$

Direct Conversion—Binary to Octal and Octal to Binary

As was previously mentioned eight is an integral power of two ($2^3 = 8$), thus

$$
\begin{aligned}
000_2 &= 0_8 \\
001_2 &= 1_8 \\
010_2 &= 2_8 \\
011_2 &= 3_8 \\
100_2 &= 4_8 \\
101_2 &= 5_8 \\
110_2 &= 6_8 \\
111_2 &= 7_8
\end{aligned}
$$

It should be obvious now, that all that is necessary to convert from binary to octal is to divide the binary number into groups of three starting at the binary point and working in either direction, then assign each group its octal equivalent. Conversely, to convert from octal to binary, all that is necessary is to assign each digit in the octal number its binary equivalent. In either case the order of the original number must be maintained.

EXAMPLE: Convert 10111010110_2 to octal.
SOLUTION:

Binary Number 010 111 010 110$_2$
Octal Equiv.   2   7   2   6$_8$

Thus $10111010110_2 = 2726_8$

EXAMPLE: Convert 1564_8 to binary.
SOLUTION:

Octal Number   1   5   6   4$_8$
Binary Equiv.   001 101 110 100$_2$

Thus $1564_8 = 1101110100_2$

NOTE: When converting, if there is an unfinished group at the beginning of a binary number, the blank spaces may be filled in with zeros.

Decimal to Binary
Coded Decimal

Although the Binary Coded Decimal (BCD) is not truly a number system, we will discuss this code because it is computer related as are certain number systems.

This code, sometimes called the 8421 code, makes use of groups of binary symbols to represent a decimal number. In the decimal system there are only ten symbols; therefore, only ten groups of binary bits (symbols) must be remembered. Each decimal digit is represented by a group of four binary bits. The ten groups to remember are as follows:

Decimal Symbol	Binary Coded Decimal (BCD)
0	0000
1	0001
2	0010
3	0011
4	0100
5	0101
6	0110
7	0111
8	1000
9	1001

Thus, to convert a decimal number into a binary coded decimal number the appropriate binary group from the above listing is substituted for each decimal symbol in the number as follows:

(1) Decimal   3   8   1
     BCD   0011   1000   0001

(2) Decimal   7   2   0   3
     BCD   0011   0010   0000   0011

(3) Decimal   4   2
     BCD   0100   0010

The separation of the BCD groups is shown for ease of reading and does not necessarily need to be written as shown. The number 381 could be written as 0011100000001. One advantage of the BCD over true binary is ease of determining the decimal value. This is shown as follows:

Decimal   9   3   4
BCD   1001   0011   0100

The number 934 in true binary is 1110100110_2. This in polynomial form is

$$1(2)^9 + 1(2)^8 + 1(2)^7 + 0(2)^6 + 1(2)^5 + 0(2)^4$$
$$+ 0(2)^3 + 1(2)^2 + 1(2)^1 + 0(2)^0$$

$$= 512 + 256 + 128 + 0 + 32 + 0 + 0 + 4 + 2 + 0$$

$$= 934$$

To convert a BCD to decimal we separate the binary bits into groups of four, working from the right, and then write the decimal digit represented by each group. Thus,

BCD   0100100110011000
= BCD   0100   1001   1001   1000
= Decimal   4   9   9   8
= 4998

One serious disadvantage of the BCD is that this code cannot provide a "decimal" carry. The following examples are given to show this.

EXAMPLE: Add the following:

Decimal		BCD
5	=	0101
+ 3	=	+ 0011
8	=	1000

Notice that the example does not have a carry in the decimal addition and the answer in BCD is equal to the answer in decimal. The BCD is in correct notation and does exist.

241

EXAMPLE: Add the following:

```
Decimal      BCD
   8     =   1000
 + 5     = + 0101
 ────        ────
  13     =   1101
```

Notice that the BCD symbol is the true binary representation of 13 but 1101 does not exist in BCD. The correct BCD answer for 13 is 0001 0011. When a carry is made in decimal the BCD system cannot indicate the correct answer in BCD form.

Excess Three Code

The excess three code is used to eliminate the inability of the decimal carry. It is really a modification of the BCD so that a carry can be made.

To change a BCD symbol to excess three add three to the BCD; that is,

```
BCD      1000
       + 0011
       ──────
excess three 1011
```

The excess three number 1011 is 8 in decimal. The following shows the correspondence between decimal, BCD, and excess three code.

Decimal	BCD	Excess Three
0	0000	0011
1	0001	0100
2	0010	0101
3	0011	0110
4	0100	0111
5	0101	1000
6	0110	1001
7	0111	1010
8	1000	1011
9	1001	1100

As previously stated, the excess three code will provide the capability of the decimal carry. The following is given for explanation.

EXAMPLE: Add 6 and 3 in excess three.
SOLUTION: Write

```
Decimal
   6   = 0011  0011  1001
 + 3   = 0011  0011  0110
 ────
   9
```

Notice that in the right-hand groups the six and three are given. In the other groups a zero (0011) is indicated.
Then,

```
  0011  0011  1001  (excess three)
+ 0011  0011  0110  (excess three)
─────────────────
  0110  0110  1111  (excess six)
```

Our answer is in excess six; therefore, we must subtract three from each group in order to return our answer to excess three; that is,

```
  0110  0110  1111
- 0011  0011  0011
─────────────────
  0011  0011  1100  (excess three)
    0     0     9   in decimal
```

When a carry is developed in any group, the following procedure is used.
EXAMPLE: Add 9 and 3 in excess three.
SOLUTION: Write

```
Decimal           Excess Three
   9      0011  0011 ⟋1100  (excess three)
                    1
 + 3      0011  0011 ⟍0110  (excess three)
 ────
  12      0110  0111  0010  (excess six)
```

NOTE: Since the right-hand group created a carry, as shown, three must be ADDED instead of subtracted in order to place this group into excess three. The other groups follow the previous example; that is,

```
  0110   0111   0010  (excess six)
- 0011  -0011  +0011
──────────────────────
  0011   0100   0101  (excess three)
    0      1      2   decimal
```

COMPLEMENT ARITHMETIC

A computer is usually designed to perform its arithmetic operations using either addition (only)

or subtraction (only). Thus in a given computer all computations are performed by adding. In another machine all operations are performed by subtracting. This is possible since algebraically:

$$a + b = s \qquad \text{eq #1}$$
$$\text{and} \quad a + (-b) = d \qquad \text{eq #2}$$

In equation #1, b is a positive quantity and a + b equals s (the sum). In equation #2, b is a negative quantity, and, when added, d represents the difference, or a - b. This procedure can be used in the computer only after a method is found for identifying and manipulating both positive and negative numbers; a process usually accomplished by using complement arithmetic.

An arithmetical complement is defined as the difference between a number and the power of the base next in series. Thus in base ten:

2 is the complement of 8;
  $(10 - 8 = 2)$
26 is the complement of 74;
  $(100 - 74 = 26)$
744 is the complement of 256;
  $(1000 - 256 = 744)$

Referring back to the definition, it is seen that complement arithmetic is not limited to base ten. Thus in base eight:

2 is the complement of 6;
  $(10 - 6 = 2)$
4 is the complement of 74;
  $(100 - 74 = 4)$
522 is the complement of 256;
  $(1000 - 256 = 522)$

The relationship between any number and its complement in any base is then redefined by the following equation.

$$C = B^D - n \qquad \text{eq #3}$$

Where: n = any number
    D = the number of digits in the number
    C = the complement
    B = the base of the system being used.

Observe that 2_{10} is the complement of 8_{10}, and at the same time 2_{10} is a number which has 8_{10} as a complement. Thus, developing a method of differentiating between 2 as a number and 2 as a complement of a number must also be accomplished before arithmetic operations can be performed using the complement method.

By modifying equation #3 as follows:

$$C = B^{D+1} - n \qquad \text{eq #4}$$

an interesting situation develops.

Considering the previously developed complements and using equation #4 yields, in base ten;

92 is the complement of 8;
  $(100 - 8 = 92)$
926 is the complement of 74;
  $(1000 - 74 = 926)$
9744 is the complement of 256;
  $(10000 - 256 = 9744)$

And in base eight:

72 is the complement of 6;
  $(100 - 6 = 72)$
704 is the complement of 74;
  $(1000 - 74 = 704)$
7522 is the complement of 256;
  $(10000 - 256 = 7522)$

Observe here that the complement is preceded in each case by the highest digit in the base used. If all numbers which are not complements are preceded by 0, the identification of complemented numbers is solved. This removes the possible ambiguity which might develop in interpreting a positive 92_{10} and complement of 8_{10}. Before considering some of the finer aspects of this system as applied in the binary computer we shall work several problems using present developments and base ten aritmetic.

R's (Radix) Complement Arithmetic

It may now be stipulated without further proof that the complement of a number represents the negative value of that number. Thus, to perform a subtraction it is only necessary to complement the subtrahend and continue as in normal addition. If at any time a resultant is a negative number this will be indicated by the appearance of the complement form in the answer. Further, the complement of a complement is the original number.

Let a = 047 and a' = 953 (read the complement of a = 953) B = 023 and b' = 977. There are eight possible combinations of a and b which can occur, these are:

$$a \pm (+b); \; a \pm (-b); \; -a \pm (+b); \; -a \pm (-b)$$

Substituting the given values for a and b into each case we have the following:

1) a + (+b)   2) a - (+b)   3) a + (-b)

```
  047         047   047      047
  023        -023  +977      977
  ───        ────  ────     ────
  070         024  1024     1024
```

4) a - (-b)   5) -a + (+b)   6) -a - (+b)

```
  047   047       953          953   953
 -977  +023       023         -023  +977
 ────  ────       ───         ────  ────
        070       976 = -24        1930 = -70
```

7) -a + (-b)   8) -a - (-b)

```
  953              953   953
  977             -977  +023
 ────             ────  ────
 1930              976 = -24
```

Consideration of the preceding examples brings to light an interesting phenomenon; that of overflow, as typified by the 1 in examples 2; 3; 6; and 7. Here the most significant digit (1) contributes nothing to the validity of the answer and is disregarded. This situation will be discussed at greater length later in this chapter.

All previously developed statements will apply to binary complement arithmetic. Thus the complement shall be indicated when the most significant digit is one, the highest permissible digit of the system, and a positive or uncomplemented number will be indicated if the most significant digit is zero. Further, the discussion is limited to the use of a finite number of digits, since this is the situation actually prevailing within the computer. For simplicity we will limit the binary number to fifteen places. Thus, looking at the binary equivalent of some numbers in the vicinity of zero:

```
 5    000   000   000   000   101
 4    000   000   000   000   100
 3    000   000   000   000   011
 2    000   000   000   000   010
 1    000   000   000   000   001
 0    000   000   000   000   000
-1    111   111   111   111   111
-2    111   111   111   111   110
-3    111   111   111   111   101
-4    111   111   111   111   100
-5    111   111   111   111   011
```

This representation is slightly awkward, but may be converted to the more conveniently handled octal form by grouping, giving:

```
 5    00005
 4    00004
 3    00003
 2    00002
 1    00001
 0    00000
-1    77777
-2    77776
-3    77775
-4    77774
-5    77773
```

Here each complement is developed by subtracting the absolute value of the number from 2^{15}.

Remembering that the only way to determine whether or not a number is positive or negative is by evaluating the most significant digit. We see that values from 00000 through 37777 (011 111 111 111 111) are positive and values from 40000 (100 000 000 000 000) through 77777 are negative (-40000 through -00001). However, since complementing a negative number gives a positive number, an undefined point exists, for complementing 40000 yields 40000, and the same number can not be both positive and negative. Thus our negative numbers are in fact limited to values from 40001 through 77777 (-37777 through -00001).

Now consider some sample problems.

1. (+5) + (-4) = 1

```
    00005
    77774
  1 ─────
    00001 ──► Once again this overflow is dropped
```

2. (-6) + (-6) = -14   3. (-12) - (-6) = -4

```
    77772               77766
    77772               00006
  1 ─────              ──────
    77764 = -14         77774 = -4
```

4. (32000) + (32000) = 64000

```
    32000                  Note here that the addition
    32000                  has produced an overflow
   ──────                  into the sign bit (highest
    64000 Complementing    order) position thereby in-
                           dicating a negative result.
```

$$100000 - 64000 = -14000$$

Once again it is emphasized that the largest possible positive number the machine can contain as the result of a computation is 37777. This problem would have to be scaled (examined to determine if the result will produce an overflow) prior to entering it into the machine.

5. (-37775) + (-37774) = -77771

```
  40003
  40004
1 00007  = 7
```

Here again is an erroneous Sum. The one is dropped and it appears that the sum is a positive seven. Once again the limitations of the machine have been exceeded since the largest negative number that it can properly define is 40001 (-37777). The solution is found in proper scaling of the factors.

The important facts about the R's complement arithmetic method to be reemphasized are:

a. The complement is indicated by the presence of the highest digit of the base being used in the most significant digit's position.

b. A positive number is indicated by the presence of a zero in the most significant digit's position.

c. The transition through zero is smooth, i.e., 00002; 00001; 00000; 77777; 77776.

d. A point of ambiguity exists at 40000 (in this case). This limits the magnitude of positive numbers to 37777 and negative numbers to -37777.

e. An erroneous result or an erroneous indication will be obtained if the sum of the factors exceed the limitations of the machine. The reader is encouraged to redo the problems above in pure binary form.

R's - 1 Complement Arithmetic

Heretofore we have considered what is most commonly termed radix complement arithmetic. In this method, the complement is formed by subtracting the number from a complete power of the base. Thus it is customarily abbreviated as R's complement. Another form of complement arithmetic, the R's -1 complement, is found to be uniquely convenient in binary arithmetic. (A comparison of the two systems is left until later in the discussion.) The R's -1 complement of a number may be obtained simply by subtracting one from the complement obtained by the previous method.

The following comparisons are made:

Number	R's Complement	R's - 1 Complement
5	00005	00005
4	00004	00004
3	00003	00003
2	00002	00002
1	00001	00001
+0	00000	00000
-0	undefined	77777
-1	77777	77776
-2	77776	77775
-3	77775	77774
-4	77774	77773
-5	77773	77772

The salient point here is the transition from positive numbers to negative numbers through zero. Note that when using R's -1 complement arithmetic there exists a positive and a negative zero. Now however, if the largest positive number is permitted to be 37777, the complement of this is 40000. Thus the previous point of ambiguity no longer exists at 40000. Once again the absolute value of the largest possible number that the machine can contain is 37777.

Consider some problems using R's -1 complement arithmetic.

1.   (+5) - (+4) = 1

```
    00005
    77773
1   00000
   └──►1
    00001
```

(Here an important difference in the two systems is demonstrated. In the R's complement system an overflow exceeding the limits of the machine had no significance and was dropped; the adder was said to be open ended. However, in R's -1 complement arithmetic, as is seen from the example, the overflow is of significance and must be added to the least significant digit of what might be termed our partial sum. Thus, an R's -1 complement adder is referred to as having end-around-carry.)

2.   (-6) + (-6) = -14

```
    77771
    77771
1   77762
   └──►1
    77763  = -14
```

245

3.   (32000) + (32000) = 64000

$$\begin{array}{r} 32000 \\ 32000 \\ \hline 64000 \end{array} = -13777$$

This is the same example that was used with the R's complement method. The capacity of the machine has been exceeded and overflowed into the sign bit yields what appears to be a negative result.

4.   (-37775) + (-37774) = -77771

$$\begin{array}{r} 40002 \\ 40003 \\ \hline 1\ \ 00005 \\ \hookrightarrow 1 \\ \hline 00006 \end{array}$$

Here is another example of exceeding the negative capacity of the machine. This also produces an erroneous answer.

5. (+5) – (+5) = 0   6.   (+0) + (+0) = +0

$$\begin{array}{r} 00005 \\ 77772 \\ \hline 777777 \end{array} = -0 \qquad \begin{array}{r} 00000 \\ 00000 \\ \hline 00000 \end{array} = +0$$

7. (-0) + (-0) no equivalent in traditional arithmetic

$$\begin{array}{r} 77777 \\ 77777 \\ \hline 1\ \ 77776 \\ \hookrightarrow 1 \\ \hline 77777 \end{array} = -0$$

8. (-0) + (+0) no equivalent in traditional arithmetic

$$\begin{array}{r} 77777 \\ 00000 \\ \hline 77777 \end{array} = -0$$

Examples 5 through 8 demonstrate the conditions resulting in a zero answer. It should be observed that only in the addition of a positive zero and a positive zero is the sum a positive zero. It is of importance for the student to realize that using either a positive or negative zero in a calculation will not affect the validity of the result. This may be readily demonstrated by working a few problems, and is left to the student for proof.

The salient facts about R's -1 complement arithmetic to be remembered are:

a. The complement is indicated by the presence of the highest digit of the base being used in the most significant digit's position.

b. A zero in the most significant digit's position indicates a positive number.

c. The transition from positive numbers to negative numbers through zero is not smooth. Zero is represented by 00000 which is termed a positive zero, and by 77777 which is termed a negative zero.

d. Use of either value for zero does not invalidate the result.

e. An erroneous resultant is once again obtained if the sum of the factors exceed the limitations of the machine.

f. End-around-carry is a necessity when using R's -1 complement arithmetic. Overflow into bit position 2^{15} (which does not physically exist in the machine) is brought around and added to the least significant digit's position.

The time has now arrived when a few comments upon the relative merits of the two systems are in order.

The main advantage of the R's complement system is found in its smooth transition through zero. However, certain difficulties are encountered in designing the circuitry necessary for complementing. While these will not be elaborated on at this time, they generally limit the use of R's complement arithmetic to countertype applications. In countertype applications there is no interest in negative resultants; further, all complements will normally be entered by the operator or as a fixed function of the machine. The great simplicity in developing the complement in the R's -1 system makes this the most frequently used in the arithmetic section of computers. To demonstrate this, consider the binary equivalent of 35452_2 and its complement in each system.

R's complement

| 011 | 101 | 100 | 101 | 010 | = | 35452 |
| 100 | 010 | 011 | 010 | 110 | = | 42326 |

R's -1 complement

| 011 | 101 | 100 | 101 | 010 | = | 35452 |
| 100 | 010 | 011 | 010 | 101 | = | 42325 |

Here it is observed that each binary digit is reversed or complemented in the R's -1 system, whereas the reversal in the R's complement system occurs only after the least significant digit, which is a one. Note that in forming the R's complement above that the first complemented digit occurs in the 2^2 bit position, and 2^1 is the first bit position containing a one.

For some applications the frequent occurrence of a negative zero in R's -1 arithmetic may be objectionable. (We may even say the normal occurrence of a negative zero, since the only time a positive zero occurs is the result of summing two positive zeros.)

Several methods have been developed to overcome this limitation. One of these is to detect for the occurrence of the negative zero and under its stimulus generate a false carry which will complement the negative zero. Thus:

101	010	111	100	011	=	52743
010	101	000	011	100	=	25034
111	111	111	111	111	=	77777

false carry   1   1

1   000   000   000   000   000   1   00000

Overflow   Overflow

In this particular instance the further carry produced by summing with the false carry is disregarded. This method will not be explored any further at this time since a true appreciation of the system requires a more complete knowledge of mechanization techniques.

Nines Complement

A further advantage of the excess three code is the ease with which the nines complement of a number indicated in excess three may be found. That is, the nines complement of seven, indicated in excess three as 1010, is found by inverting each digit in 1010 to read 0101. This 0101, in excess three represents decimal two which is the nines complement of seven.

The following shows the nines complement of the decimal digits.

Decimal	Excess Three	Excess Three Nines Complement	Decimal Nines Complement
0	0011	1100	9
1	0100	1011	8
2	0101	1010	7
3	0110	1001	6
4	0111	1000	5
5	1000	0111	4
6	1001	0110	3
7	1010	0101	2
8	1011	0100	1
9	1100	0011	0

Subtractive Adders

The other method for circumventing the production of negative zeros is through the use of a subtractive adder, i.e., an arithmetic unit which can only subtract. Just as it is possible to subtract using a unit capable only of addition so is it possible to add while limited to the process of subtraction. This is clearly demonstrated through the use of the following equation:

$$a + b = a - (-b)$$

There is only one new concept to consider when using a subtractive adder, and that is end-around-borrow. This is demonstrated by the next problem.

$$(00047) + (00023) = 00072$$

```
    1    00047
    ↑    77754
         00073
    └──    -1
         00072
```

This is a demonstration of end-around-borrow. Note that one is subtracted from the difference and added in the most significant digit's position of the minuend whenever the subtrahend is greater than the minuend. It is to be appreciated naturally that the actual circuitry using this technique does not step through the problem in precisely the manner indicated above. The important thing is that the above example can faithfully represent the action of the machine.

It now remains to consider the resultants obtained with problems which will produce a zero difference.

1. (00054) - (00054) = 0
```
   00054
   00054
   00000
```

2. (-0) - (-0) = 0
```
   77777
   77777
   00000
```

3. 0 - (-0) = 0
```
 1  00000
    77777
    00001
    →  -1
    00000
```

4. (-0) - (+0) = -0
```
   77777
   00000
   77777
```

Here note that the only way a negative zero is obtained is by performing the operation indicated in step four. Thus to ensure the normal appearance of a positive zero it is only necessary to use a subtractive adder. All other previously developed rules as applied to R's -1 arithmetic still apply.

As important remainders:

a. The absolute value of a number cannot exceed 37777...7.

b. Addition of two or more factors which exceed this limit will give either an erroneous answer or one that is interpreted erroneously.

c. All positive numbers will enter the machine preceded by a most significant digit of zero; all negative numbers will enter the machine in complement form.

d. Whereas, with the additive adder a subtract instruction caused the addend to be complemented it is shown that in the subtractive adder an add command will cause the subtrahend to be complemented.

e. Using R's - 1 arithmetic an end-around-borrow must be performed whenever the subtrahend is greater than the minuend.

f. The end-around-borrow is eliminated using R's complement arithmetic and a subtractive adder, however, the problem of mechanizing to form the complement is a limitation which precludes the use of this form.

BOOLEAN ALGEBRA

The father of Boolean algebra was the English logician and mathematician George Boole (1815-1864). In the spring of 1847 he wrote a pamphlet entitled "A Mathematical Analysis of Logic." Later (1854), he wrote a more exhaustive treatise on this subject which was entitled "An Investigation of the Laws of Thought." It is this later work which forms the basis for our present day mathematical theories used for the analysis of logical processes.

Although conceived in the 18th century, little practical application was found for Boolean algebra until 1938 when it was found that Boolean algebra could be adapted for the analysis of telephone relay and switching circuits. Since that time, the extent of its use has expanded rapidly, roughly paralleling the development and use of more complex switching circuits such as found in present day automatic telephone dialing systems and digital computers. Thus Boolean algebra has become an important subject which must be learned, if the operation of digital computers and other devices using complex switching circuits is to be understood.

This section discusses the basic symbols, laws, axioms, and theories of Boolean algebra in sufficient detail to enable you to analyze and simplify Boolean equations. While equations with more than four variables are not covered in this text, the information given here will aid you in the analysis and simplification of such operations.

CLASSES AND ELEMENTS

In our universe, it is logical to visualize two divisions; all things of interest in a discussion are in one division and all things which are not of interest in the discussion are in the other division. These two divisions comprise a set or class which is designated the UNIVERSAL CLASS, and all things contained in the universal class are called ELEMENTS. One may also visualize another set or class; this class contains no elements and has been designated the NULL CLASS.

In a particular discussion certain elements of the universal class may be grouped together to form combinations which are known as classes. However, these classes are actually subclasses of the universal class and thus should not be confused with the universal class or the null class. Each subclass of the universal class is dependent on its elements and the possible states (stable, unstable, or both) that these elements may have.

Boolean algebra is limited to the use of elements which have only two possible states, both of which are stable. These states are usually designated TRUE (1) or FALSE (0). Thus, to determine the number of classes or combinations of elements in Boolean algebra, we solve for the numerical value of 2^n where n equals the number of elements. If we have two elements (each element has two possible states) then we have 2^n or 2^2 possible classes. If the elements are designated A and B, then A may be true or false and B may be true or false. Using the connective word "and" the classes which could be formed are as follows:

A true and B false
A true and B true
A false and B true
A false and B false

However, if the connective work "or" is used then four additional classes are formed. The differences between the two class, groups, or forms are discussed below.

Venn Diagrams

Since the Venn diagram is a topographical picture of logic, composed of the universal class divided into classes depending on the n number of elements, Venn diagrams may be used to illustrate Boolean logic as follows:

Consider the universal class as containing submarines and atomic powered sound sources. Let A equal submarines and B equal atomic powered sound sources. Therefore, we have four classes which are:

(1) Submarines and not atomic
(2) Submarines and atomic
(3) Atomic and not submarines
(4) Not submarines and not atomic

These four classes are called MINTERMS since they represent the four minimum classes of elements. The opposite of the minterms are the MAXTERMS which are stated as follows:

(1) Atomic or not submarines
(2) Not submarines or not atomic
(3) Submarines or not atomic
(4) Submarines or atomic

The various relationships which may exist are represented by the Venn diagrams (fig. 13-3).

CONNECTIVES AND VARIABLES

Before proceeding it will be necessary to identify and define the symbology used in Boolean algebra. As may be seen, most of these symbols are common to other branches of mathematics, however in Boolean algebra they may have a slightly different meaning or application.

=   The equal sign, just as in conventional mathematics, represents a relationship of equivalence between the expressions so connected.

. or x   The dot or small x indicates the logical product, or conjunction of the terms so connected. The operation is also frequently indicated with no symbol used, i.e., $A \cdot B = A \times B = AB$. Most generally referred to as the AND operation, the terms so related are said to be "ANDed."

+   The plus sign indicates the logical sum operation, a disjunction of the terms so connected. Usually called the OR operation and the terms so connected are said to be ORed.

———   The vinculum serves a dual purpose. It is at the same time a symbol of grouping and of operation. As a sign of operation it indicates that the term(s) so overlined are to be complemented. As a symbol of grouping it collects all terms to be complemented together. Terms so overlined are often said to be NEGATED, the process of taking the complement is then called NEGATION.

() [] { }   These familiar signs of grouping are used in the customary fashion to indicate that all terms so contained are to be treated as a unit.

A, B, etc.   Various letters are used to represent the variables under consideration, generally starting with A. Since the variables are capable of being in only one of two states the numerals 0 and 1 are the only numbers used in a Boolean expression.

APPLICATIONS TO SWITCHING CIRCUITS

Since Boolean algebra is based upon elements having two possible stable states, it becomes very useful for analyzing switching circuits. The reason for this is that a switching circuit can be in only one of two possible states. That is, it is either open or it is closed. We may represent these two states as 0 and 1, respectively. The basic switching operations are discussed below, all other switching operations (even the most complex) are merely combinations of these basic operations.

The And Operation

Let us consider the Venn diagram in figure 13-4A. Its classes are labeled using the basic expressions of Boolean algebra. Note that there are two elements, or variables, A and B. The shaded area represents the class of elements that are $A \cdot B$ in Boolean notation and is expressed as:

$$f(A,B) = A \cdot B$$

The other three classes are also indicated in figure 13-4A. This expression is called an AND operation because it represents one of the four minterms previously discussed. Recall that AND indicates class intersection and both A and B must be considered simultaneously.

We can conclude then that a minterm of n variables is a logical product of these n variables with each variable present in either its noncomplemented or its complemented form, and is considered an AND operation.

For any Boolean function there is a corresponding truth table which shows, in tabular form, the true conditions of the function for each way in which conditions can be assigned its variables. In Boolean algebra, 0 and 1 are the symbols assigned to the variables of any function. Figure 13-4B shows the AND operation function of two variables and its corresponding truth table.

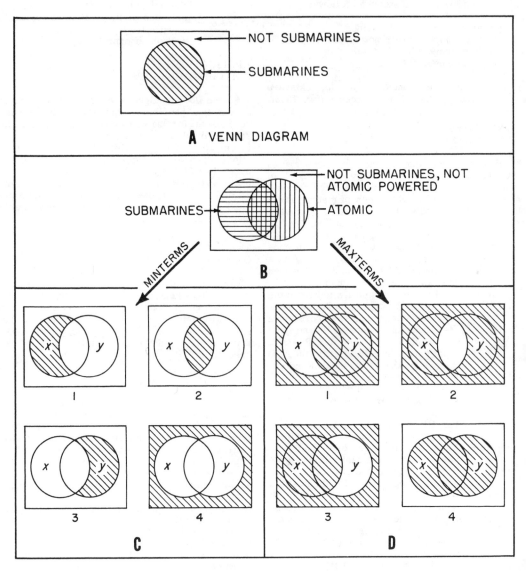

164.4(179)
Figure 13-3. — The Venn diagram.

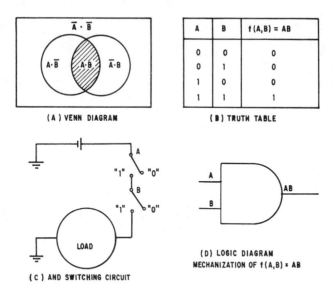

(A) VENN DIAGRAM

A	B	f (A,B) = AB
0	0	0
0	1	0
1	0	0
1	1	1

(B) TRUTH TABLE

(C) AND SWITCHING CIRCUIT

(D) LOGIC DIAGRAM
MECHANIZATION OF f(A,B) = AB

12.133
Figure 13-4.—The AND operation.

This function can be seen to be true if one thinks of the logic involved: AB is equal to A and B which is the function f(A,B). Thus, if either A or B takes the condition of 0, or both take this condition, then the functions f(A,B) equals AB is equal to 0. But if both A and B take the condition of 1 then the AND operation function has the condition of 1.

Figure 13-4C shows a switching circuit for the function f(A,B) equals AB in that there will be an output only if both A and B are closed. An output in this case equals 1. If either switch is open, 0 condition, then there will be no output or 0.

In any digital equipment, there will be many circuits like the one shown in figure 13-4C. In order to analyze circuit operation, it is necessary to refer frequently to these circuits without looking at their switch arrangements. This is done by logic diagram mechanization as shown in figure 13-4D. This indicates that there are two inputs, A and B, into an AND operation circuit producing the function in Boolean algebra form of AB. These diagrams simplify equipment circuit

diagrams by indicating operations without drawing all the circuit details.

It should be understood that while the previous discussion concerning the AND operation dealt with only two variables that any number of variables will fit the discussion. For example, in figure 13-5 three variables are shown along with their Venn diagram, truth table, switching circuit, and logic diagram mechanization.

The OR Operation

Consider the Venn diagram in figure 13-6A; note that there are two elements, or variables, A and B. The shaded area represents the class of elements that are A+B in Boolean notation and is expressed in Boolean algebra as:

$$f(A,B) = A + B$$

This expression is called an OR operation for it represents one of the four maxterms previously discussed. Recall that OR indicates class union, and either A or B or both must be considered.

251

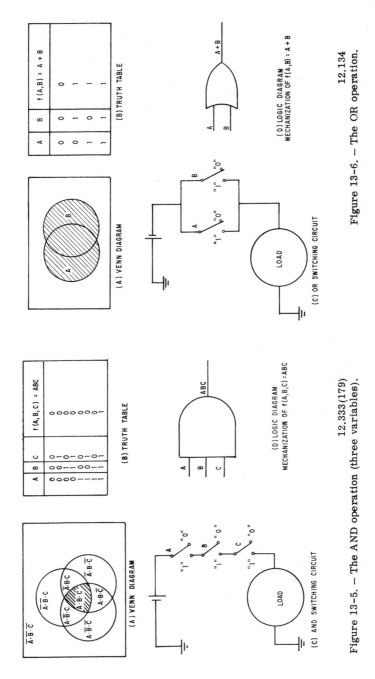

12.134

Figure 13-6. – The OR operation.

12.333(179)

Figure 13-5. – The AND operation (three variables).

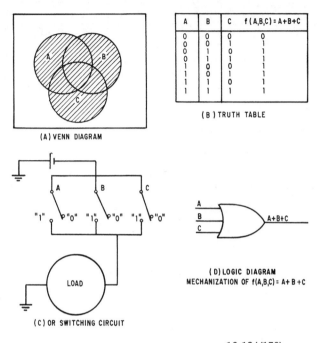

A	B	C	f (A,B,C) = A+B+C
0	0	0	0
0	0	1	1
0	1	0	1
0	1	1	1
1	0	0	1
1	0	1	1
1	1	0	1
1	1	1	1

(B) TRUTH TABLE

(A) VENN DIAGRAM

(D) LOGIC DIAGRAM
MECHANIZATION OF f(A,B,C) = A+ B +C

(C) OR SWITCHING CIRCUIT

12.134(179)
Figure 13-7.—The OR operation (three variables).

The conclusion is that a maxterm of n variables is a logical sum of these n variables where each variable is present in either its noncomplemented or its complemented form.

In figure 13-6B the truth table of an OR operation is shown. This truth table can be seen to be true if one thinks of A+B being equal to A or B which is the function f(A,B). Thus if A or B takes the value 1, then f(A,B) must equal 1. If not, then the function equals zero.

Figure 13-6C shows a switching circuit for the OR operation which is two or more switches in parallel. It is apparent that the circuit will transmit if either A or B is in a closed position; that is, equal to 1. If, and only if, both A and B are open, equal to 0, the circuit will not transmit.

The logic diagram for the OR operation is given in figure 13-6D. This means that there are two inputs, A and B, into an OR operation circuit producing the function in Boolean form of A+B. Note the difference in the diagram from that of figure 13-4D.

As in the discussion of the AND operation the OR operation may also be used with more than two inputs. Figure 13-7 shows the OR operation with three inputs.

The Not Operation

The shaded area in figure 13-8A represents the complement of A which in Boolean algebra is $\overline{A}$ and read as "NOT A." The expression f(A) equals $\overline{A}$ is called a NOT operation. The truth table for the NOT operation (fig. 13-8B) is explained by the NOT switching circuit. The requirement of a NOT circuit is that a signal injected at the input produce the complement of this signal at the output. Thus, in figure 13-8C it can be seen that when switch A is closed, that is, equal to 1, the relay opens the circuit to the load. When switch A is open, that is, equal to 0, the relay completes a closed circuit to the load. The logic diagram for the NOT operation is given in figure 13-8D. This means that A is the input to a

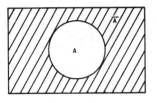

(A) Venn Diagram

A	f(A) = $\overline{A}$
1	0
0	1

(B) Truth Table

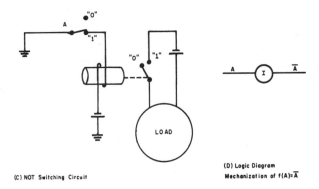

(C) NOT Switching Circuit

(D) Logic Diagram
Mechanization of f(A)=$\overline{A}$

12.135

Figure 13-8.—The NOT operation.

NOT operation circuit and gives an output of $\overline{A}$. The NOT operation may be applied to any operation circuit such as AND or OR. This is discussed in the following section.

The NOR Operation

The shaded area in figure 13-9A represents the quantity, A OR B, negated. This figure represents the minterm expression $\overline{AB}$; that is, A OR B negated is $\overline{A\ OR\ B}$ and by application of DeMorgan's theorem (to be discussed later) is equal to $\overline{AB}$.

The truth table for the NOR operation is shown in figure 13-9B. The table shows that if either A or B is equal to 1, then f(A,B) is equal to 0. Furthermore, if A and B equal 0, then f(A,B) equals 1.

The NOR operation is a combination of the OR operation and the NOT operation. The NOR switching circuit in figure 13-9C is the OR circuit placed in series with the NOT circuit. If either switch A, switch B, or both are in the closed position, equal to 1, then there is no transmission to the load. If both switches A and B are open, equal to 0, then current is transmitted to the load.

The logic diagram mechanization of f(A,B) equals $\overline{A+B}$ (NOR operation) is shown in figure 13-9D. It uses both the OR logic diagrams and the NOT logic diagrams. The NOR logic diagram mechanization shows there are two inputs, A and B, into an OR circuit producing the function in Boolean form of A+B. This function is the input to the NOT (inverter) which gives the output, in Boolean form, of $\overline{A+B}$. Note that the whole quantity of A + B is complemented and not the separate variables.

The NAND Operation

The shaded area in figure 13-10A represents the quantity A AND B negated (NOT), and is a maxterm expression. Notice that $\overline{AB}$ is equal to the maxterm expression $\overline{A}+\overline{B}$.

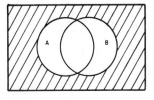

(A) Venn Diagram

A	B	A + B	f(A,B) = $\overline{A+B}$
0	0	0	1
0	1	1	0
1	0	1	0
1	1	1	0

(B) Truth Table

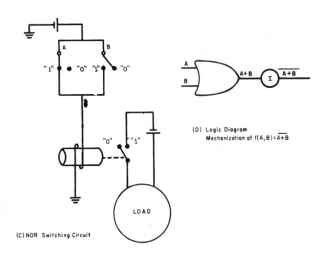

(C) NOR Switching Circuit

(D) Logic Diagram
Mechanization of f(A,B) = $\overline{A+B}$

67.89
Figure 13-9.—The NOR operation.

The truth table is shown for the NAND operation in figure 13-10B. When A and B equal 1, then f(A,B) is equal to 0. In all other cases, the function is equal to 1.

The NAND operation is a combination of the AND operation and the NOT operation. The NAND switching circuit is shown in figure 13-10C. Note that the AND circuit is in series with the NOT circuit. If either switch A or B is open, equal to 0, then current is transmitted to the load. If both switches A and B are closed, equal to 1, then there is no transmission to the load.

The logic diagram mechanization of f(A,B) equals $\overline{AB}$ (NAND operation) is shown in figure

13-10D. The AND operation logic diagram and the NOT logic diagram mechanization show that there are two inputs, A and B, into the AND circuit producing the function in Boolean form of AB. This function is the input to the NOT circuit which gives the output, in Boolean form, of $\overline{AB}$. Note that the entire quantity AB is complemented and not the separate variables.

NOTE: The logic diagrams in figures 13-9 and 13-10 do not conform to the American Standard Logic Symbology (MIL-STD). They were drawn in this fashion merely to illustrate a point in the text, normally they will be drawn as shown in figure 13-11.

255

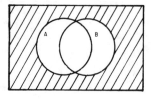

(A) Venn Diagram

A	B	AB	f(A,B)= $\overline{AB}$
0	0	0	1
0	1	0	1
1	0	0	1
1	1	1	0

(B) Truth Table

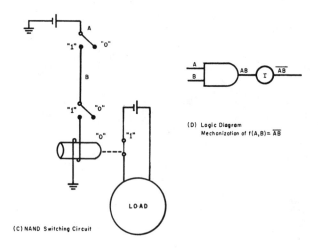

(C) NAND Switching Circuit

(D) Logic Diagram
Mechanization of f(A,B)= $\overline{AB}$

67.90
Figure 13-10.—The NAND operation.

Exclusive OR Operation

The exclusive OR operation is actually a special application of the OR operation discussed previously. In this operation either A or B must be true in order for the function to be true, however, if both are true at the same time the function will be false.

As can be seen by the Venn diagram (fig. 13-12A) this operation must be assigned a special class since it does not conform to any of the minterm or maxterm classes previously discussed.

In the mechanization of this operation (fig. 13-12C) the switches are mechanically linked to-

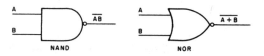

NAND   NOR

67.89:.90(179)
Figure 13-11.—American Standard logic symbology for the NOR and the NAND operations.

gether so that one or the other, but not both, may be closed at a time. The truth table for this operation is shown in figure 13-12B and the logic symbol in figure 13-12D.

256

A. VENN DIAGRAM

A	B	$f(A,B) = A\bar{B} + B\bar{A}$
0	0	0
0	1	1
1	0	1
1	1	1

B. TRUTH TABLE

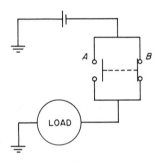

LOAD

C. EXCLUSIVE "OR" SWITCHING CIRCUIT

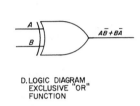

D. LOGIC DIAGRAM
EXCLUSIVE "OR"
FUNCTION

12.134(179)

Figure 13-12.—The EXCLUSIVE OR operation.

FUNDAMENTAL LAWS AND AXIOMS OF BOOLEAN ALGEBRA

This section lists the fundamental laws and axioms of Boolean algebra which should be memorized since they may be used to simplify Boolean expressions. While these laws and axioms may be used to simplify any Boolean expression, it is difficult to determine when the expression is in its simplest form. To overcome this difficulty, other methods of simplification have been devised and will be discussed later.

The laws and axioms are listed below and figures 13-13 through 13-22 show the truth tables, logic diagrams, and mechanization for these laws and axioms.

I. Law of Identity
   A = A
II. Law of Complementarity
   1. $\overline{A}A = 0$
   2. $A + \overline{A} = 1$
III. Idempotent Law
   1. AA = A
   2. A + A = A

IV. Commutative Law
   1. AB = BA
   2. A + B = B + A
V. Associative Law
   1. (AB)C = A(BC)
   2. (A + B) + C = A + (B + C)
VI. Distributive Law
   1. A(B + C) = (AB) + (AC)
   2. A + (BC) = (A + B)(A + C)
VII. Law of Dualization (DeMorgan's Theorem)
   1. $\overline{(A + B)} = \overline{A}\overline{B}$
   2. $\overline{(AB)} = \overline{A} + \overline{B}$
VIII. Law of Double Negation
   $\overline{\overline{A}} = A$
IX. Law of Absorption
   1. A(A + B) = A
   2. A + (AB) = A

AXIOMS

   1. A + 0 = A
   2. A · 0 = 0   (The variable A may
   3. A + 1 = 1   be 1 or 0.)
   4. A · 1 = A

257

BOOLEAN EQUATION SIMPLIFICATION AND MECHANIZATION

In this section various means of simplifying logic equations and the mechanization of the resultant will be discussed.

Applications Of Theorems

Boolean algebra comprises a set of axioms and theorems (discussed earlier) which are useful in describing logic equations such as those used in computer technology. Likewise, these laws and

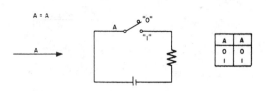

179.791
Figure 13-13.—Law of identity.

(A)   (B)

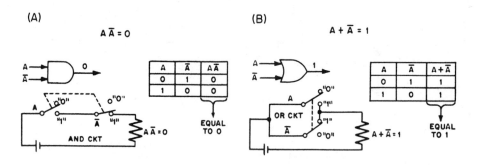

179.792
Figure 13-14.—Complementary law.

(A)   (B)

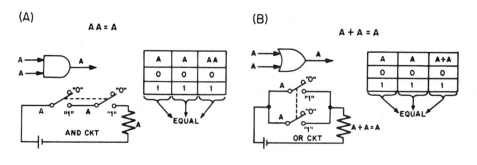

179.793
Figure 13-15.—Idempotent law.

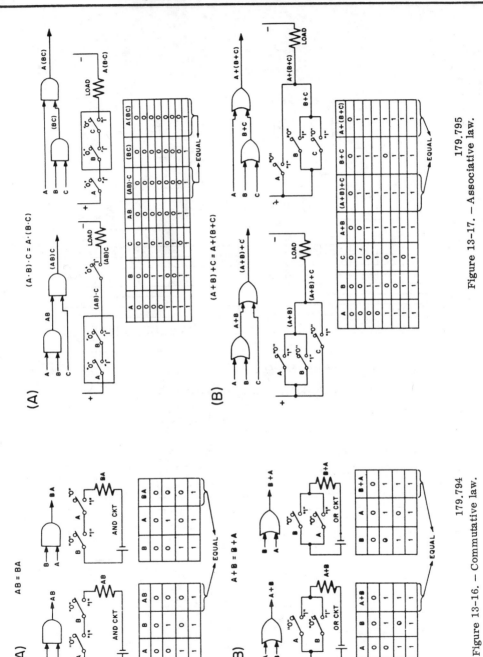

Figure 13-17. – Associative law.

179.795

Figure 13-16. – Commutative law.

179.794

259

(A) A(B+C) = (AB) + (AC)

(B) A + BC = (A+B)(A+C)

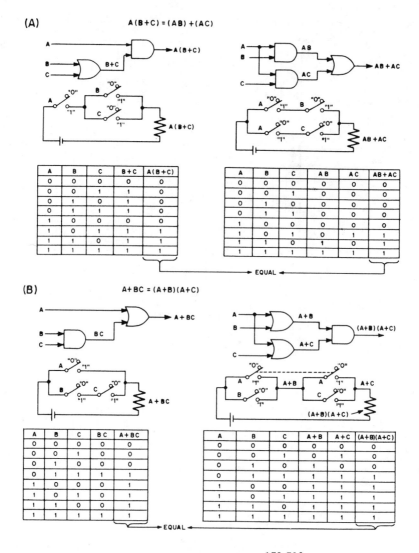

A	B	C	B+C	A(B+C)
0	0	0	0	0
0	0	1	1	0
0	1	0	1	0
0	1	1	1	0
1	0	0	0	0
1	0	1	1	1
1	1	0	1	1
1	1	1	1	1

A	B	C	AB	AC	AB+AC
0	0	0	0	0	0
0	0	1	0	0	0
0	1	0	0	0	0
0	1	1	0	0	0
1	0	0	0	0	0
1	0	1	0	1	1
1	1	0	1	0	1
1	1	1	1	1	1

→ EQUAL ←

A	B	C	BC	A+BC
0	0	0	0	0
0	0	1	0	0
0	1	0	0	0
0	1	1	1	1
1	0	0	0	1
1	0	1	0	1
1	1	0	0	1
1	1	1	1	1

A	B	C	A+B	A+C	(A+B)(A+C)
0	0	0	0	0	0
0	0	1	0	1	0
0	1	0	1	0	0
0	1	1	1	1	1
1	0	0	1	1	1
1	0	1	1	1	1
1	1	0	1	1	1
1	1	1	1	1	1

→ EQUAL ←

179.796
Figure 13-18. — Distributive law.

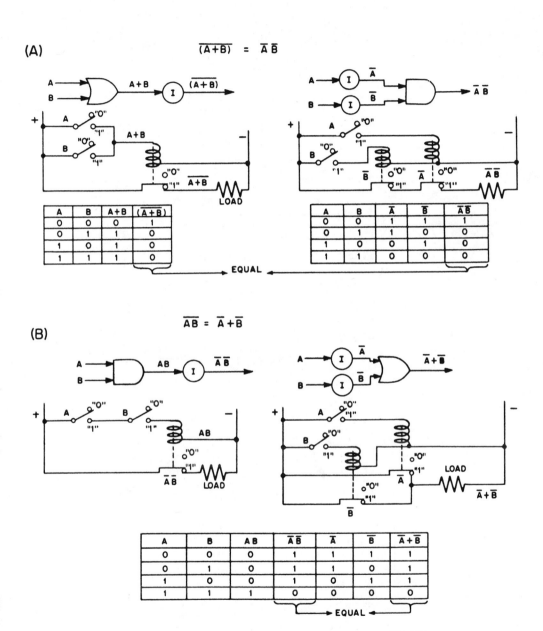

179.797
Figure 13-19.— Law of dualization (De Morgan's theorem).

261

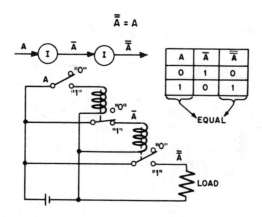

179.798
Figure 13-20. — Law of double negation.

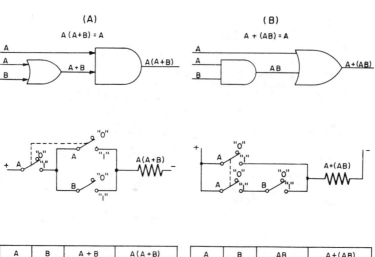

179.799
Figure 13-21. — Absorption law.

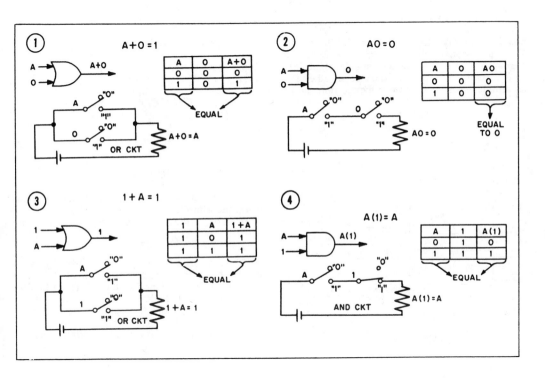

179.800
Figure 13-22.—Axiomatic expressions.

axioms are used to simplify logic equations so
that logic circuits can be designed in their simplest
and most economic form. For example, the equation below is a logic equation which describes
the logic circuit (fig. 13-23A) in Boolean terms.

$$F = AC + AD + BC + BD$$

If Boolean algebra is used to simplify the
logic equation,

$$F = AC + AD + BC + BD$$
$$= A(C + D) + B(C + D) \qquad \text{Dist. Law}$$
$$= (A + B)(C + D)$$

then the logic circuit arrangement for the simplified expression is shown in figure 13-23B. Factors such as the loading and standardization of
logic circuits may dictate the use of other than
the simplest possible Boolean expression. In this

discussion the only concern is with the equation
simplification without regard to other design
considerations.

Consider the following as a second example:
EXAMPLE: Simplify the logic equation,

$$f = ABC + AB\overline{D} + A\overline{C} + \overline{ABCD} + \overline{A}C$$

SOLUTION: Rearrange terms and factor as
follows:

$$f = ABC + A\overline{C} + \overline{ABCD} + \overline{A}C + AB\overline{D}$$
$$= A(BC + \overline{C}) + \overline{A}(\overline{BCD} + C) + AB\overline{D}$$

Applying the complementary law to $(BC+\overline{C})$ and
$(\overline{BCD}+C)$
Then:

$$f = A(B+\overline{C}) + \overline{A}(\overline{BD}+C) + AB\overline{D}$$

Apply distributive law
Then:

$$f = AB + A\overline{C} + \overline{A}\,\overline{BD} + \overline{A}C + AB\overline{D}$$
$$f = (AB+AB\overline{D}) + A\overline{C} + \overline{A}C + \overline{A}\,\overline{BD}$$

Apply the law of absorption to $(AB+AB\overline{D})$ and
rearrange terms. Then:

$$f = AB + A\overline{C} + \overline{A}C + \overline{A}\,\overline{BD}$$

This equation is the easiest to mechanize; however, the simplification process could be carried
one step further by factoring in which case

$$f = A(B+\overline{C})+\overline{A}(C+\overline{BD})$$

The foregoing examples of simplification show
the process to be rather difficult at first with
no positive indication (for the beginner) that the
simplest possible logic equation has been reached.
Repeated use of these theorems is the only solution. Simplification theorems are of greatest
value in the preliminary stages of simplification,
or in the simplification of elementary functions.

Veitch Diagrams

A second approach to equation simplification
is the Veitch diagram. These diagrams provide
a very quick and easy way for finding the simplest
logic equation needed to express a given function.
Veitch diagrams for two, three, or four variables
are readily constructed (fig. 13-24). Any number
of variables may be plotted on a Veitch diagram,
though the diagrams are difficult to construct and
use when more than four variables are involved.

Because each variable has two possible states
(true or false), the number of squares needed is

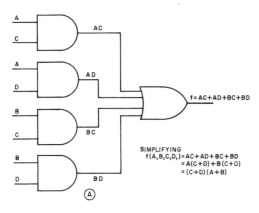
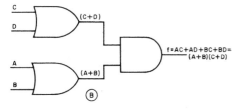

164.8
Figure 13-23.—Simplified logic circuitry resulting from simplifying logic equations.

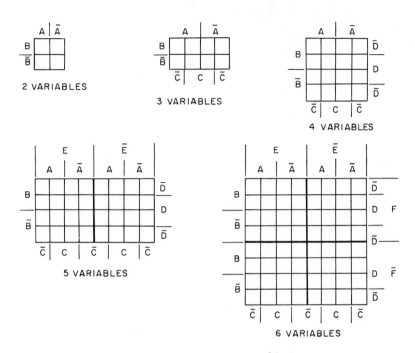

124.17
Figure 13-24.—Veitch diagrams.

the number of possible states (two) raised to a power dictated by the number of variables. Thus, for four variables the Veitch diagram must contain 2^4 or 16 squares. Five variables require 2^5 or 32 squares. An eight variable Veitch diagram needs 2^8 or 256 squares—a rather unwieldy diagram. If it becomes necessary to simplify logic equations containing more than six variables, other methods of simplification should be used.

An exploded view of a four variable Veitch diagram is shown in figure 13-25. Note the division of the diagram into labeled columns and rows. The entries into the diagram are placed in these columns and rows in accordance with the function values for a given Boolean expression.

Looking at the square in the upper left corner of the main Veitch diagram in figure 13-25, and using the extensions, it is seen that is contains the variables $AB\overline{C}\overline{D}$; the next lower block contains $AB\overline{C}D$; the next lower block contains $A\overline{B}\overline{C}D$; and the block in the lower left corner contains the variables $A\overline{B}\overline{C}\overline{D}$. All of the squares in the diagram are similarly identified. Note that

the term $A\overline{C}$ is contained in each of the four terms just discussed; and, by the distributive law, since the variables B and D appear in both asserted and complemented form they can be dropped. Thus, the left vertical column identifies the term $A\overline{C}$.

This is proven below by equation using the laws considered earlier.

$$\begin{aligned}
A\overline{C} &= A\overline{C} \ (1) \\
&= A\overline{C} \ (B + \overline{B}) \\
&= AB\overline{C} + A\overline{B}\overline{C} \\
&= AB\overline{C} \ (1) + A\overline{B}\overline{C} \ (1) \\
&= AB\overline{C} \ (D + \overline{D}) + A\overline{B}\overline{C} \ (D + \overline{D}) \\
&= AB\overline{C}D + AB\overline{C}\overline{D} + A\overline{B}\overline{C}D + A\overline{B}\overline{C}\overline{D}
\end{aligned}$$

The final expression represents four of the maxterms of the term $A\overline{C}$. Also note that a two-variable term is represented by four squares. A study of the diagram will reveal that a term with one variable is represented by eight squares, a three-variable term by 2 squares, and a four-variable term by 1 square.

265

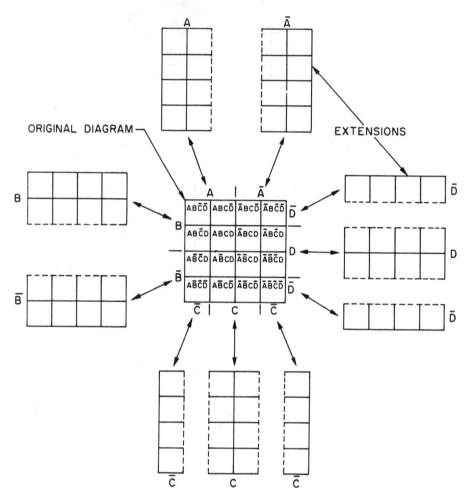

124.21(164)
Figure 13-25. — Exploded Veitch diagrams.

To illustrate the use of the Veitch diagram, the logic equation

$$f = ABC + AB\overline{D} + A\overline{C} + \overline{A}\,\overline{B}\overline{C}\overline{D} + \overline{A}C$$

will be used. (This is one of the same equations used earlier to show some applications of the simplification theorems.) Because there are four variables, a four variable Veitch diagram is needed. The step-by-step process is as follows:

Step 1. Draw the appropriate Veitch diagram (table 13-11).

Step 2. Plot the logic function on the Veitch diagram, term by term. This is accomplished by placing a "1" in each square representative of the term. (Use table 13-11 to identify the squares and table 13-12 to understand the plotting of the terms on the diagram.)

Table 13-11.—Identifying the squares of a Veitch diagram

	A		$\overline{A}$		
B	$AB\overline{C}\overline{D}$ 1	$ABC\overline{D}$ 2	$\overline{A}BC\overline{D}$ 3	$\overline{A}B\overline{C}\overline{D}$ 4	$\overline{D}$
	$AB\overline{C}D$ 5	$ABCD$ 6	$\overline{A}BCD$ 7	$\overline{A}B\overline{C}D$ 8	D
$\overline{B}$	$A\overline{B}\overline{C}D$ 9	$A\overline{B}CD$ 10	$\overline{A}\overline{B}CD$ 11	$\overline{A}\overline{B}\overline{C}D$ 12	D
	$A\overline{B}\overline{C}\overline{D}$ 13	$A\overline{B}C\overline{D}$ 14	$\overline{A}\overline{B}C\overline{D}$ 15	$\overline{A}\overline{B}\overline{C}\overline{D}$ 16	$\overline{D}$
	$\overline{C}$	C	$\overline{C}$		

NOTE: The numbers in each square are for the purpose of illustration only.

164.66

Table 13-12.—Plotting of the logic function

	A		$\overline{A}$		
B	$AB\overline{C}\overline{D}$ 1	$ABC\overline{D}$ 1	$\overline{A}BC\overline{D}$ 1	$\overline{A}B\overline{C}\overline{D}$	$\overline{D}$
	$AB\overline{C}D$ 1	$ABCD$ 1	$\overline{A}BCD$ 1	$\overline{A}B\overline{C}D$	D
$\overline{B}$	$A\overline{B}\overline{C}D$ 1	$A\overline{B}CD$	$\overline{A}\overline{B}CD$ 1	$\overline{A}\overline{B}\overline{C}D$	D
	$A\overline{B}\overline{C}\overline{D}$ 1	$A\overline{B}C\overline{D}$	$\overline{A}\overline{B}C\overline{D}$ 1	$\overline{A}\overline{B}\overline{C}\overline{D}$ 1	$\overline{D}$
	$\overline{C}$	C	$\overline{C}$		

165.67

$$f = ABC + AB\overline{D} + A\overline{C} + \overline{A}\overline{B}C\overline{D} + \overline{A}C$$

The term ABC of the equation is identified in the Veitch diagram by squares 2 and 6. The derivation is as follows:

$$ABC\overline{D} + ABCD = ABC\,(\overline{D} + D) = ABC\,(1) = ABC$$

The term $AB\overline{D}$, identified by squares 1 and 2 as follows:

$$AB\overline{C}\overline{D} + ABC\overline{D} + AB\overline{D}\,(\overline{C} + C) = AB\overline{D}\,(1) = AB\overline{D}$$

For the term $A\overline{C}$, squares 1, 5, 9, and 13:

$$AB\overline{C}\overline{D} + AB\overline{C}D + A\overline{B}\overline{C}D + A\overline{B}\overline{C}\overline{D} = AB\overline{C}\,(D+\overline{D})$$
$$+ A\overline{B}\overline{C}\,(D + \overline{D}) = AB\overline{C}\,(1) + A\overline{B}\overline{C}\,(1) + A\overline{C}\,(B +$$
$$\overline{B}) = A\overline{C}\,(1) = A\overline{C}$$

The term $\overline{A}\overline{B}\overline{C}\overline{D}$, identified by square 16 is self-explanatory.

The term $\overline{A}C$, squares 3, 7, 11, and 15 as follows:

$$\overline{A}BC\overline{D} + \overline{A}BCD + \overline{A}\overline{B}CD + \overline{A}\overline{B}C\overline{D} = \overline{A}BC$$
$$(\overline{D} + D) + \overline{A}\overline{B}C\,(D + \overline{D}) = \overline{A}BC\,(1) + \overline{A}\overline{B}C\,(1) =$$
$$\overline{A}C\,(B + \overline{B}) = \overline{A}C\,(1) = \overline{A}C$$

Step 3. Obtain the simplified logic equation by using figure 13-26 and observing the following rules:

a. If 1's are located in adjacent squares or at opposite ends of any row or column, one of the variables may be dropped.
b. If any row or column of squares, any block of four squares, or the four end squares of any adjacent rows or columns, or the four corner squares are filled with 1's, two of the variables may be dropped.
c. If any two adjacent rows or columns, the top and bottom rows, or the right and left columns are completely filled with 1's, three of the variables may be dropped.
d. To reduce the original equation to its simplest form, sufficient simplification must be made until all 1's have been included in the final equation. The digit "1" may be used more than once, and the largest possible combination of 1's in groups of 8, 4, 2, or as a single 1 (block) should be used.

Squares 1, 5, 9, and 13 are combined (fig. 13-26) using rule (b) to yield $A\overline{C}$ (1's in B and $\overline{B}$ cancel).
Squares 3, 7, 11, and 15 are combined using rule (b) to yield $\overline{A}C$.
Squares 1, 2, 5, and 6 are combined using rule (b) to yield AB.

267

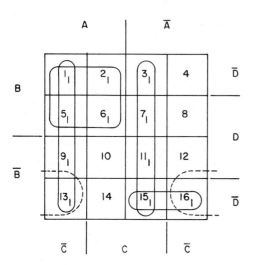

Figure 13-26.—Derivation of resultant.

Squares 15 and 16 are combined using rule (a) to yield $\overline{A}\overline{B}D$.

To keep track of the squares combined, draw loops around the combined squares. In doing this, the Veitch diagram takes on the appearance shown in figure 13-26.

All 1's have been used, therefore, a logic equation can now be written

$$f = AB + A\overline{C} + \overline{A}C + \overline{A}\overline{B}D$$

which agrees with the simplified logic equation obtained by the use of the simplifying theorems.

A second simplification for the equation, $f = ABC + AB\overline{D} + A\overline{C} + \overline{A}\overline{B}\overline{C}D + \overline{A}C$, the equation just discussed, is derived by grouping blocks 13 and 16. Although shown in the form of a table, the Veitch diagram (fig. 13-26) is to be considered as a cylinder. (This permits continuity of the variable, i.e., the rows containing $\overline{D}$ and $\overline{C}$ are contiguous though it is not readily possible to show this.) As such, the $\overline{D}$ blocks become adjacent when the bottom end of the table is folded back and up and the top is folded back and down until the ends meet. The $\overline{C}$ blocks become adjacent when the right end is folded back and to the left and the left end is folded back and to the right. Thus, blocks 13 and 16 are adjacent. The simplification of the equation is therefore,

$$f = AB + A\overline{C} + \overline{A}C + \overline{B}C\overline{D}.$$

It is to be recognized from the above example that a single equation in Boolean can be represented in more than one simplified form.

A Veitch diagram provides a convenient means of finding the complement of a logic equation. This is done by plotting the original equation on a Veitch diagram, and then plotting ones on another Veitch diagram everywhere except where the original diagram has ones. An example will illustrate the procedure.

EXAMPLE:

If: $f = ABC$

What is $\overline{f}$?

A three variable Veitch should give the answer. The original equation is first plotted as shown in table 13-13.

On another Veitch diagram, all squares which do not have a "one" in the original diagram, are assigned a "one" (fig. 13-27). Now the equation for $\overline{f}$ can be written. Squares 3, 4, 7, and 8 combine to form $\overline{A}$; squares 5, 6, 7, and 8 combine to form $\overline{B}$; and squares 1, 5, 4, and 8 combine to form $\overline{C}$. Therefore, the equation for $\overline{f}$ is,

$$\overline{f} = \overline{A} + \overline{B} + \overline{C}$$

which agrees with the result obtained by directly applying DeMorgan's theorem.

In the earlier discussion of logic operations, switching circuits were used to illustrate the various operations. These switching circuits were actually the mechanization of the logic operations

Table 13-13.—Three-variable Veitch diagram showing statement True

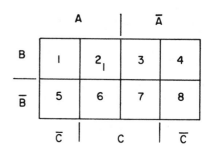

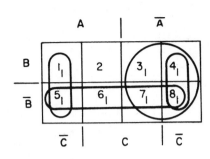

164.70
Figure 13-27. — Three-variable Veitch diagram
showing statement complemented.

using conventional single-pole double-throw
switches. Before using the actual logic symbols,
the conventional switches are again used to mechanize an equation. The equation:

$$f = AB + A\overline{C} + \overline{A}C + \overline{A}B\overline{D}$$

will be mechanized. (This is the equation used
in the discussion on simplification.)

It will be recalled that the AND function used
a series switching circuit, and the OR function
used a parallel switching circuit. Therefore, the
mechanization of the above equation is as illustrated in figure 13-28. This diagram illustrates
the AND and OR functions. The AND functions
are each series connected switch grouping of the
four possible parallel paths—the OR function. The
above equation and mechanization are shown in
figure 13-29, using the logic symbols for AND
and OR gates. A logic equation can always be
mechanized by a switching network. This involves
the following four steps:

1. Construct a truth table.
2. Write the logic equation.
3. Simplify the equation, if possible.
4. Draw the required switching network.

The reader is encouraged to apply these four
steps to several hypothetical problems.

Minterms And
Maxterms

In this discussion, consider the following
equations:

eq 1 $f = AB + A\overline{B} + \overline{A}B$
eq 2 $f = A + \overline{B}$

Equation 1 is the sum of three Boolean terms,
each of which is the product of two variables
(A and B), each variable of which is represented in either its true or complemented form.
Equation 2 represents the sum of the two variables A and B with B complemented. Equation
1 is a minterm expression of the two variables,
and equation 2 is a maxterm expression of these
variables. (The proof of the equality of the two
equations is left to the reader.)

In general then, a minterm expression of n
variables is defined as the product of these n
variables where each variable is expressed in
either its true or complemented form, and a
maxterm of n variables is the sum of these n
variables where each variable is added in either
its true or complemented form. Consequently,
there are four minterms of two variables, and
they are ($\overline{AB}$, $\overline{A}B$, $A\overline{B}$ and AB). Likewise, there
are four maxterms of two variables. They are
($\overline{A} + \overline{B}$), ($\overline{A} + B$), ($A + \overline{B}$), and ($A + B$).

There are eight minterms and eight maxterms
of three variables as shown in table 13-14. As
might be expected, there are 2^n minterms and
2^n maxterms where n variables are considered.

Observe that the minterm identified by an odd
number in the upper left corner of each block
in the left column relates to the next higher even
number in the right column in accordance with
DeMorgan's Theorem. Example. The maxterm in
block #2 is the complement of the minterm in
block #1. Using table 13-15, the unshaded area
represents the minterm while the shaded areas
represent the maxterms.

The Harvard Chart

One other technique of interest is the Harvard
Chart. Its use is demonstrated in table 13-16.

Simplify $f = AB\overline{C} + ABC + \overline{A}BC + \overline{A}B\overline{C} + A\overline{B}C$

1. Draw a line through all rows whose terms
are not contained in the expression being simplified (Rows 1, 2, and 5).

2. Starting with the left column (column 1)
cross out all terms which have been lined out in
step 1. ($\overline{A}$ is lined out in rows 1 and 2 and A in
row 5. Thus, all terms are lined out in the left
column of this example.)

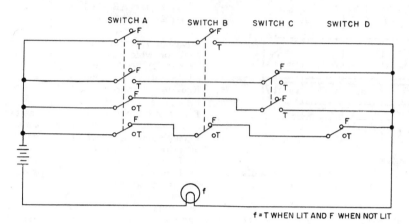

164.9

Figure 13-28.—Mechanization of a logic equation.

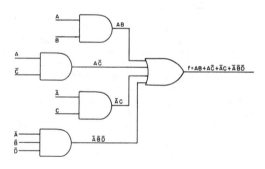

164.10

Figure 13-29.—Mechanization of a logic equation using logic symbols.

Table 13-14.—Minterms and maxterms of variables A, B, and C

MINTERM		MAXTERM	
1	$\bar{A}\,\bar{B}\,\bar{C}$	16	$\bar{A}+\bar{B}+\bar{C}$
3	$\bar{A}\,\bar{B}\,C$	14	$\bar{A}+\bar{B}+C$
5	$\bar{A}\,B\,\bar{C}$	12	$\bar{A}+B+\bar{C}$
7	$A\,\bar{B}\,\bar{C}$	10	$A+\bar{B}+\bar{C}$
9	$\bar{A}\,B\,C$	8	$\bar{A}+B+C$
11	$A\,\bar{B}\,C$	6	$A+\bar{B}+C$
13	$A\,B\,\bar{C}$	4	$A+B+\bar{C}$
15	$A\,B\,C$	2	$A+B+C$

3. In column 2 only $\bar{B}$ is eliminated. Circle all B's for easy identification as a part of the final answer.

4. Going to the right, cross out all terms containing B in all rows with a circled B. (For example, in row 4 the terms $\bar{A}B$, BC, and $\bar{A}BC$ are lined out.)

5. Continue with column C, etc.

6. In column AC the term AC is not crossed out, circle it.

7. Going to the right in the rows containing AC, cross out all other terms containing AC.

164.71

270

Table 13-15.—Representation of minterm and maxterm on a Veitch diagram

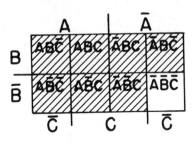

164.72

Table 13-16.—Harvard Chart Columns

1	2	3	4	5	6	7	
A	B	C	AB	AC	BC	ABC	
A̅	B̅	C̅	A̅B̅	A̅C̅	B̅C̅	A̅B̅C̅	Row 1
A̅	B̅	C	A̅B̅	A̅C	B̅C	A̅B̅C	Row 2
A̅	(B)	C̅	A̅B	A̅C̅	BC̅	A̅BC̅	Row 3
A̅	(B)	C	A̅B	A̅C	BC	A̅BC	Row 4
A	B̅	C̅	AB̅	AC̅	B̅C̅	AB̅C̅	Row 5
A	B̅	C	AB̅	(AC)	B̅C	AB̅C	Row 6
A	(B)	C̅	AB	AC̅	BC̅	ABC̅	Row 7
A	(B)	C	AB	(AC)	BC	ABC	Row 8

179.812

All terms in columns BC and ABC (columns 6 and 7) are now crossed out and the process ends. Only B and AC are left. The final answer is, therefore, f = B + AC.

Proof of this solution by use of the simplification theorems or Veitch diagrams is left to the reader. While it is easier to use Veitch diagrams and simplification theorems with problems containing four variables or less, this is not the case when more variables are involved. Harvard charts are more suitable for solving problems with five or more variables.

Inhibitor Circuit

There are three possible subtraction combinations of the two binary digits, 0 and 1. These are as follows:

$$\begin{array}{ccc} M & S & \\ 0 & - 0 & = 0 \\ 1 & - 0 & = 1 \\ 1 & - 1 & = 0 \end{array}$$

(The fourth, which is not possible without a borrow, is $0-1$, which is not considered at this time.)

A circuit that will produce the correct result for each of the three possible subtractions is the INHIBITOR (fig. 13-30). The inhibitor symbol is shown in figure 13-30A. A truth table, showing the inputs and outputs from the circuit is shown in figure 13-30B.

The inhibitor circuit produces an output when there is a signal on M but not S. An INVERTER (small circle) connected in the S input path causes the S input to the AND-element to be inverted. (The inverter output to the AND-element will be 1 when the S input is 0, and 0 when the S input is 1.) Thus the inhibitor output is 1 when (and only when) the voltage pulses at the input represents $M\overline{S}$. For all other input combinations the inhibitor output will be 0. The actions of this circuit thus satisfy the basic requirements of a binary subtractor. Although this is a relatively simple illustration of the relationship between mathematics and logic, it is representative of logic principles used to perform more complex arithmetic operations.

MIXED LOGIC

If all signal lines on a logic diagram of a system or device are assumed to have the same

INPUT		OUTPUT
MINUEND (M) (A)	SUBTRAHEND (S) (B)	DIFFERENCE (M$\overline{S}$) (A$\overline{B}$)
0	0	0
1	0	1
1	1	0

B TRUTH TABLE

124.38.1(164)
Figure 13-30.—Inhibitor.

271

state when active, and if all are electrical potentials, and if the more positive potential is consistently selected as the 1-state, the resultant system or device is said to have positive logic. If the less positive potential is consistently selected as the 1-state, the resultant system or device is said to have a negative logic. In either case, the system or device uses a FIXED LOGIC.

Consider the symbols shown at A and B in figure 13-31 and the truth table shown at C. In positive logic, the -3 volt level is the 0-state and the +2 volt level is the 1-state. Note that only the last combinations of input levels in the table at C produce a high output, and the circuit performs the AND operation. The table at D shows the logic values substituted for the voltage levels represented in the table at C.

In negative logic, the -3 volt level is the 1-state (high) and the +2 volt level is the 0-state. With this logic, the circuit at B produces the -3 volt level output for all input combinations except the last, as shown in the table at C. This circuit performs the OR function. The table at E shows the substitution of the logic values for the voltage levels using negative logic. This is the truth table for the OR function, and the circuit performs the OR operation.

A more recent method of assigning logic levels to binary logic elements uses MIXED LOGIC. The details of this method are delineated in the American Standard, Graphic Symbols for Logic Diagrams, Y32.14 of September 26, 1962. A brief description of this method is treated below.

Consider again the symbol shown in figure 13-31A and the truth table shown at C. The filled-in right triangle at the point where the signal line joins the symbol indicates that the 1-state (or activating signal) with respect to that particular logic symbol is the more positive potential. An open right triangle indicates that the 1-state for that particular logic symbol is the less positive potential. (Either of the two kinds of right triangles, open or filled, may be omitted provided the convention is suitably noted on the diagram.)

Again, the output (F) of each circuit is a function of two variables (A,B). The output and input levels may be either +2 volts or -3 volts. The circuit at A produces the AND function only when the +2 volt level is taken as the activating level. The same circuit produces the OR function if the -3 volt level is the active level. Thus, a method is available for using both positive and negative logic on the same diagram. Inverters are frequently used with mixed logic to

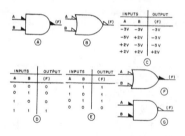

20.472(164)
Figure 13-31.—Mixed logic.

ensure that the desired function is applied to a subsequent circuit at the selected activating level.

Recall now that with fixed logic (as heretofore considered) the basic symbols in figure 13-31A and B represent the AND and OR functions respectively. When mixed logic is used, a single circuit can perform either the AND operation or the OR operation. Thus, given a physical on-off device and a table of logic combinations, the logic function performed by the device is determined by the specified choices of the 1-state at its inputs and outputs.

It is also possible, using mixed logic, to have a circuit whose active output level is the opposite of the activating input level as shown in figure 13-31F and G. When both inputs to the circuit at F are low, the output is high. However, with fixed logic, the circuit can be thought of as combining the OR and NOT functions, and is sometimes referred to as the NOR circuit. Likewise, the circuit at G produces a low output at (F) when both inputs are high. This circuit combines the AND and NOT functions and is referred to as a NAND operation.

LOGIC CIRCUITS

Logic circuits are used to perform mathematical operations such as addition, subtraction, multiplication, division, counting, etc. They are designed to express logic equations, i.e., to produce either a true or false condition at the output dependent upon the value or condition of the binary variables.

FLIP-FLOPS

The flip-flop (bistable multivibrator) is used extensively in logic equipment either as a storage device or for counting and scaling purposes. Since it is extremely stable in either of its two states, one state may be used to represent the true condition of a variable and the other state the false.

Flip-flops differ widely in the manner in which information is applied. The number of inputs to a given flip-flop may vary from as few as one to more than seven. Of course only two outputs are available.

Flip-flops may or may not be clocked, i.e., in synchronization or in time with other flip-flops in that equipment. When they are clocked, timing or clock pulses will be derived from a stable oscillator called a MASTER CLOCK. Clocked flip-flops may react to information on the leading or trailing edge of the clock pulse.

The specific operation to be performed by a given flip-flop is usually set forth in a concise form by the truth table. The logic symbols for two types of flip-flops are shown in figure 13-32.

Set-Reset
Flip-Flop

A truth table for a basic set-reset flip-flop is shown in table 13-17. In figure 13-32A the set output, Q, is obtained from the "1" output terminal and the $\bar{Q}$ from the "0" output terminal. When both inputs are at a logical zero, there will be no change in state (Qn). A "0" in the S and a "1" into the R will cause the flip-flop to go to the "0" or clear state or to stay in the "0" or clear state. A "1" into the S and a "0" into the R will put or keep the circuit in the logical "1" or set state. Logical "1's" into both S and R will result in an undetermined condition (U); i.e., it may go to either the "1" or "0" state.

As was noted before, flip-flops may be clocked. When clocked the flip-flop can be designed to change state on either a positive or negative going clock pulse. The truth tables are still valid but the

Table 13-17. — Truth table for basic set-reset flip-flop

S	R	Q	$\bar{Q}$
0	0	Qn	$\overline{Qn}$
0	1	0	1
1	0	1	0
1	1	U	U

179.811

flip-flop will respond to the conditions only during the specified clock transition.

Toggle Flip-Flop

The basic toggle flip-flop has only one input terminal (fig. 13-32B) and will change state each time it is pulsed. Most toggle flip-flops are designed to respond only to the negative going pulses.

COUNTERS

A flip-flop can hold a single bit of data (a single variable) or a single digit in the binary system. A group of variables or a number of digits may be stored in several flip-flops which make up a register. For example, the number 1010 would require a four flip-flop register for storage.

Information may be fed into or removed from a register by either of two methods. One method is SEQUENTIAL TRANSMISSION or SERIAL FORM. In this method, pulses, representing the digits of a number, are fed into the register one at a time, usually beginning with the least significant digit. This is illustrated in figure 13-33. The pulses are shifted through the register until each flip-flop is set to the proper number. The second method is SIMULTANEOUS TRANSMISSION or PARALLEL FORM. In this method, the pulses representing the digits of a number are applied simultaneously to each flip-flop via a separate line. This is illustrated in figure 13-34.

A. SET-RESET   B. TOGGLE

124.26(179)

Figure 13-32. — Flip-flop logic symbols.

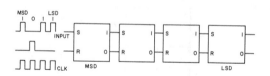

179.815
Figure 13-33. — Sequential transmission.

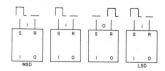

179.816
Figure 13-34. — Simultaneous transmission.

Ring Counter

Figure 13-35 illustrates a RING COUNTER. This circuit will continually cycle a one or pattern of one's as long as clock pulses are applied. The S and R inputs to a flip-flop are determined by the state of the flip-flop which precedes it; therefore, with each clock pulse the flip-flop will be set to the state which previously had been exhibited by the flip-flop which precedes it. Since the output of the last flip-flop is fed to the input of the first, the pattern of digits originally set into the ring counter will be circulated repeatedly.

Binary Up Counter

Figure 13-36 depicts a BINARY UP COUNTER. The input to this circuit is a series of pulses, which represent events which are to be counted. The counter illustrated is a modulus 16 binary up counter. That is, it will represent 16 numbers — 0000 to 1111. The number of digits a counter may represent is determined by the number of flip-flops it contains. For example, a counter made up of five flip-flops can represent 32 numbers — 00000 to 11111.

The operation of the circuit of figure 13-36 is as follows. Referring to figure 13-37, with all the flip-flops set at 0, the first input pulse will set FF1 to the one state. The counter now contains 0001. The second pulse will reset or

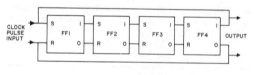

124.31.(179)
Figure 13-35. — Ring counter.

clear FF1 (return it to the zero state). The negative (or positive) going signal generated by FF1 changing from the one to the zero state will set FF2 and advance the count. The counter will now contain 0010. This action will continue until the counter contains 1111. The next pulse, the sixteenth, will reset all the flip-flops to zero.

Down Counter

In order to perform certain functions, digital equipment may be required to cycle through a predetermined number of operations and then stop. A circuit used to count down from a preset number is the DECREMENTING or DOWN COUNTER. Such a circuit is illustrated in figure 13-38. This circuit is a modulus 16 down counter. It will count down from 1111 to 0000. The operation of this circuit may be understood by considering all the flip-flops to be set to the one state (1111) and referring to the waveforms depicted in figure 13-39.

COMPUTATIONAL CIRCUITS

In most digital equipments, arithmetic operations are necessary. Addition is one of the most basic functions. Recall the basic binary rules which are:

augend	0	0	1	1
addend	+0	+1	+0	+1
sum	0	1	1	0 with a carry

These conditions are illustrated on the truth table (table 13-18).

Half Adder

The HALF ADDER is a logic circuit which produces a sum output and a carry output for addend and augend inputs. The block diagram symbol for the half adder is shown in figure

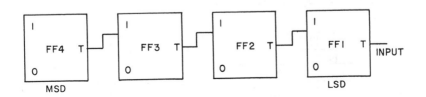

179.817
Figure 13-36. — Binary up counter.

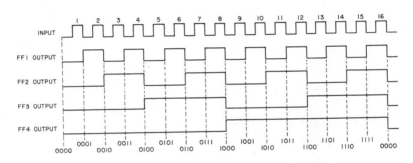

179.818
Figure 13-37. — Operation of figure 13-36.

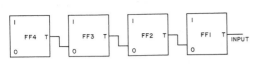

179.819
Figure 13-38. — Down counter.

13-40. From the truth table a logic equation may be derived for both the sum and carry:

$$\text{SUM} = \overline{X} Y + X \overline{Y}$$
$$\text{CARRY} = X Y$$

The logic diagram in figure 13-41 is the mechanization of these equations. The reader at this point should follow all possible input combinations through the diagram.

Full Adder

A carry may be generated from a previous order, as can occur when numbers having more than one digit are added. The input combinations are greater and the circuitry is more involved than with half adders. The circuit which can handle the input carries (carries from a previous order) as well as the addend and augend inputs is called a FULL ADDER. The block diagram symbol and the truth table for the full adder are shown in figure 13-42.

275

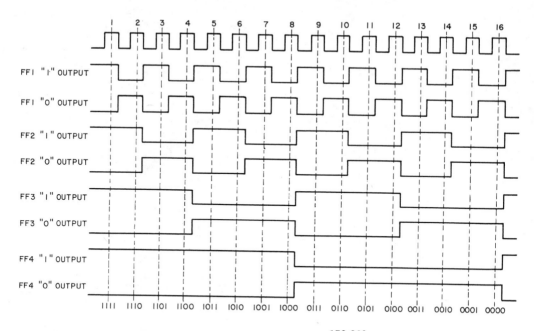

179.820
Figure 13-39. — Operation of figure 13-38.

Table 13-18. — Truth table for ADD operations

AUGEND CONDITIONS (X)	ADDEND CONDITIONS (Y)	SUM	CARRY INPUT (C)
0	0	0	0
0	1	1	0
1	0	1	0
1	1	0	1

124.259(179)

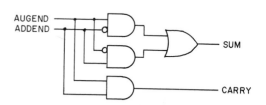

124.40(179)
Figure 13-41. — Half adder logic diagram.

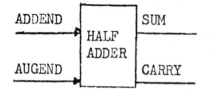

179.821
Figure 13-40.—Half adder block diagram symbol.

The logic equations derived from the full adder truth table are:

$$\text{SUM} = \bar{X}\bar{Y}C + \bar{X}Y\bar{C} + X\bar{Y}\bar{C} + XYC$$
$$\text{CARRY} = \bar{X}YC + X\bar{Y}C + XY\bar{C} + XYC$$

276

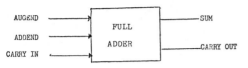

A. BLOCK DIAGRAM SYMBOL

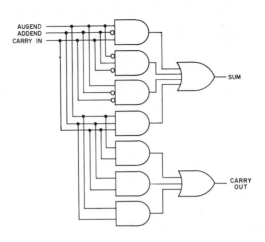

179.822

Figure 13-43. — Full adder logic diagram.

X	Y	C	SUM OUT	CARRY OUT
0	0	0	0	0
0	0	1	1	0
0	1	0	1	0
0	1	1	0	1
1	0	0	1	0
1	0	1	0	1
1	1	0	0	1
1	1	1	1	1

B. TRUTH TABLE

124.260(179)

Figure 13-42. — Full adder block diagram symbol and truth table.

The carry can be simplified using Boolean algebra or one of the simplification tables to:

$$CARRY = CX + CY + XY$$

These equations may be mechanized without change as shown in figure 13-43.

Again it would behoove the reader to follow each input combination through the diagram. There are many ways to achieve this same function. It can be done with OR and AND circuits, NOR circuits, NAND circuits, or any combination of these circuits.

Serial Addition

The serial adder makes use of the full adder in combination with several shift registers. Serial addition processes each digit in a number individually in a sequential order. The shift registers must have as many flip-flops as there are digits in the numbers to be added and the flip-flops must be clocked. Since the information is added in sequential fashion, one clock pulse is required for each addition. Therefore, the number of flip-flops and the amount of time required increase with the magnitude of numbers which are to be added. A block diagram of a serial adder is shown in figure 13-44.

The LSD in the addend and augend registers are added as soon as the numbers are transferred into these registers. The sum and carry outputs of the adder are connected to the sum register and the carry flip-flop, respectively. For each clock pulse, the contents of the addend and augend register are shifted one position to the right; the sum output for the LSD is shifted into the sum register; and the carry output is shifted into the carry flip-flop. It must be remembered that the outputs may be one or zero depending on the numbers being added. The second least significant digits are now added with the carry, if one was generated by the lower

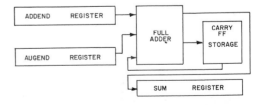

179.823

Figure 13-44. — Serial adder block diagram.

277

order digits. Succeeding clock pulses shift higher and higher order digits into the adder, while simultaneously shifting the new sum digits into the sum register. With the correct number of clock pulses, the LSD of the sum will be contained in the rightmost flip-flop in the sum register. The numbers which were in the addend and augend registers are now gone. New numbers could be shifted into these registers as others are shifted into the adder. However, complex timing signal patterns are required.

If the sum of the serial adder was shifted back into the addend register, the circuit could function as an accumulator; i.e., contents of this register could be added to. The sum of a series of numbers could be totaled in this manner. The subtotals would be added in serial fashion to the next number in the series to form a new accumulated sum.

Multiplication

Another slight modification will allow this configuration to perform the function of multiplication by repeated addition. The augend register would be connected in a ring, and a down counter controlling addition cycles would be required as shown in figure 13-45. Since this diagram is shown with eight flip-flops, it will take eight clock pulses to add two numbers. At the end of one addition cycle (the eighth clock pulse) a pulse is applied to the down counter.

Initially, the multiplicand is placed in the augend register, the multiplier into the down counter and zero's in the accumulator register. After the first addition cycle, the multiplicand will be in the accumulator register as well as in the augend register, and the down counter is decremented by one. Each succeeding series of eight pulses decrements the down counter by one and the accumulator adds the magnitude of the multiplicand to its contents. When the down counter reaches a zero count in each of its flip-flops, gating circuitry will disable clock pulses to the serial adder. The contents of the accumulator are now equal to the product.

DIGITAL TO ANALOG CONVERSION

Digital information, as previously explained, is expressed in digits of a particular number system. Analog data is in physical and often electrical form. The term ANALOG stems from the term ANALOGY. In analog representations, voltage, current, and shaft rotation can be used to

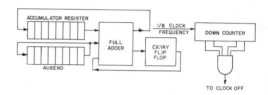

124.45(179)
Figure 13-45. — Multiplication by repeated addition.

represent variables such as temperature, fluid flow, and angular displacement. Analog data expressed as a physical representation bears an exact relationship to the original information; as a result, it is a continuous variable, whereas digital is discrete and discontinuous.

ANALOG VERSUS DIGITAL INFORMATION

Just as problems in digital form may be handled by digital computers or other digital equipment, many problems can be solved in an analog computer. To further illuminate the differences between digital and analog, consider the following problem. A missile is to be launched from a cruiser and it is necessary to determine acceleration. Force divided by mass will yield acceleration according to Newton's second law. For example, if F (force) is 150 units and M (mass) is 2 units, then A (acceleration) equals 75 units. The differences in the two methods should be made apparent by analyzing the approach to this problem by each method.

A digital equipment would process data in binary form. The mass will be relatively constant so it would be a stored value, in this case 00000010, assuming the equipment will handle eight digits. The force would be fed into the device through a radio link in digital form; in this example it would be 10010110. The values would be sent to a binary divider which would yield 01001011. Notice that this example uses only whole integers of a positive power; therefore, a binary point with negative power digits would need to be added to increase accuracy.

An analog device would handle the information by analogy. One method frequently used is electrical analogy. For example, since A = F/M and I = E/R, a simple resistive circuit could be used to determine the answer.

The potential in figure 13-46 is radio controlled so that the voltage, E, is directly related to F. The resistance, R, is a fixed value representing mass. Thus, acceleration can be obtained from the current meter in the circuit.

There are many familiar applications of analog. The gas gauge in automobiles is one which uses a simple electrical circuit. In a synchro system, various voltages are proportional to angular displacement. An antenna may go to a position indicator through a gear arrangement, in which shaft rotation is proportional to antenna orientation.

Note that digital data is discrete; i.e., in the example shown only whole digits could be used. The analog output could be any value along a continuum within the range of the device. The digital example above could be increased in accuracy simply by using negative power digits.

DIGITAL-TO-ANALOG CONVERTER

In many digital equipments the input data is in analog form (physical or electrical). The data to be read into the digital equipment may be from a gun turret. The turret may be connected by a system of gears to a shaft. The position of the shaft is related to the gun turret position. Therefore, shaft position must be translated into binary form. Conversely, digital outputs must be translated to analog form. For example, a d.c. control voltage to a magnetic amplifier may have to be determined from digital information. One way of converting digital information to an analog voltage will now be presented.

Flip-Flop Register

A digital to analog converter is sometimes called a decoder. A simple version could be used

which contains a single resistor for each flip-flop in a register containing the binary number. Figure 13-47 shows such an arrangement. The values of the resistors are inversely proportional to the weight of the digit in the flip-flop to which each resistor is connected. Consequently, the smaller resistors will contribute more to development of the output voltage than those which are larger.

An electrical analysis will illustrate the operation. For convenience, the flip-flops will be assumed to operate with positive logic between zero volts and a positive sixteen volts. When 0000 is contained in the register the top of each resistor is at ground potential (0 volts); therefore, the output is at zero volts. A 0001 in the register would connect R4 to a plus sixteen volts and the other resistors to zero volts. This results in an equivalent circuit as shown in figure 13-48A. When a 0011 is contained in the register, the resulting equivalent circuit is shown in figure 13-48B. The output, in this example, will increase approximately one volt for each binary increment.

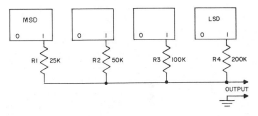

179.825
Figure 13-47. — Simple D-to-A converter.

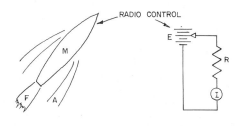

179.824
Figure 13-46. — Simple analog computation.

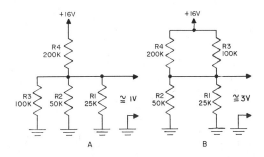

179.826
Figure 13-48. — Equivalent circuits of figure 13-47.

Level Amplifier

There are several shortcomings manifested by this simple configuration. For example, loading conditions will affect the accuracy of this simple resistive network; thus, the output levels obtained from the divider may not be an exact analog representation of the digital input. Figure 13-49 is a block diagram of a more sophisticated D-to-A converter.

In most applications the level amplifiers are IC chips which produce precise outputs from inputs within a wide range. Due to circuit variations flip-flop output levels will vary and level amplifiers are placed between the flip-flops and the divider network to equalize the outputs of each flip-flop. When triggered by the flip-flop amplifiers will connect the divider network either to ground or to a reference voltage supplied by a precision supply.

Operational Amplifier

Operational amplifier is a general term given to a device which can perform a multitude of functions. Technically, the title operational amplifier is commonly reserved for a two or more stage assembly which has been designed for insertion into other equipment. This assembly can be designed to perform almost any mathematical function. The behavior of the operational amplifier is influenced very strongly by externally connected circuits; i.e., a single amplifier can be externally connected to invert, add, subtract, multiply by a constant, integrate, or differentiate, dependent on external circuitry. The operation is primarily determined by feedback circuitry (which can be external). In the previous diagram (fig. 13-49), the device can be used to multiply by a constant.

ANALOG-TO-DIGITAL CONVERTERS

As previously stated, the information to be acted upon by digital equipment is often in the form of analog data that must be converted to digital form with as much accuracy and in as little time as possible. The device that makes the conversion is the analog-to-digital converter.

The development of analog-to-digital converters was accelerated during and immediately after World War II. Previously, analog computers had been in extensive use for solving a wide variety of experimental and simulative problems. Their answers, however, had to be

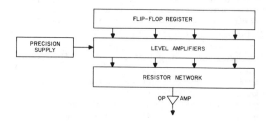

179.827
Figure 13-49.—D-to-A converter block diagram.

converted, by hand, from analog to digital form, and these results tabulated on desk-type digital machines. Determination of final results, from the vast amounts of data produced by some of these analog computations, required months or even years. The obvious need was for analog-to-digital converters that could operate at speeds compatible with the computers available.

With the development of high-speed digital equipments the need became even more acute. Instead of doing the hand-machine computations, data could be processed on automatic machines. Analog-to-digital converters now had the additional problem of converting analog information to digital data in a code suitable for digital computers and other digital equipment. With the advent of analog-to-digital and digital-to-analog converters it became feasible to manufacture hybrid digital-analog computers.

ELECTROMECHANICAL A-TO-D CONVERTERS

Analog-to-digital converters are commonly known as encoders. They can be categorized into two major types, those in which the input information is in the form of a physical position (usually involving shaft rotation, but in some cases involving linear motion) and those in which the input information is in the form of a voltage. Encoders that receive shaft-position information are electromechanical. Types of electromechanical converters are coded surface, optical interference, and digital gear. The voltage-input encoders are electronic and will be discussed later in this chapter. It should be noted that the type of encoding method selected is limited by the original source of the data. For example, a shaft position can be converted to a voltage by connecting the shaft to the arm of a potentiometer. By the same token, a voltage can be converted to a shaft position by a servo system.

The CODED SURFACE CONVERTER is one method of changing shaft position into digital representation. It is shown in figure 13-50A. This disk is attached to a shaft, with which it rotates. The shaded areas of the disk are of conducting materials while the unshaded areas are nonconductive. A set of brushes makes contact with the concentric circular patterns of the disk and for a particular position of the disk (i.e., position of the shaft) a particular binary output is produced. The illustration shows the brushes in a position to yield an output of 101. This device is usually called a binary wheel encoder.

Another method which can be used that is very similar to the binary wheel has a binary pattern of conducting and nonconducting areas wrapped around a drum. As the drum is rotated, various binary codes are picked up by the brushes. These binary codes are indicative of the position of the drum and also the position of the shaft. Such a drum is shown in figure 13-50B. If a translatory motion is to be digitized, a similar binary pattern can be used also. In this case, the coded surface is flat with the brushes translating over the binary combinations. Figure 13-50C shows this type of encoder.

To increase the degree of RESOLUTION (the degree of accuracy to which the position of the shaft is measured) of the wheel, additional concentric patterns can be added to the wheel converter. The figures shown previously used three channels representing three bits. With the three bits, eight different codes can be obtained which yield an indication of the wheel's position to within 45 degrees of its actual position. If one more channel were used (i.e., a 4-bit code), 16 code combinations could be obtained enabling the position to be measured to within 22 1/2 degrees of its actual position. If a wheel with 10 digits and 1,024 code combinations were used, the position of the wheel could be measured to within about 1/3 degree of its actual position.

Another method of sensing the position of a binary wheel or drum encoder uses a light source with photocell pickups. The shaded areas of the wheel as shown in figure 13-51 are opaque and the unshaded areas are transparent. As the

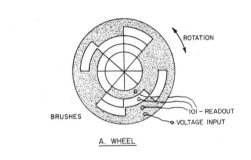

A. WHEEL

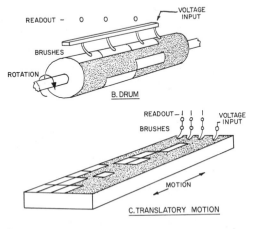

B. DRUM

C. TRANSLATORY MOTION

179.828
Figure 13-50. — Types of 3 bit binary encoders.

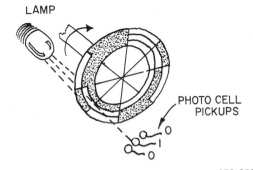

LAMP

PHOTO CELL PICKUPS

179.829
Figure 13-51. — Photocell pickup binary wheel encoder.

wheel rotates, various photocells are exposed to the light while others are unexposed due to the opaque portion of the wheel.

Gray Code

When a natural binary code is used on the wheel or drum converter an unexpected problem arises. For example, the case in which a binary wheel encoder is moving from 011 to 100 will be analyzed. At the boundary, each of the output signals should reverse itself. However, if the wheel stops on or very near the boundary or passes through the boundary at a relatively slow speed, the output from each channel depends on the precision of the markings on the wheel, the accuracy of the alignment of the brushes or photocells, and the width of the contactive surface of the brushes or the sensitivity of the photocells. Depending upon these factors, the output could be any of the codes from 000 to 111. It is obvious that this is an undesirable condition. As the digital equipment samples the output configuration of the converter, and its output is incorrect due to the reasons previously discussed an equipment will, in many applications, initiate a corrective action or some other action when it is not needed. This would cause an erratic operation of the system. There are means of eliminating this problem.

One method is to use a CYCLIC or GRAY CODE. The best method of understanding this code is to examine its derivation. The following is an example of deriving a four-digit reflected binary code.

The first two natural binary bits are put down. The reflected binary bits are identical with these:

```
0 0 0 0
0 0 0 1
```

These two bits are repeated, followed by their mirror image, but with the second digit changed from 0 to 1:

This results in the first four bits. These bits are repeated again, followed by their mirror

images, but this time the third digit is switched from 0 to 1:

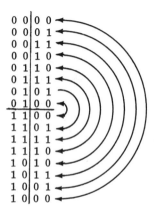

This gives the first eight bits. Finally, these bits are repeated, followed by their mirror images, but the last digit is switched from 0 to 1:

We now have the full four-digit code, consisting of sixteen bits extending from 0 to 15. An examination of the code will reveal that no more than one digit ever switches at a given instant. Brush misalignment now can produce an error of only one angular position.

By increasing the number of digits and repeating the same process a code as large as may be desired can be derived. Five digits will yield 32 bits, six digits will yield 65 bits, etc. A large number of coded positions increases the accuracy of the readout, since smaller angles of rotation will cause the readout to change.

In addition to the possible error of one count, another problem is introduced when the reflected binary code is used in the form of a decade readout. The code that is used to convert a three-digit decimal counter into a four-digit reflected binary counter is as follows:

HUNDREDS	TENS	UNITS
0 0 0 0	0 0 0 0	0 0 0 0
0 0 0 1	0 0 0 1	0 0 0 1
0 0 1 1	0 0 1 1	0 0 1 1
0 0 1 0	0 0 1 0	0 0 1 0
0 1 1 0	0 1 1 0	0 1 1 0
0 1 1 1	0 1 1 1	0 1 1 1
0 1 0 1	0 1 0 1	0 1 0 1
0 1 0 0	0 1 0 0	0 1 0 0
1 1 0 0	1 1 0 0	1 1 0 0
1 1 0 1	1 1 0 1	1 1 0 1

Examination of the first and last bit of each of the decades reveals the nature of the problem. Four digits must switch when going from 9 to 10. Three of these are in the units decade, and one is in the tens decade. Similarly, seven digits must switch when going from 99 to 100. Each time that a decade must return from 9 to 0, three digits must switch. The use of another code solves this problem.

Unit-Distance Codes

The term UNIT-DISTANCE CODE is not used to denote one specific code, but rather a family of codes, each of which is designed to eliminate multiple switching problems. A typical unit-distance code is as follows:

```
1 0 0 0
1 1 0 0
1 1 0 1
0 1 0 1
0 0 0 1
0 0 1 1
0 1 1 1
1 1 1 1
1 1 1 0
1 0 1 0
```

An examination of the first and last bits of this code reveals that only one digit must change when returning from 9 to 0. However, when it is used as a decade counter, there is the additional switch required in the tens column, and in the hundreds column, etc. In order to eliminate this problem, the techniques of the unit-distance code and the reflected binary code are combined. Two codes are used representing the bits from 0 to 9. Odd-numbered decades use the unit-distance

code that has been shown, while even-numbered decades employ a mirror image of this code. Thus, only the switch in the next higher column is required, that is, the tens column when going from 9 to 10, or 19 to 20, etc., and the hundreds column when going from 99 to 100, or 199 to 200, etc. Such a code would be as in figure 13-52.

By means of this code, we can proceed from 0 to 999, employing only one, basic, four-digit code that is reflected and repeated in accordance with a relatively simple pattern. Figure 13-52 represents a typical ten-bit coded disc of the reflected binary type.

By rearranging the conducting and nonconducting areas on the wheel or drum converter into the proper pattern for the Gray code the converter becomes a cyclic code generator. Since a negative zero is not generated and there is only a single digit change each time the code is advanced the problems of ambiguity and erratic readout are reduced or eliminated. Ambiguity does not exist since only a positive zero will be generated. The problem of erratic readout is greatly reduced since only a single digit will be affected in event of erratic brush contact when the digit is changing from a 0 to a 1 or vice versa. Since a continuous sampling is being made as the wheel or drum rotates the correct code will eventually be recorded even should an error be generated as a result of erratic brush contact. Converting from Gray code to natural binary is a relatively simple process requiring only the proper arrangement of logic circuitry.

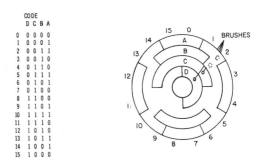

179.830

Figure 13-52. — A Gray coded wheel analog to digital converter.

ELECTRONIC A-TO-D CONVERTERS

In a number of instances, the analog input to digital equipment is in the form of an electrical signal. The electrical signal is of a voltage level proportional to and varying with the quantity it represents, such as shown in figure 13-53. These electrical analog signals may be representative of temperature, pressure, position, etc., and must be converted to a digital representation for use. Several methods for converting these analog inputs to digital form can be used.

One fairly simple method of converting an analog signal to a digital representation is shown by the block diagram in figure 13-54. Basically the unit consists of a counter, a digital-to-analog converter, and a comparator. The counter is used to hold the digital representation of the analog signal input. The output of the counter forms the input to a digital-to-analog converter, whose output is a voltage level proportional to the digital input from the counter. The analog output of the digital-to-analog converter is compared to the input analog signal by a voltage comparator. If the input signal is greater than the signal from the digital-to-analog converter, the comparator produces a voltage output representative of a binary 1. If the input signal is less than the signal developed by the digital-to-analog converter, a 0 level output is produced. As long as the signal from the digital-to-analog converter is less than the magnitude of the input analog signal, the voltage comparator has a 1 output signal. This causes the counter to count up or increase the magnitude of its contents. As the counter steps, the magnitude of its content increases, and, therefore, the output of the digital-to-analog converter increases. This continues until the output of the digital-to-analog converter becomes minutely greater than the input analog signal at which time the comparator output goes to a 0 level. The counter now contains a digital value representative of

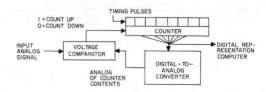

179.832

Figure 13-54. — An electronic analog to digital converter.

the analog value. The system is designed so that the digital number indicates the information represented by the analog voltage.

The comparator unit shown on the analog-to-digital converter block can be a simple circuit or an extremely complex network depending on the type of application. Usually these are modular units having characteristics which produce a logical "one" when one of the inputs is less than or equal to the other input voltage amplitude.

DIGITAL VOLTMETER

Due to the rapid advance of test equipment in the Navy, the digital voltmeter is becoming more widely used. The following covers only the general principles behind the digital voltmeter. The circuitry is not considered, nor is any specific voltmeter shown since the emphasis is on the analog-to-digital concept and the digital voltmeter as an encoder.

The digital voltmeter is essentially an analog-to-digital converter since it takes an analog voltage (continuous voltage) input and displays a digital representation of this voltage. The following discussion covers only one of the many ways this can be done.

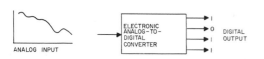

ANALOG INPUT

179.831

Figure 13-53. — Electrical signals converted to digital form.

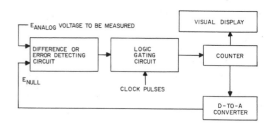

179.833

Figure 13-55. — Digital voltmeter block diagram.

Figure 13-55 is a basic block diagram of a digital voltmeter employing a feedback encoder. The voltage to be measured is applied to the difference amplifier or error detection circuit. As long as there is an error (E_{NULL} less than E_{ANALOG}) the gating circuit will be held open allowing the clock pulses to be fed to the counter. The counter output is applied to a digital-to-analog converter where the digital information is converted into an analog null voltage. The error detection circuitry produces a logical zero when the contents of the counter are proportional to the analog voltage being measured. This stops the count. The counter contents will normally be converted into a decimal equivalent (for ease of reading) prior to display by the readout device.

Another electronic A-to-D converter is the voltage-to-frequency type. The output is made proportional to the analog voltage magnitude. The binary pulses are then sent to a binary coded decimal counter for readout.

APPENDIX I

PERIODIC TABLE OF THE ELEMENTS

ATOMIC NUMBER — 1
ELEMENT SYMBOL — H
ATOMIC WEIGHT — 1.002

LIGHT METALS — HEAVY METALS — NONMETALS — INERT GASES

IA	IIA	IIIB	IVB	VB	VIB	VIIB	VIII	VIII	VIII	IB	IIB	IIIA	IVA	VA	VIA	VIIA	INERT GASES
1 H 1.008																	2 He 4.003
3 Li 6.94	4 Be 9.012											5 B 10.81	6 C 12.011	7 N 14.007	8 O 15.999	9 F 18.998	10 Ne 20.18
11 Na 22.990	12 Mg 24.305											13 Al 26.982	14 Si 28.086	15 P 30.974	16 S 32.06	17 Cl 35.453	18 Ar 39.95
19 K 39.102	20 Ca 40.08	21 Sc 44.956	22 Ti 47.90	23 V 50.942	24 Cr 51.996	25 Mn 54.938	26 Fe 55.847	27 Co 58.933	28 Ni 58.71	29 Cu 63.546	30 Zn 65.37	31 Ga 69.72	32 Ge 72.59	33 As 74.922	34 Se 78.96	35 Br 79.904	36 Kr 83.80
37 Rb 85.47	38 Sr 87.62	39 Y 88.905	40 Zr 91.22	41 Nb 92.906	42 Mo 95.94	43 • Tc (99)	44 Ru 101.07	45 Rh 102.905	46 Pd 106.4	47 Ag 107.868	48 Cd 112.40	49 In 114.82	50 Sn 118.69	51 Sb 121.75	52 Te 127.60	53 I 126.904	54 Xe 131.30
55 Cs 132.905	56 Ba 137.34	57 La 138.91	72 Hf 178.49	73 Ta 180.948	74 W 183.85	75 Re 186.2	76 Os 190.2	77 Ir 192.2	78 Pt 195.09	79 Au 196.967	80 Hg 200.59	81 Tl 204.37	82 Pb 207.2	83 Bi 208.980	84 • Po (210)	85 • At (210)	86 • Rn (222)
87 • Fr (223)	88 • Ra (226)	89 • Ac (227)	104 • Ku (257)														

LANTHANUM SERIES

58 Ce 140.12	59 Pr 140.907	60 Nd 144.24	61 Pm (147)	62 Sm 150.35	63 Eu 151.96	64 Gd 157.25	65 Tb 158.924	66 Dy 162.50	67 Ho 164.930	68 Er 167.26	69 Tm 168.934	70 Yb 173.04	71 Lu 174.97

ACTINIUM SERIES

90 • Th 232.038	91 • Pa (231)	92 • U 238.03	93 • Np (237)	94 • Pu (242)	95 • Am (243)	96 • Cm (247)	97 • Bk (247)	98 • Cf (249)	99 • Es (254)	100 • Fm (257)	101 • Md (258)	102 • No (255)	103 • Lr (256)	105 • (258-261)	106 • (258-261)

• — INDICATES PRINCIPAL RADIOACTIVE ELEMENTS

SEE NEXT PAGE FOR INTERPRETATION OF SYMBOLS

Figure A1-1. – Periodic table of the elements. 5.36(179)A

Symbol	Name	Atomic Number	Atomic Weight
Ac	Actinium	89	1(227)
Ag	Silver	47	107.868
Al	Aluminum	13	26.982
Am	Americium	95	(243)
Ar	Argon	18	39.95
As	Arsenic	33	74.922
At	Astatine	85	(210)
Au	Gold	79	196.967
B	Boron	5	10.81
Ba	Barium	56	137.34
Be	Beryllium	4	9.012
Bi	Bismuth	83	208.980
Bk	Berkelium	97	(247)
Br	Bromine	35	79.904
C	Carbon	6	12.011
Ca	Calcium	20	40.08
Cd	Cadmium	48	112.40
Ce	Cerium	58	140.12
Cf	Californium	98	(249)
Cl	Chlorine	17	35.453
Cm	Curium	96	(247)
Co	Cobalt	27	58.933
Cr	Chromium	24	51.996
Cs	Cesium	55	132.905
Cu	Copper	29	63.546
Dy	Dysprosium	66	162.50
Es	Einsteinium	99	(254)
Er	Erbium	68	167.26
Eu	Europium	63	151.96
F	Fluorine	9	18.998
Fe	Iron	26	55.847
Fm	Fermium	100	(257)
Fr	Francium	87	(223)
Ga	Gallium	31	69.72
Gd	Gadolinium	64	157.25
Ge	Germanium	32	72.59
H	Hydrogen	1	1.008
He	Helium	2	4.003
Hf	Hafnium	72	178.49
Hg	Mercury	80	200.59
Ho	Holmium	67	164.930
I	Iodine	53	126.904
In	Indium	49	114.82
Ir	Iridium	77	192.2
K	Potassium	19	39.102
Kr	Krypton	36	83.80
*Ku	Kurchatovium	104	(257)
La	Lanthanum	57	138.91
Li	Lithium	3	6.94
Lr	Lawrencium	103	(256)
Lu	Lutetium	71	174.97
Md	Mendelevium	101	(258)
Mg	Magnesium	12	24.305

Figure A1-1. — Periodic table of the elements — continued.   5.36(179)B

Symbol	Name	Atomic Number	Atomic Weight
Mn	Manganese	25	54.938
Mo	Molybdenum	42	95.94
N	Nitrogen	7	14.007
Na	Sodium	11	22.990
Nb	Niobium	41	92.906
Nd	Neodymium	60	144.24
Ne	Neon	10	20.18
Ni	Nickel	28	58.71
No	Nobelium	102	(255)
Np	Neptunium	93	(237)
O	Oxygen	8	15.999
Os	Osmium	76	190.2
P	Phosphorus	15	30.974
Pa	Protactinium	91	(231)
Pb	Lead	82	207.2
Pd	Palladium	46	106.4
Pm	Promethium	61	(147)
Po	Polonium	84	(210)
Pr	Praseodymium	59	140.907
Pt	Platinum	78	195.09
Pu	Plutonium	94	(242)
Ra	Radium	88	(226)
Rb	Rubidium	37	85.47
Re	Rhenium	75	186.2
Rh	Rhodium	45	102.905
Rn	Radon	86	(222)
Ru	Ruthenium	44	101.07
S	Sulfur	16	32.06
Sb	Antimony	51	121.75
Sc	Scandium	21	44.956
Se	Selenium	34	78.96
Si	Silicon	14	28.086
Sm	Samarium	62	150.35
Sn	Tin	50	118.69
Sr	Strontium	38	87.62
Ta	Tantalum	73	180.948
Tb	Terbium	65	158.924
Tc	Technetium	43	(99)
Te	Tellurium	52	127.60
Th	Thorium	90	232.038
Ti	Titanium	22	47.90
Tl	Thallium	81	204.37
Tm	Thulium	69	158.934
U	Uranium	92	238.03
V	Vanadium	23	50.942
W	Tungsten	74	183.85
Xe	Xenon	54	131.30
Y	Yttrium	39	88.905
Yb	Ytterbium	70	173.04
Zn	Zinc	30	65.37
Zr	Zirconium	40	91.22

* Note: Element proposed but not confirmed.

Figure A1-1. — Periodic table of the elements — continued.   5.36(179)B

TRIGONOMETRIC FUNCTIONS

0°–14.9°

Degs.	Function	0.0°	0.1°	0.2°	0.3°	0.4°	0.5°	0.6°	0.7°	0.8°	0.9°
0	sin	0.0000	0.0017	0.0035	0.0052	0.0070	0.0087	0.0105	0.0122	0.0140	0.0157
	cos	1.0000	1.0000	1.0000	1.0000	1.0000	1.0000	0.9999	0.9999	0.9999	0.9999
	tan	0.0000	0.0017	0.0035	0.0052	0.0070	0.0087	0.0105	0.0122	0.0140	0.0157
1	sin	0.0175	0.0192	0.0209	0.0227	0.0244	0.0262	0.0279	0.0297	0.0314	0.0332
	cos	0.9998	0.9998	0.9998	0.9997	0.9997	0.9997	0.9996	0.9996	0.9995	0.9995
	tan	0.0175	0.0192	0.0209	0.0227	0.0244	0.0262	0.0279	0.0297	0.0314	0.0332
2	sin	0.0349	0.0366	0.0384	0.0401	0.0419	0.0436	0.0454	0.0471	0.0488	0.0506
	cos	0.9994	0.9993	0.9993	0.9992	0.9991	0.9990	0.9990	0.9989	0.9988	0.9987
	tan	0.0349	0.0367	0.0384	0.0402	0.0419	0.0437	0.0454	0.0472	0.0489	0.0507
3	sin	0.0523	0.0541	0.0558	0.0576	0.0593	0.0610	0.0628	0.0645	0.0663	0.0680
	cos	0.9986	0.9985	0.9984	0.9983	0.9982	0.9981	0.9980	0.9979	0.9978	0.9977
	tan	0.0524	0.0542	0.0559	0.0577	0.0594	0.0612	0.0629	0.0647	0.0664	0.0682
4	sin	0.0698	0.0715	0.0732	0.0750	0.0767	0.0785	0.0802	0.0819	0.0837	0.0854
	cos	0.9976	0.9974	0.9973	0.9972	0.9971	0.9969	0.9968	0.9966	0.9965	0.9963
	tan	0.0699	0.0717	0.0734	0.0752	0.0769	0.0787	0.0805	0.0822	0.0840	0.0857
5	sin	0.0872	0.0889	0.0906	0.0924	0.0941	0.0958	0.0976	0.0993	0.1011	0.1028
	cos	0.9962	0.9960	0.9959	0.9957	0.9956	0.9954	0.9952	0.9951	0.9949	0.9947
	tan	0.0875	0.0892	0.0910	0.0928	0.0945	0.0963	0.0981	0.0998	0.1016	0.1033
6	sin	0.1045	0.1063	0.1080	0.1097	0.1115	0.1132	0.1149	0.1167	0.1184	0.1201
	cos	0.9945	0.9943	0.9942	0.9940	0.9938	0.9936	0.9934	0.9932	0.9930	0.9928
	tan	0.1051	0.1069	0.1086	0.1104	0.1122	0.1139	0.1157	0.1175	0.1192	0.1210
7	sin	0.1219	0.1236	0.1253	0.1271	0.1288	0.1305	0.1323	0.1340	0.1357	0.1374
	cos	0.9925	0.9923	0.9921	0.9919	0.9917	0.9914	0.9912	0.9910	0.9907	0.9905
	tan	0.1228	0.1246	0.1263	0.1281	0.1299	0.1317	0.1334	0.1352	0.1370	0.1388
8	sin	0.1392	0.1409	0.1426	0.1444	0.1461	0.1478	0.1495	0.1513	0.1530	0.1547
	cos	0.9903	0.9900	0.9898	0.9895	0.9893	0.9890	0.9888	0.9885	0.9882	0.9880
	tan	0.1405	0.1423	0.1441	0.1459	0.1477	0.1495	0.1512	0.1530	0.1548	0.1566
9	sin	0.1564	0.1582	0.1599	0.1616	0.1633	0.1650	0.1668	0.1685	0.1702	0.1719
	cos	0.9877	0.9874	0.9871	0.9869	0.9866	0.9863	0.9860	0.9857	0.9854	0.9851
	tan	0.1584	0.1602	0.1620	0.1638	0.1655	0.1673	0.1691	0.1709	0.1727	0.1745
10	sin	0.1736	0.1754	0.1771	0.1788	0.1805	0.1822	0.1840	0.1857	0.1874	0.1891
	cos	0.9848	0.9845	0.9842	0.9839	0.9836	0.9833	0.9829	0.9826	0.9823	0.9820
	tan	0.1763	0.1781	0.1799	0.1817	0.1835	0.1853	0.1871	0.1890	0.1908	0.1926
11	sin	0.1908	0.1925	0.1942	0.1959	0.1977	0.1994	0.2011	0.2028	0.2045	0.2062
	cos	0.9816	0.9813	0.9810	0.9806	0.9803	0.9799	0.9796	0.9792	0.9789	0.9785
	tan	0.1944	0.1962	0.1980	0.1998	0.2016	0.2035	0.2053	0.2071	0.2089	0.2107
12	sin	0.2079	0.2096	0.2113	0.2130	0.2147	0.2164	0.2181	0.2198	0.2215	0.2232
	cos	0.9781	0.9778	0.9774	0.9770	0.9767	0.9763	0.9759	0.9755	0.9751	0.9748
	tan	0.2126	0.2144	0.2162	0.2180	0.2199	0.2217	0.2235	0.2254	0.2272	0.2290
13	sin	0.2250	0.2267	0.2284	0.2300	0.2318	0.2334	0.2351	0.2368	0.2385	0.2402
	cos	0.9744	0.9740	0.9736	0.9732	0.9728	0.9724	0.9720	0.9715	0.9711	0.9707
	tan	0.2309	0.2327	0.2345	0.2364	0.2382	0.2401	0.2419	0.2438	0.2456	0.2475
14	sin	0.2419	0.2436	0.2453	0.2470	0.2487	0.2504	0.2521	0.2538	0.2554	0.2571
	cos	0.9703	0.9699	0.9694	0.9690	0.9686	0.9681	0.9677	0.9673	0.9668	0.9664
	tan	0.2493	0.2512	0.2530	0.2549	0.2568	0.2586	0.2605	0.2623	0.2642	0.2661
Degs.	Function	0′	6′	12′	18′	24′	30′	36′	42′	48′	54′

Figure A2-1.— Natural Sines, Cosines, and Tangents.   45.45.2(179)A

15°-29.9°

Degs.	Function	0.0°	0.1°	0.2°	0.3°	0.4°	0.5°	0.6°	0.7°	0.8°	0.9°
15	sin	0.2588	0.2605	0.2622	0.2639	0.2656	0.2672	0.2689	0.2706	0.2723	0.2740
	cos	0.9659	0.9655	0.9650	0.9646	0.9641	0.9636	0.9632	0.9627	0.9622	0.9617
	tan	0.2679	0.2698	0.2717	0.2736	0.2754	0.2773	0.2792	0.2811	0.2830	0.2849
16	sin	0.2756	0.2773	0.2790	0.2807	0.2823	0.2840	0.2857	0.2874	0.2890	0.2907
	cos	0.9613	0.9608	0.9603	0.9598	0.9593	0.9588	0.9583	0.9578	0.9573	0.9568
	tan	0.2867	0.2886	0.2905	0.2924	0.2943	0.2962	0.2981	0.3000	0.3019	0.3038
17	sin	0.2924	0.2940	0.2957	0.2974	0.2990	0.3007	0.3024	0.3040	0.3057	0.3074
	cos	0.9563	0.9558	0.9553	0.9548	0.9542	0.9537	0.9532	0.9527	0.9521	0.9516
	tan	0.3057	0.3076	0.3096	0.3115	0.3134	0.3153	0.3172	0.3191	0.3211	0.3230
18	sin	0.3090	0.3107	0.3123	0.3140	0.3156	0.3173	0.3190	0.3206	0.3223	0.3239
	cos	0.9511	0.9505	0.9500	0.9494	0.9489	0.9483	0.9478	0.9472	0.9466	0.9461
	tan	0.3249	0.3269	0.3288	0.3307	0.3327	0.3346	0.3365	0.3385	0.3404	0.3424
19	sin	0.3256	0.3272	0.3289	0.3305	0.3322	0.3338	0.3355	0.3371	0.3387	0.3404
	cos	0.9455	0.9449	0.9444	0.9438	0.9432	0.9426	0.9421	0.9415	0.9409	0.9403
	tan	0.3443	0.3463	0.3482	0.3502	0.3522	0.3541	0.3561	0.3581	0.3600	0.3620
20	sin	0.3420	0.3437	0.3453	0.3469	0.3486	0.3502	0.3518	0.3535	0.3551	0.3567
	cos	0.9397	0.9391	0.9385	0.9379	0.9373	0.9367	0.9361	0.9354	0.9348	0.9342
	tan	0.3640	0.3659	0.3679	0.3699	0.3719	0.3739	0.3759	0.3779	0.3799	0.3819
21	sin	0.3584	0.3600	0.3616	0.3633	0.3649	0.3665	0.3681	0.3697	0.3714	0.3730
	cos	0.9336	0.9330	0.9323	0.9317	0.9311	0.9304	0.9298	0.9291	0.9285	0.9278
	tan	0.3839	0.3859	0.3879	0.3899	0.3919	0.3939	0.3959	0.3979	0.4000	0.4020
22	sin	0.3746	0.3762	0.3778	0.3795	0.3811	0.3827	0.3843	0.3859	0.3875	0.3891
	cos	0.9272	0.9265	0.9259	0.9252	0.9245	0.9239	0.9232	0.9225	0.9219	0.9212
	tan	0.4040	0.4061	0.4051	0.4101	0.4122	0.4142	0.4163	0.4183	0.4204	0.4224
23	sin	0.3907	0.3923	0.3939	0.3955	0.3971	0.3987	0.4003	0.4019	0.4035	0.4051
	cos	0.9205	0.9198	0.9191	0.9184	0.9178	0.9171	0.9164	0.9157	0.9150	0.9143
	tan	0.4245	0.4265	0.4286	0.4307	0.4327	0.4348	0.4369	0.4390	0.4411	0.4431
24	sin	0.4067	0.4083	0.4099	0.4115	0.4131	0.4147	0.4163	0.4179	0.4195	0.4210
	cos	0.9135	0.9128	0.9121	0.9114	0.9107	0.9100	0.9092	0.9085	0.9078	0.9070
	tan	0.4452	0.4473	0.4494	0.4515	0.4536	0.4557	0.4578	0.4599	0.4621	0.4642
25	sin	0.4226	0.4242	0.4258	0.4274	0.4289	0.4305	0.4321	0.4337	0.4352	0.4368
	cos	0.9063	0.9056	0.9048	0.9041	0.9033	0.9026	0.9018	0.9011	0.9003	0.8996
	tan	0.4663	0.4684	0.4706	0.4727	0.4748	0.4770	0.4791	0.4813	0.4834	0.4856
26	sin	0.4384	0.4399	0.4415	0.4431	0.4446	0.4462	0.4478	0.4493	0.4509	0.4524
	cos	0.8988	0.8980	0.8973	0.8965	0.8957	0.8949	0.8942	0.8934	0.8926	0.8918
	tan	0.4877	0.4899	0.4921	0.4942	0.4964	0.4986	0.5008	0.5029	0.5051	0.5073
27	sin	0.4540	0.4555	0.4571	0.4586	0.4602	0.4617	0.4633	0.4648	0.4664	0.4679
	cos	0.8910	0.8902	0.8894	0.8886	0.8878	0.8870	0.8862	0.8854	0.8846	0.8838
	tan	0.5095	0.5117	0.5139	0.5161	0.5184	0.5206	0.5228	0.5250	0.5272	0.5295
28	sin	0.4695	0.4710	0.4726	0.4741	0.4756	0.4772	0.4787	0.4802	0.4818	0.4833
	cos	0.8829	0.8821	0.8813	0.8805	0.8796	0.8788	0.8780	0.8771	0.8763	0.8755
	tan	0.5317	0.5340	0.5362	0.5384	0.5407	0.5430	0.5452	0.5475	0.5498	0.5520
29	sin	0.4848	0.4863	0.4879	0.4894	0.4909	0.4924	0.4939	0.4955	0.4970	0.4985
	cos	0.8746	0.8738	0.8729	0.8721	0.8712	0.8704	0.8695	0.8686	0.8678	0.8669
	tan	0.5543	0.5566	0.5589	0.5612	0.5635	0.5658	0.5681	0.5704	0.5727	0.5750
Degs.	Function	0'	6'	12'	18'	24'	30'	36'	42'	48'	54'

Figure A2-1.— Natural Sines, Cosines, and Tangents — continued.   45.45.2(179)B

Degs.	Function	0.0°	0.1°	0.2°	0.3°	0.4°	0.5°	0.6°	0.7°	0.8°	0.9°
30	sin	0.5000	0.5015	0.5030	0.5045	0.5060	0.5075	0.5090	0.5105	0.5120	0.5135
	cos	0.8660	0.8652	0.8643	0.8634	0.8625	0.8616	0.8607	0.8599	0.8590	0.8581
	tan	0.5774	0.5797	0.5820	0.5844	0.5867	0.5890	0.5914	0.5938	0.5961	0.5985
31	sin	0.5150	0.5165	0.5180	0.5195	0.5210	0.5225	0.5240	0.5255	0.5270	0.5284
	cos	0.8572	0.8563	0.8554	0.8545	0.8536	0.8526	0.8517	0.8508	0.8499	0.8490
	tan	0.6009	0.6032	0.6056	0.6080	0.6104	0.6128	0.6152	0.6176	0.6200	0.6224
32	sin	0.5299	0.5314	0.5329	0.5344	0.5358	0.5373	0.5388	0.5402	0.5417	0.5432
	cos	0.8480	0.8471	0.8462	0.8453	0.8443	0.8434	0.8425	0.8415	0.8406	0.8396
	tan	0.6249	0.6273	0.6297	0.6322	0.6346	0.6371	0.6395	0.6420	0.6445	0.6469
33	sin	0.5446	0.5461	0.5476	0.5490	0.5505	0.5519	0.5534	0.5548	0.5563	0.5577
	cos	0.8387	0.8377	0.8368	0.8358	0.8348	0.8339	0.8329	0.8320	0.8310	0.8300
	tan	0.6494	0.6519	0.6544	0.6569	0.6594	0.6619	0.6644	0.6669	0.6694	0.6720
34	sin	0.5592	0.5606	0.5621	0.5635	0.5650	0.5664	0.5678	0.5693	0.5707	0.5721
	cos	0.8290	0.8281	0.8271	0.8261	0.8251	0.8241	0.8231	0.8221	0.8211	0.8202
	tan	0.6745	0.6771	0.6796	0.6822	0.6847	0.6873	0.6899	0.6924	0.6950	0.6976
35	sin	0.5736	0.5750	0.5764	0.5779	0.5793	0.5807	0.5821	0.5835	0.5850	0.5864
	cos	0.8192	0.8181	0.8171	0.8161	0.8151	0.8141	0.8131	0.8121	0.8111	0.8100
	tan	0.7002	0.7028	0.7054	0.7080	0.7107	0.7133	0.7159	0.7186	0.7212	0.7239
36	sin	0.5878	0.5892	0.5906	0.5920	0.5934	0.5948	0.5962	0.5976	0.5990	0.6004
	cos	0.8090	0.8080	0.8070	0.8059	0.8049	0.8039	0.8028	0.8018	0.8007	0.7997
	tan	0.7265	0.7292	0.7319	0.7346	0.7373	0.7400	0.7427	0.7454	0.7481	0.7508
37	sin	0.6018	0.6032	0.6046	0.6060	0.6074	0.6088	0.6101	0.6115	0.6129	0.6143
	cos	0.7986	0.7976	0.7965	0.7955	0.7944	0.7934	0.7923	0.7912	0.7902	0.7891
	tan	0.7536	0.7563	0.7590	0.7618	0.7646	0.7673	0.7701	0.7729	0.7757	0.7785
38	sin	0.6157	0.6170	0.6184	0.6198	0.6211	0.6225	0.6239	0.6252	0.6266	0.6280
	cos	0.7880	0.7869	0.7859	0.7848	0.7837	0.7826	0.7815	0.7804	0.7793	0.7782
	tan	0.7813	0.7841	0.7869	0.7898	0.7926	0.7954	0.7983	0.8012	0.8040	0.8069
39	sin	0.6293	0.6307	0.6320	0.6334	0.6347	0.6361	0.6374	0.6388	0.6401	0.6414
	cos	0.7771	0.7760	0.7749	0.7738	0.7727	0.7716	0.7705	0.7694	0.7683	0.7672
	tan	0.8098	0.8127	0.8156	0.8185	0.8214	0.8243	0.8273	0.8302	0.8332	0.8361
40	sin	0.6428	0.6441	0.6455	0.6468	0.6481	0.6494	0.6508	0.6521	0.6534	0.6547
	cos	0.7660	0.7649	0.7638	0.7627	0.7615	0.7604	0.7593	0.7581	0.7570	0.7559
	tan	0.8391	0.8421	0.8451	0.8481	0.8511	0.8541	0.8571	0.8601	0.8632	0.8662
41	sin	0.6561	0.6574	0.6587	0.6600	0.6613	0.6626	0.6639	0.6652	0.6665	0.6678
	cos	0.7547	0.7536	0.7524	0.7513	0.7501	0.7490	0.7478	0.7466	0.7455	0.7443
	tan	0.8693	0.8724	0.8754	0.8785	0.8816	0.8847	0.8878	0.8910	0.8941	0.8972
42	sin	0.6691	0.6704	0.6717	0.6730	0.6743	0.6756	0.6769	0.6782	0.6794	0.6807
	cos	0.7431	0.7420	0.7408	0.7396	0.7385	0.7373	0.7361	0.7349	0.7337	0.7325
	tan	0.9004	0.9036	0.9067	0.9099	0.9131	0.9163	0.9195	0.9228	0.9260	0.9293
43	sin	0.6820	0.6833	0.6845	0.6858	0.6871	0.6884	0.6896	0.6909	0.6921	0.6934
	cos	0.7314	0.7302	0.7290	0.7278	0.7266	0.7254	0.7242	0.7230	0.7218	0.7206
	tan	0.9325	0.9358	0.9391	0.9424	0.9457	0.9490	0.9523	0.9556	0.9590	0.9623
44	sin	0.6947	0.6959	0.6972	0.6984	0.6997	0.7009	0.7022	0.7034	0.7046	0.7059
	cos	0.7193	0.7181	0.7169	0.7157	0.7145	0.7133	0.7120	0.7108	0.7096	0.7083
	tan	0.9657	0.9691	0.9725	0.9759	0.9793	0.9827	0.9861	0.9896	0.9930	0.9965
Degs.	Function	0′	6′	12′	18′	24′	30′	36′	42′	48′	54′

Figure A2-1. — Natural Sines, Cosines, and Tangents — continued.   45.45.2(179)C

45°–59.9°

Degs.	Function	0.0°	0.1°	0.2°	0.3°	0.4°	0.5°	0.6°	0.7°	0.8°	0.9°
45	sin	0.7071	0.7063	0.7096	0.7108	0.7120	0.7133	0.7145	0.7157	0.7169	0.7181
	cos	0.7071	0.7059	0.7046	0.7034	0.7022	0.7009	0.6997	0.6984	0.6972	0.6959
	tan	1.0000	1.0035	1.0070	1.0105	1.0141	1.0176	1.0212	1.0247	1.0283	1.0319
46	sin	0.7193	0.7206	0.7218	0.7230	0.7242	0.7254	0.7266	0.7278	0.7290	0.7302
	cos	0.6947	0.6934	0.6921	0.6909	0.6896	0.6884	0.6871	0.6858	0.6845	0.6833
	tan	1.0355	1.0392	1.0428	1.0464	1.0501	1.0538	1.0575	1.0612	1.0649	1.0686
47	sin	0.7314	0.7325	0.7337	0.7349	0.7361	0.7373	0.7385	0.7396	0.7408	0.7420
	cos	0.6820	0.6807	0.6794	0.6782	0.6769	0.6756	0.6743	0.6730	0.6717	0.6704
	tan	1.0724	1.0761	1.0799	1.0837	1.0875	1.0913	1.0951	1.0990	1.1028	1.1067
48	sin	0.7431	0.7443	0.7455	0.7466	0.7478	0.7490	0.7501	0.7513	0.7524	0.7536
	cos	0.6691	0.6678	0.6665	0.6652	0.6639	0.6626	0.6613	0.6600	0.6587	0.6574
	tan	1.1106	1.1145	1.1184	1.1224	1.1263	1.1303	1.1343	1.1383	1.1423	1.1463
49	sin	0.7547	0.7559	0.7570	0.7581	0.7593	0.7604	0.7615	0.7627	0.7638	0.7649
	cos	0.6561	0.6547	0.6534	0.6521	0.6508	0.6494	0.6481	0.6468	0.6455	0.6441
	tan	1.1504	1.1544	1.1585	1.1626	1.1667	1.1708	1.1750	1.1792	1.1833	1.1875
50	sin	0.7660	0.7672	0.7683	0.7694	0.7705	0.7716	0.7727	0.7738	0.7749	0.7760
	cos	0.6428	0.6414	0.6401	0.6388	0.6374	0.6361	0.6347	0.6334	0.6320	0.6307
	tan	1.1918	1.1960	1.2002	1.2045	1.2088	1.2131	1.2174	1.2218	1.2261	1.2305
51	sin	0.7771	0.7782	0.7793	0.7804	0.7815	0.7826	0.7837	0.7848	0.7859	0.7869
	cos	0.6293	0.6280	0.6266	0.6252	0.6239	0.6225	0.6211	0.6198	0.6184	0.6170
	tan	1.2349	1.2393	1.2437	1.2482	1.2527	1.2572	1.2617	1.2662	1.2708	1.2753
52	sin	0.7880	0.7891	0.7902	0.7912	0.7923	0.7934	0.7944	0.7955	0.7965	0.7976
	cos	0.6157	0.6143	0.6129	0.6115	0.6101	0.6088	0.6074	0.6060	0.6046	0.6032
	tan	1.2799	1.2846	1.2892	1.2938	1.2985	1.3032	1.3079	1.3127	1.3175	1.3222
53	sin	0.7986	0.7997	0.8007	0.8018	0.8028	0.8039	0.8049	0.8059	0.8070	0.8080
	cos	0.6018	0.6004	0.5990	0.5976	0.5962	0.5948	0.5934	0.5920	0.5906	0.5892
	tan	1.3270	1.3319	1.3367	1.3416	1.3465	1.3514	1.3564	1.3613	1.3663	1.3713
54	sin	0.8090	0.8100	0.8111	0.8121	0.8131	0.8141	0.8151	0.8161	0.8171	0.8181
	cos	0.5878	0.5864	0.5850	0.5835	0.5821	0.5807	0.5793	0.5779	0.5764	0.5750
	tan	1.3764	1.3814	1.3865	1.3916	1.3968	1.4019	1.4071	1.4124	1.4176	1.4229
55	sin	0.8192	0.8202	0.8211	0.8221	0.8231	0.8241	0.8251	0.8261	0.8271	0.8281
	cos	0.5736	0.5721	0.5707	0.5693	0.5678	0.5664	0.5650	0.5635	0.5621	0.5606
	tan	1.4281	1.4335	1.4388	1.4442	1.4496	1.4550	1.4605	1.4659	1.4715	1.4770
56	sin	0.8290	0.8300	0.8310	0.8320	0.8329	0.8339	0.8348	0.8355	0.8368	0.8377
	cos	0.5592	0.5577	0.5563	0.5548	0.5534	0.5519	0.5505	0.5490	0.5476	0.5461
	tan	1.4826	1.4882	1.4938	1.4994	1.5051	1.5108	1.5166	1.5224	1.5282	1.5340
57	sin	0.8387	0.8396	0.8406	0.8415	0.8425	0.8434	0.8443	0.8453	0.8462	0.8471
	cos	0.5446	0.5432	0.5417	0.5402	0.5388	0.5373	0.5358	0.5344	0.5329	0.5314
	tan	1.5399	1.5458	1.5517	1.5577	1.5637	1.5697	1.5757	1.5818	1.5880	1.5941
58	sin	0.8480	0.8490	0.8499	0.8508	0.8517	0.8526	0.8536	0.8545	0.8551	0.8563
	cos	0.5299	0.5284	0.5270	0.5255	0.5240	0.5225	0.5210	0.5195	0.5180	0.5165
	tan	1.6003	1.6066	1.6128	1.6191	1.6255	1.6319	1.6383	1.6447	1.6512	1.6577
59	sin	0.8572	0.8581	0.8590	0.8599	0.8607	0.8616	0.8625	0.8634	0.8643	0.8652
	cos	0.5150	0.5135	0.5120	0.5105	0.5090	0.5075	0.5060	0.5045	0.5030	0.5015
	tan	1.6643	1.6709	1.6775	1.6842	1.6909	1.6977	1.7045	1.7113	1.7182	1.7251
Degs.	Function	0'	6'	12'	18'	24'	30'	36'	42'	48'	54'

Figure A2-1.—Natural Sines, Cosines, and Tangents—continued.   45.45.2(179)D

60°–74.9°

Degs.	Function	0.0°	0.1°	0.2°	0.3°	0.4°	0.5°	0.6°	0.7°	0.8°	0.9°
60	sin	0.8660	0.8669	0.8678	0.8686	0.8695	0.8704	0.8712	0.8721	0.8729	0.8738
	cos	0.5000	0.4985	0.4970	0.4955	0.4939	0.4924	0.4909	0.4894	0.4879	0.4863
	tan	1.7321	1.7391	1.7461	1.7532	1.7603	1.7675	1.7747	1.7820	1.7893	1.7966
61	sin	0.8746	0.8755	0.8763	0.8771	0.8780	0.8788	0.8796	0.8805	0.8813	0.8821
	cos	0.4848	0.4833	0.4818	0.4802	0.4787	0.4772	0.4756	0.4741	0.4726	0.4710
	tan	1.8040	1.8115	1.8190	1.8265	1.8341	1.8418	1.8495	1.8572	1.8650	1.8728
62	sin	0.8829	0.8838	0.8846	0.8854	0.8862	0.8870	0.8878	0.8886	0.8894	0.8902
	cos	0.4695	0.4679	0.4664	0.4648	0.4633	0.4617	0.4602	0.4586	0.4571	0.4555
	tan	1.8807	1.8887	1.8967	1.9047	1.9128	1.9210	1.9292	1.9375	1.9458	1.9542
63	sin	0.8910	0.8918	0.8926	0.8934	0.8942	0.8949	0.8957	0.8965	0.8973	0.8980
	cos	0.4540	0.4524	0.4509	0.4493	0.4478	0.4462	0.4446	0.4431	0.4415	0.4399
	tan	1.9626	1.9711	1.9797	1.9883	1.9970	2.0057	2.0145	2.0233	2.0323	2.0413
64	sin	0.8988	0.8996	0.9003	0.9011	0.9018	0.9026	0.9033	0.9041	0.9048	0.9056
	cos	0.4384	0.4368	0.4352	0.4337	0.4321	0.4305	0.4289	0.4274	0.4258	0.4242
	tan	2.0503	2.0594	2.0686	2.0778	2.0872	2.0965	2.1060	2.1155	2.1251	2.1348
65	sin	0.9063	0.9070	0.9078	0.9085	0.9092	0.9100	0.9107	0.9114	0.9121	0.9128
	cos	0.4226	0.4210	0.4195	0.4179	0.4163	0.4147	0.4131	0.4115	0.4099	0.4083
	tan	2.1445	2.1543	2.1642	2.1742	2.1842	2.1943	2.2045	2.2148	2.2251	2.2355
66	sin	0.9135	0.9143	0.9150	0.9157	0.9164	0.9171	0.9178	0.9184	0.9191	0.9198
	cos	0.4067	0.4051	0.4035	0.4019	0.4003	0.3987	0.3971	0.3955	0.3939	0.3923
	tan	2.2460	2.2566	2.2673	2.2781	2.2889	2.2998	2.3109	2.3220	2.3332	2.3445
67	sin	0.9205	0.9212	0.9219	0.9225	0.9232	0.9239	0.9245	0.9252	0.9259	0.9265
	cos	0.3907	0.3891	0.3875	0.3859	0.3843	0.3827	0.3811	0.3795	0.3778	0.3762
	tan	2.3559	2.3673	2.3789	2.3906	2.4023	2.4142	2.4262	2.4383	2.4504	2.4627
68	sin	0.9272	0.9278	0.9285	0.9291	0.9298	0.9304	0.9311	0.9317	0.9323	0.9330
	cos	0.3746	0.3730	0.3714	0.3697	0.3681	0.3665	0.3649	0.3633	0.3616	0.3600
	tan	2.4751	2.4876	2.5002	2.5129	2.5257	2.5386	2.5517	2.5649	2.5782	2.5916
69	sin	0.9336	0.9342	0.9348	0.9354	0.9361	0.9367	0.9373	0.9379	0.9385	0.9391
	cos	0.3584	0.3567	0.3551	0.3535	0.3518	0.3502	0.3486	0.3469	0.3453	0.3437
	tan	2.6051	2.6187	2.6325	2.6464	2.6605	2.6746	2.6889	2.7034	2.7179	2.7326
70	sin	0.9397	0.9403	0.9409	0.9415	0.9421	0.9426	0.9432	0.9438	0.9444	0.9449
	cos	0.3420	0.3404	0.3387	0.3371	0.3355	0.3338	0.3322	0.3305	0.3289	0.3272
	tan	2.7475	2.7625	2.7776	2.7929	2.8083	2.8239	2.8397	2.8556	2.8716	2.8878
71	sin	0.9455	0.9461	0.9466	0.9472	0.9478	0.9483	0.9489	0.9494	0.9500	0.9505
	cos	0.3256	0.3239	0.3223	0.3206	0.3190	0.3173	0.3156	0.3140	0.3123	0.3107
	tan	2.9042	2.9208	2.9375	2.9544	2.9714	2.9887	3.0061	3.0237	3.0415	3.0595
72	sin	0.9511	0.9516	0.9521	0.9527	0.9532	0.9537	0.9542	0.9548	0.9553	0.9558
	cos	0.3090	0.3074	0.3057	0.3040	0.3024	0.3007	0.2990	0.2974	0.2957	0.2940
	tan	3.0777	3.0961	3.1146	3.1334	3.1524	3.1716	3.1910	3.2106	3.2305	3.2506
73	sin	0.9563	0.9568	0.9573	0.9578	0.9583	0.9588	0.9593	0.9598	0.9603	0.9608
	cos	0.2924	0.2907	0.2890	0.2874	0.2857	0.2840	0.2823	0.2807	0.2790	0.2773
	tan	3.2709	3.2914	3.3122	3.3332	3.3544	3.3759	3.3977	3.4197	3.4420	3.4646
74	sin	0.9613	0.9617	0.9622	0.9627	0.9632	0.9636	0.9641	0.9646	0.9650	0.9655
	cos	0.2756	0.2740	0.2723	0.2706	0.2689	0.2672	0.2656	0.2639	0.2622	0.2605
	tan	3.4874	3.5105	3.5339	3.5576	3.5816	3.6059	3.6305	3.6554	3.6806	3.7062
Degs.	Function	0'	6'	12'	18'	24'	30'	36'	42'	48'	54'

Figure A2-1. — Natural Sines, Cosines, and Tangents — continued.   45.45.2(179) E

75°–89.9°

Degs.	Function	0.0°	0.1°	0.2°	0.3°	0.4°	0.5°	0.6°	0.7°	0.8°	0.9°
75	sin	0.9659	0.9664	0.9668	0.9673	0.9677	0.9681	0.9686	0.9690	0.9694	0.9699
	cos	0.2588	0.2571	0.2554	0.2538	0.2521	0.2504	0.2487	0.2470	0.2453	0.2436
	tan	3.7321	3.7583	3.7848	3.8118	3.8391	3.8667	3.8947	3.9232	3.9520	3.9812
76	sin	0.9703	0.9707	0.9711	0.9715	0.9720	0.9724	0.9728	0.9732	0.9736	0.9740
	cos	0.2419	0.2402	0.2385	0.2368	0.2351	0.2334	0.2317	0.2300	0.2284	0.2267
	tan	4.0108	4.0408	4.0713	4.1022	4.1335	4.1653	4.1976	4.2303	4.2635	4.2972
77	sin	0.9744	0.9748	0.9751	0.9755	0.9759	0.9763	0.9767	0.9770	0.9774	0.9778
	cos	0.2250	0.2232	0.2215	0.2198	0.2181	0.2164	0.2147	0.2130	0.2113	0.2096
	tan	4.3315	4.3662	4.4015	4.4374	4.4737	4.5107	4.5483	4.5864	4.6252	4.6646
78	sin	0.9781	0.9785	0.9789	0.9792	0.9796	0.9799	0.9803	0.9806	0.9810	0.9813
	cos	0.2079	0.2062	0.2045	0.2328	0.2011	0.1994	0.1977	0.1959	0.1942	0.1925
	tan	4.7046	4.7453	4.7867	4.8288	4.8716	4.9152	4.9594	5.0045	5.0504	5.0970
79	sin	0.9816	0.9820	0.9823	0.9826	0.9829	0.9833	0.9836	0.9839	0.9842	0.9845
	cos	0.1908	0.1891	0.1874	0.1857	0.1840	0.1822	0.1805	0.1788	0.1771	0.1754
	tan	5.1446	5.1929	5.2422	5.2924	5.3435	5.3955	5.4486	5.5026	5.5578	5.6140
80	sin	0.9848	0.9851	0.9854	0.9857	0.9860	0.9863	0.9866	0.9869	0.9871	0.9874
	cos	0.1736	0.1719	0.1702	0.1685	0.1668	0.1650	0.1633	0.1616	0.1599	0.1582
	tan	5.6713	5.7297	5.7894	5.8502	5.9124	5.9758	6.0405	6.1066	6.1742	6.2432
81	sin	0.9877	0.9880	0.9882	0.9885	0.9888	0.9890	0.9893	0.9895	0.9898	0.9900
	cos	0.1564	0.1547	0.1530	0.1513	0.1495	0.1478	0.1461	0.1444	0.1426	0.1409
	tan	6.3138	6.3859	6.4596	6.5350	6.6122	6.6912	6.7720	6.8548	6.9395	7.0264
82	sin	0.9903	0.9905	0.9907	0.9910	0.9912	0.9914	0.9917	0.9919	0.9921	0.9923
	cos	0.1392	0.1374	0.1357	0.1340	0.1323	0.1305	0.1288	0.1271	0.1253	0.1236
	tan	7.1154	7.2066	7.3002	7.3962	7.4947	7.5958	7.6996	7.8062	7.9158	8.0285
83	sin	0.9925	0.9928	0.9930	0.9932	0.9934	0.9936	0.9938	0.9940	0.9942	0.9943
	cos	0.1219	0.1201	0.1184	0.1167	0.1149	0.1132	0.1115	0.1097	0.1080	0.1063
	tan	8.1443	8.2636	8.3863	8.5126	8.6427	8.7769	8.9152	9.0579	9.2052	9.3572
84	sin	0.9945	0.9947	0.9949	0.9951	0.9952	0.9954	0.9956	0.9957	0.9959	0.9960
	cos	0.1045	0.1028	0.1011	0.0993	0.0976	0.0958	0.0941	0.0924	0.0906	0.0889
	tan	9.5144	9.6768	9.8448	10.02	10.20	10.39	10.58	10.78	10.99	11.20
85	sin	0.9962	0.9963	0.9965	0.9966	0.9968	0.9969	0.9971	0.9972	0.9973	0.9974
	cos	0.0872	0.0854	0.0837	0.0819	0.0802	0.0785	0.0767	0.0750	0.0732	0.0715
	tan	11.43	11.66	11.91	12.16	12.43	12.71	13.00	13.30	13.62	13.95
86	sin	0.9976	0.9977	0.9978	0.9979	0.9980	0.9981	0.9982	0.9983	0.9984	0.9985
	cos	0.0698	0.0680	0.0663	0.0645	0.0628	0.0610	0.0593	0.0576	0.0558	0.0541
	tan	14.30	14.67	15.06	15.46	15.89	16.35	16.83	17.34	17.89	18.46
87	sin	0.9986	0.9987	0.9988	0.9989	0.9990	0.9990	0.9991	0.9992	0.9993	0.9993
	cos	0.0523	0.0506	0.0488	0.0471	0.0454	0.0436	0.0419	0.0401	0.0384	0.0366
	tan	19.08	19.74	20.45	21.20	22.02	22.90	23.86	24.90	26.03	27.27
88	sin	0.9994	0.9995	0.9995	0.9996	0.9996	0.9997	0.9997	0.9997	0.9998	0.9998
	cos	0.0349	0.0332	0.0314	0.0297	0.0279	0.0262	0.0244	0.0227	0.0209	0.0192
	tan	28.64	30.14	31.82	33.69	35.80	38.19	40.92	44.07	47.74	52.08
89	sin	0.9998	0.9999	0.9999	0.9999	0.9999	1.000	1.000	1.000	1.000	1.000
	cos	0.0175	0.0157	0.0140	0.0122	0.0105	0.0087	0.0070	0.0052	0.0035	0.0017
	tan	57.29	63.66	71.62	81.85	95.49	114.6	143.2	191.0	286.5	573.0
Degs.	Function	0′	6′	12′	18′	24′	30′	36′	42′	48′	54′

Figure A2-1.— Natural Sines, Cosines, and Tangents — continued.   45.45.2(179)F

APPENDIX III

TABLE OF LOGARITHMS

N	0	1	2	3	4	5	6	7	8	9
0	....	0000	3010	4771	6021	6990	7782	8451	9031	9542
1	0000	0414	0792	1139	1461	1761	2041	2304	2553	2788
2	3010	3222	3424	3617	3802	3979	4150	4314	4472	4624
3	4771	4914	5051	5185	5315	5441	5563	5682	5798	5911
4	6021	6128	6232	6335	6435	6532	6628	6721	6812	6902
5	6990	7076	7160	7243	7324	7404	7482	7559	7634	7709
6	7782	7853	7924	7993	8062	8129	8195	8261	8325	8388
7	8451	8513	8573	8633	8692	8751	8808	8865	8921	8976
8	9031	9085	9138	9191	9243	9294	9345	9395	9445	9494
9	9542	9590	9638	9685	9731	9777	9823	9868	9912	9956
10	0000	0043	0086	0128	0170	0212	0253	0294	0334	0374
11	0414	0453	0492	0531	0569	0607	0645	0682	0719	0755
12	0792	0828	0864	0899	0934	0969	1004	1038	1072	1106
13	1139	1173	1206	1239	1271	1303	1335	1367	1399	1430
14	1461	1492	1523	1553	1584	1614	1644	1673	1703	1732
15	1761	1790	1818	1847	1875	1903	1931	1959	1987	2014
16	2041	2068	2095	2122	2148	2175	2201	2227	2253	2279
17	2304	2330	2355	2380	2405	2430	2455	2480	2504	2529
18	2553	2577	2601	2625	2648	2672	2695	2718	2742	2765
19	2788	2810	2833	2856	2878	2900	2923	2945	2967	2989
20	3010	3032	3054	3075	3096	3118	3139	3160	3181	3201
21	3222	3243	3263	3284	3304	3324	3345	3365	3385	3404
22	3424	3444	3464	3483	3502	3522	3541	3560	3579	3598
23	3617	3636	3655	3674	3692	3711	3729	3747	3766	3784
24	3802	3820	3838	3856	3874	3892	3909	3927	3945	3962
25	3979	3997	4014	4031	4048	4065	4082	4099	4116	4133
26	4150	4166	4183	4200	4216	4232	4249	4265	4281	4298
27	4314	4330	4346	4362	4378	4393	4409	4425	4440	4456
28	4472	4487	4502	4518	4533	4548	4564	4579	4594	4609
29	4624	4639	4654	4669	4683	4698	4713	4728	4742	4757
30	4771	4786	4800	4814	4829	4843	4857	4871	4886	4900
31	4914	4928	4942	4955	4969	4983	4997	5011	5024	5038
32	5051	5065	5079	5092	5105	5119	5132	5145	5159	5172
33	5185	5198	5211	5224	5237	5250	5263	5276	5289	5302
34	5315	5328	5340	5353	5366	5378	5391	5403	5416	5428
35	5441	5453	5465	5478	5490	5502	5514	5527	5539	5551
36	5563	5575	5587	5599	5611	5623	5635	5647	5658	5670
37	5682	5694	5705	5717	5729	5740	5752	5763	5775	5786
38	5798	5809	5821	5832	5843	5855	5866	5877	5888	5899
39	5911	5922	5933	5944	5955	5966	5977	5988	5999	6010
40	6021	6031	6042	6053	6064	6075	6085	6096	6107	6117
41	6128	6138	6149	6160	6170	6180	6191	6201	6212	6222
42	6232	6243	6253	6263	6274	6284	6294	6304	6314	6325
43	6335	6345	6355	6365	6375	6385	6395	6405	6415	6425
44	6435	6444	6454	6464	6474	6484	6493	6503	6513	6522
45	6532	6542	6551	6561	6571	6580	6590	6599	6609	6618
46	6628	6637	6646	6656	6665	6675	6684	6693	6702	6712
47	6721	6730	6739	6749	6758	6767	6776	6785	6794	6803
48	6812	6821	6830	6839	6848	6857	6866	6875	6884	6893
49	6902	6911	6920	6928	6937	6946	6955	6964	6972	6981
50	6990	6998	7007	7016	7024	7033	7042	7050	7059	7067
N	0	1	2	3	4	5	6	7	8	9

Figure A3-1. — Table of Logarithms.

45.45.1(179)A

N	0	1	2	3	4	5	6	7	8	9
50	6990	6998	7007	7016	7024	7033	7042	7050	7059	7067
51	7076	7084	7093	7101	7110	7118	7126	7135	7143	7152
52	7160	7168	7177	7185	7193	7202	7210	7218	7226	7235
53	7243	7251	7259	7267	7275	7284	7292	7300	7308	7316
54	7324	7332	7340	7348	7356	7364	7372	7380	7388	7396
55	7404	7412	7419	7427	7435	7443	7451	7459	7466	7474
56	7482	7490	7497	7505	7513	7520	7528	7536	7543	7551
57	7559	7566	7574	7582	7589	7597	7604	7612	7619	7627
58	7634	7642	7649	7657	7664	7672	7679	7686	7694	7701
59	7709	7716	7723	7731	7738	7745	7752	7760	7767	7774
60	7782	7789	7796	7803	7810	7818	7825	7832	7839	7846
61	7853	7860	7868	7875	7882	7889	7896	7903	7910	7917
62	7924	7931	7938	7945	7952	7959	7966	7973	7980	7987
63	7993	8000	8007	8014	8021	8028	8035	8041	8048	8055
64	8062	8069	8075	8082	8089	8096	8102	8109	8116	8122
65	8129	8136	8142	8149	8156	8162	8169	8176	8182	8189
66	8195	8202	8209	8215	8222	8228	8235	8241	8248	8254
67	8261	8267	8274	8280	8287	8293	8299	8306	8312	8319
68	8325	8331	8338	8344	8351	8357	8363	8370	8376	8382
69	8388	8395	8401	8407	8414	8420	8426	8432	8439	8445
70	8451	8457	8463	8470	8476	8482	8488	8494	8500	8506
71	8513	8519	8525	8531	8537	8543	8549	8555	8561	8567
72	8573	8579	8585	8591	8597	8603	8609	8615	8621	8627
73	8633	8639	8645	8651	8657	8663	8669	8675	8681	8686
74	8692	8698	8704	8710	8716	8722	8727	8733	8739	8745
75	8751	8756	8762	8768	8774	8779	8785	8791	8797	8802
76	8808	8814	8820	8825	8831	8837	8842	8848	8854	8859
77	8865	8871	8876	8882	8887	8893	8899	8904	8910	8915
78	8921	8927	8932	8938	8943	8949	8954	8960	8965	8971
79	8976	8982	8987	8993	8998	9004	9009	9015	9020	9025
80	9031	9036	9042	9047	9053	9058	9063	9069	9074	9079
81	9085	9090	9096	9101	9106	9112	9117	9122	9128	9133
82	9138	9143	9149	9154	9159	9165	9170	9175	9180	9186
83	9191	9196	9201	9206	9212	9217	9222	9227	9232	9238
84	9243	9248	9253	9258	9263	9269	9274	9279	9284	9289
85	9294	9299	9304	9309	9315	9320	9325	9330	9335	9340
86	9345	9350	9355	9360	9365	9370	9375	9380	9385	9390
87	9395	9400	9405	9410	9415	9420	9425	9430	9435	9440
88	9445	9450	9455	9460	9465	9469	9474	9479	9484	9489
89	9494	9499	9504	9509	9513	9518	9523	9528	9533	9538
90	9542	9547	9552	9557	9562	9566	9571	9576	9581	9586
91	9590	9595	9600	9605	9609	9614	9619	9624	9628	9633
92	9638	9643	9647	9652	9657	9661	9666	9671	9675	9680
93	9685	9689	9694	9699	9703	9708	9713	9717	9722	9727
94	9731	9736	9741	9745	9750	9754	9759	9763	9768	9773
95	9777	9782	9786	9791	9795	9800	9805	9809	9814	9818
96	9823	9827	9832	9836	9841	9845	9850	9854	9859	9863
97	9868	9872	9877	9881	9886	9890	9894	9899	9903	9908
98	9912	9917	9921	9926	9930	9934	9939	9943	9948	9952
99	9956	9961	9965	9969	9974	9978	9983	9987	9991	9996
100	0000	0004	0009	0013	0017	0022	0026	0030	0035	0039
N	0	1	2	3	4	5	6	7	8	9

Figure A3-1. — Table of Logarithms — continued.

45.45.1(179)B

SEMICONDUCTOR LETTER SYMBOLS

Letter symbols used in solid state circuits are those proposed as standard for industry, or are special symbols not included as standard. Semiconductor symbols consist of a basic letter with subscripts, either alphabetical or numerical, or both, in accordance with the following rules:

1. A capital (upper case) letter designates external circuit parameters and components, large-signal device parameters, and maximum (peak), average (d.c.), or root-mean-square values of current, voltage, and power (I, V, P, etc.)

2. Instantaneous values of current, voltage, and power, which vary with time, and small-signal values are represented by the lower case (small) letter of the proper symbol (i, v, p, i_e v_{eb}, etc.).

3. D.c. values, instantaneous total values, and large-signal values, are indicated by capital subscripts (i_C, I_C, v_{EB}, V_{EB}, P_C, etc.).

4. Alternating component values are indicated by using lower case subscripts: note the examples i_c, I_c, v_{eb}, V_{eb}, P_c, p_c.

5. When it is necessary to distinguish between maximum, average, or root-mean-square values, maximum or average values may be represented by addition of a subscript m or av; examples are i_{cm}, I_{CM}, I_{cav}, i_{CAV}.

6. For electrical quantities, the first subscript designates the electrode at which the measurement is made.

7. For device parameters, the first subscript designates the element of the four-pole matrix; examples are I or i for input, O or o for output, F or f for forward transfer, and R or r for reverse transfer.

8. The second subscript normally designates the reference electrode.

9. Supply voltages are indicated by repeating the associated device electrode subscript, in which case, the reference terminal is then designated by the third subscript; note the cases V_{EE}, V_{CC}, V_{EEB}, V_{CCB}.

10. In devices having more than one terminal of the same type (say two bases), the terminal subscripts are modified by adding a number following the subscript and placed on the same line, for example, V_{B1-B2}.

11. In multiple-unit devices the terminal subscripts are modified by a number preceding the electrode subscript; note the example, V_{1B-2B}.

Semiconductor symbols change, and new symbols are developed to cover new devices as the art changes; an alphabetical list of the complex symbols is presented below for easy reference.

The list is divided into six sections. These sections are signal and rectifier diodes, zener diodes, thyristors and SCR's, transistors, unijunction transistors, and field effect transistors.

SIGNAL AND RECTIFIER DIODES

PRV	Peak Reverse Voltage
I_o	Average Rectifier Forward Current
I_r	Average Reverse Current
I_{surge}	Peak Surge Current
V_F	Average Forward Voltage Drop
V_R	D.c. Blocking Voltage

ZENER DIODES

I_F	Forward current
I_Z	Zener current
I_{ZK}	Zener current near breakdown knee
I_{ZM}	Maximum D.c. zener current (limited by power dissipation)

Figure A4a-1.—Semiconductor Letter Symbols.

I_{ZT}	Zener test current	I_{HO}	Holding current. That value of forward anode current below which the controlled rectifier switches from the conducting state to the forward blocking condition with the gate open, at stated conditions.
V_f	Forward voltage		
V_Z	Nominal zener voltage		
Z_Z	Zener impedance		
Z_{ZK}	Zener impedance near breakdown knee	I_{HX}	Holding current (gate connected). The value of forward anode current below which the controlled rectifier switches from the conducting state to the forward blocking condition with the gate terminal returned to the cathode terminal through specified impedance and/or bias voltage.
Z_{ZT}	Zener impedance at zener test current		
I_R	Reverse current		
V_R	Reverse test voltage		

THYRISTORS AND SCRs

I_f	Forward current, r.m.s. value of forward anode current during the "on" state.	$P_{F(AV)}$	Average forward power. Average value of power dissipation between anode and cathode.
$I_{FM(pulse)}$	Repetitive pulse current. Repetitive peak forward anode current after application of gate signal for specified pulse conditions.	P_{GFM}	Peak gate power. The maximum instantaneous value of gate power dissipation permitted.
$I_{FM(surge)}$	Peak forward surge current. The maximum forward current having a single forward cycle in a 60 Hz single-phase resistive load system.	I_{ROM}	Peak reverse blocking current. The maximum current through the thyristor when the device is in the reverse blocking state (anode negative) for a stated anode-to-cathode voltage and junction temperature with the gate open.
I_{FOM}	Peak forward blocking current, gate open. The maximum current through the thyristor when the device is in the "off" state for a stated anode-to-cathode voltage (anode positive) and junction temperature with the gate open.	I_{RXM}	Peak reverse blocking current. Same as I_{ROM} except that the gate terminal is returned to the cathode through a stated impedance and/or bias voltage.
I_{FXM}	Peak forward blocking current. Same as I_{FOM} except that the gate terminal is returned to the cathode through a stated impedance and/or bias voltage.	$P_{GF(AV)}$	Average forward gate power. The value of maximum allowable gate power dissipation averaged over a full cycle.
I_{GFM}	Peak forward gate current. The maximum instantaneous value of current which may flow between gate and cathode.	V_F	Forward "on" voltage. The voltage measured between anode and cathode during the "on" condition for specified conditions of anode and temperature.
I_{GT}	Gate trigger current (continuous d.c.). The minimum d.c. gate current required to cause switching from the "off" state at a stated condition.	$V_{F(on)}$	Dynamic forward "on" voltage. The voltage measured between anode and cathode at a specified time after turn-on function has been initiated at stated conditions.

Figure A4a-1.—Semiconductor Letter Symbols—continued.

BASIC ELECTRONICS VOLUME II

V_{FOM} — Peak forward blocking voltage, gate open. The peak repetitive forward voltage which may be applied to the thyristor between anode and cathode (anode positive) with the gate open at stated conditions.

V_{FXM} — Peak forward blocking voltage. Same as V_{FOM} except that the gate terminal is returned to the cathode through a stated impedance and/or voltage.

V_{GFM} — Peak forward gate voltage. The maximum instantaneous voltage between the gate terminal and the cathode terminal resulting from the flow of forward gate current.

V_{GRM} — Peak reverse gate voltage. The maximum instantaneous voltage which may be applied between the gate terminal and the cathode terminal when the junction between the gate region and the adjacent cathode region is reverse biased.

V_{GT} — Gate trigger voltage (continuous d.c.). The d.c. voltage between the gate and the cathode required to produce the d.c. gate trigger current.

$V_{ROM(rep)}$ — Peak reverse blocking voltage, gate open. The maximum allowable value of reverse voltage (repetitive or continuous d.c.) which can be applied between anode and cathode (anode negative) with the gate open for stated conditions.

V_{RXM} — Peak reverse blocking voltage. Same as V_{ROM} except that the gate terminal is returned to the cathode through a stated impedance and/or bias voltage.

TRANSISTORS

A_G — Available gain

A_P — Power gain

A_I — Current gain

B or b — Base electrode

BV_{BCO} — D.c. base-to-collector breakdown voltage, base reverse-biased with respect to collector, emitter open.

BV_{BEO} — D.c. base-to-emitter breakdown voltage, base reverse-biased with respect to emitter, collector open.

BV_{CBO} — D.c. collector-to-base breakdown voltage, collector reverse-biased with respect to base, emitter open.

BV_{CEO} — D.c. collector-to-emitter breakdown voltage, collector reverse-biased with respect to emitter, base open.

BV_{EBO} — D.c. emitter-to-base breakdown voltage, emitter reverse-biased with respect to base, collector open.

BV_{ECO} — D.c. emitter-to-collector breakdown voltage, emitter reverse-biased with respect to collector, base open.

C or c — Collector electrode

C_c — Collector junction capacitance

C_e — Emitter junction capacitance

C_{ib}, C_{ic}, C_{ie} — Input capacitance for common base, collector, and emitter, respectively.

C_{ob}, C_{oc}, C_{oe} — Output terminal capacitance, a.c. input open, for common base, collector and emitter, respectively.

D — Distortion

E or e — Emitter electrode

$f_{\alpha b}, f_{\alpha c}, f_{\alpha e}$ — Alpha cutoff frequency for common base, collector, and emitter, respectively.

f_{co} — Cutoff frequency

f_{max} — Maximum frequency of oscillation

GC (CB), GC (CC), GE (CE) — Grounded (or common) base, collector, and emitter, respectively.

G_b, G_c, G_e — Power gain for common base, collector, and emitter, respectively.

Figure A4a-1.—Semiconductor Letter Symbols—continued.

300

h	Hybrid parameter	t_s	Storage time (switching applications).
h_{fe}, h_{fb}, h_{fc}	Small signal forward current transfer ratio, a.c. output shorted, common emitter, common base, common collector, respectively.	V_{BE}	Base-to-emitter d.c. voltage
		V_{CE}	Collector-to-base d.c. voltage
		V_{CE}	Collector-to-emitter d.c. voltage
h_{ib}	Small-signal input impedance, a.c. output shorted, common base.	V_{CEO}	D.c. collector-to-emitter voltage with collector junction reverse-biased, zero base current.
h_{ob}	Small-signal output admittance, a.c. input open, common base.	V_{CER}	Similar to V_{CEO}, except with a resistor (of value R) between base and emitter.
I	Direct current (d.c.).		
I_B, I_C, I_E	D.c. current for base, collector, and emitter, respectively.	V_{CES}	Similar to V_{CEO}, except with base shorted to emitter.
I_{CBO}	D.c. collector current, collector reverse-biased with respect to base, emitter-to-base open.	V_{CEV}	D.c. collector-to-emitter voltage, used when only voltage bias is used.
I_{CES}	D.c. collector current, collector reverse-biased with respect to emitter, base shorted to emitter.	V_{CEX}	D.c. collector-to-emitter voltage, base-emitter back biased.
I_{EBO}	D.c. emitter current, emitter reverse-biased with respect to base, collector-to-base open.	V_{EB}	Emitter-to-base d.c. voltage
		V_{pt}	Punch-through voltage
NF	Noise Figure		
P_D	Total average power dissipation of all electrodes of a semiconductor device.		UNIJUNCTION TRANSISTORS
P_G	Power gain	I_E	Emitter current
P_{Go}	Over-all power gain	I_{EO}	Emitter reverse current. Measured between emitter and base-two at a specified voltage, and base-one open-circuited.
P_{in}	Input power		
P_{out}	Output power	I_p	Peak point emitter current. The maximum emitter current that can flow without allowing the UJT to go into the negative resistance region.
r'b	Equivalent base resistance, high frequencies		
T_j	Junction temperature	I_V	Valley point emitter. The current flowing in the emitter when the device is biased to the valley point.
T_{stg}	Storage temperature		
t_f	Fall time, from 90 percent to 10 percent of pulse (switching applications).	r_{BB}	Interbase resistance. Resistance between base-two and base-one measured at a specified interbase voltage.
t_r	Rise time, from 10 percent to 90 percent pulse (switching applications).	V_{B2B1}	Voltage between base-two and base-one. Positive at base-two.

Figure A4a-1. — Semiconductor Letter Symbols — continued.

V_p — Peak point emitter voltage. The maximum voltage seen at the emitter before the UJT goes into the negative resistance region.

V_D — Forward voltage drop of the emitter junction.

V_{EB1} — Emitter to base-one voltage

$V_{EB1(SAT)}$ — Emitter saturation voltage. Forward voltage drop from emitter to base-one at a specified emitter current (larger than I_V) and specified interbase voltage.

V_V — Valley point emitter voltage. The voltage at which the valley point occurs with a specified V_{B2B1}.

V_{OB1} — Base-one peak pulse voltage. The peak voltage measured across a resistor in series with base-one when the UJT is operated as a relaxation oscillator in a specified circuit.

α_{rBB} — Interbase resistance temperature coefficient. Variation of resistance between B2 and B1 over the specified temperature range and measured at the specific interbase voltage and temperature with emitter open circuited.

$I_{B2(mod)}$ — Interbase modulation current. B2 current modulation due to firing. Measured at a specified interbase voltage, emitter and temperature.

FIELD EFFECT TRANSISTORS

I_D — Drain current

I_{DGO} — Maximum leakage from drain to gate with source open

I_{DSS} — Drain current with gate connected to source

I_G — Gate current

I_{GSS} — Maximum gate current (leakage) with drain connected to source

$V_{(BR)DGO}$ — Drain to gate, source open

V_D — D.c. drain voltage

$V_{(BR)DGS}$ — Drain to gate, source connected to drain

$V_{(BR)DS}$ — Drain to source, gate connection not specified

$V_{(BR)DSX}$ — Drain to source, gate biased to cutoff or beyond

$V_{(BR)GS}$ — Gate to source, drain connection not specified

$V_{(BR)GSS}$ — Gate to source, drain connected to source

$V_{(BR)GD}$ — Gate to drain, source connection not specified

$V_{(BR)GDS}$ — Gate to drain, source connected to drain

V_G — D.c. gate voltage

$V_{G1S(OFF)}$ — Gate 1-source cutoff voltage (with gate 2 connected to source)

$V_{G2S(OFF)}$ — Gate 2-source cutoff voltage (with gate 1 connected to source)

$V_{GS(OFF)}$ — Cutoff

Figure A4a-1.—Semiconductor Letter Symbols—continued.

ELECTRON TUBE LETTER SYMBOLS

A number of letter symbols which are used as a form of shorthand notation in technical literature when designating electron-tube operating conditions are explained and listed below.

1. Maximum, average, and root-mean-square values are represented by capital (upper case) letters, for example: I, E, P.

2. Where needed to distinguish between values in item 1 above, the maximum value may be represented by the subscript "m", for example: E_m, I_m, P_m.

3. Average values may be represented by the subscript "av," for example: E_{av}, I_{av}, P_{av}. (When items 2 and 3 above are used, then item 1 indicates r-m-s, or effective, values.)

4. Instantaneous values of current, voltage, and power which vary with time are represented by the small (lower case) letter of the proper symbol, for example: i, e, p.

5. External resistance, impedance, etc, in the circuit external to an electron tube electrode may be represented by the upper case symbol with the proper electrode subscripts, for example: R_g, R_{sc}, Z_g, Z_{sc}.

6. Values of resistance, impedance, etc, inherent within the electron tube are represented by the lower case symbol with the proper electrode subscripts, for example: r_g, Z_g, r_p, Z_p, C_{gp}.

7. The symbols "g" and "p" are used as subscripts to identify a.c. values of electrode currents and voltages, for example: e_g, e_p, i_g, i_p.

8. The total instantaneous values of electrode currents and voltages (d.c. plus a.c. components) are indicated by the lower case symbol and the subscripts "b" for plate and "c" for grid, for example: i_b, e_c, i_c, e_b.

9. No-signal or static currents and voltages are indicated by upper case symbol and lower case subscripts "b" for plate and "c" for grid, for example: E_c, I_b, E_b, I_c.

10. R.m.s. and maximum values of a varying component are indicated by the upper case letter and the subscripts "g" and "p", for example: E_g, I_p, E_p, I_g.

11. Average values of current and voltage for the with-signal condition are indicated by adding the subscript "s" to the proper symbol and subscript, for example: I_{bs}, E_{bs}.

12. Supply voltages are indicated by the upper case symbol and double subscript "bb" for plate, "cc" for grid, "ff" for filament, for example: E_{ff}, E_{cc}, E_{bb}.

An alphabetical list of electron tube symbols follows for easy reference.

C_{gk}	Grid-cathode capacitance
C_{gp}	Grid-plate capacitance
C_{pk}	Plate-cathode capacitance
E_b	Plate voltage, d.c. value
E_c	Grid voltage, d.c. value
E_{cc}	Grid bias supply voltage
E_{co}	Negative tube cutoff voltage
E_f	Filament voltage
E_{ff}	Filament supply voltage
E_k	D.c. cathode voltage
e_b	Instantaneous plate voltage
e_c	Instantaneous grid voltage
e_g	A.c. component of grid voltage
e_p	A.c. component of plate voltage

Figure A4b-1.— Electron tube letter symbols.

I_b, I_o	D.c. plate current	P_p	Plate dissipation power
I_c	D.c. grid current	R_g	Grid resistance
I_f	Filament current	R_L	Load resistance
I_k	Cathode current		
i_b	Instantaneous plate current	R_k	Cathode resistance
i_c	Instantaneous grid current	R_p	Plate resistance, d.c.
i_g	A.c. component of grid current	R_{sc}	Screen resistance
i_p	A.c. component of plate current	r_L	A.c. load resistance
P_g	Grid dissipation power	r_p	Plate resistance, a.c.
P_o	Output power	t_k	Cathode heating time

Figure A4b-1.— Electron tube letter symbols — continued.

APPENDIX V

JOINT ELECTRONICS TYPE DESIGNATION
(AN) SYSTEM

THE *AN* NOMENCLATURE WAS DESIGNED SO THAT A COMMON DESIGNATION COULD BE USED FOR ARMY, NAVY, AND AIR FORCE EQUIPMENT. THE SYSTEM INDICATOR *AN* DOES NOT MEAN THAT THE ARMY, NAVY, AND AIR FORCE USE THE EQUIPMENT, BUT MEANS THAT THE TYPE NUMBER WAS ASSIGNED IN THE *AN* SYSTEM.

AN NOMENCLATURE IS ASSIGNED TO COMPLETE SETS OF EQUIPMENT AND MAJOR COMPONENTS OF MILITARY DESIGN; GROUPS OF ARTICLES OF EITHER COMMERCIAL OR MILITARY DESIGN WHICH ARE GROUPED FOR MILITARY PURPOSES; MAJOR ARTICLES OF MILITARY DESIGN WHICH ARE NOT PART OF OR USED WITH A SET; AND COMMERCIAL ARTICLES WHEN NOMENCLATURE WILL NOT FACILITATE MILITARY IDENTIFICATION AND/OR PROCEDURES.

AN NOMENCLATURE IS NOT ASSIGNED TO ARTICLES CATALOGED COMMERCIALLY EXCEPT AS STATED ABOVE; MINOR COMPONENTS OF MILITARY DESIGN FOR WHICH OTHER ADEQUATE MEANS OF IDENTIFICATION ARE AVAILABLE; SMALL PARTS SUCH AS CAPACITORS AND RESISTORS; AND ARTICLES HAVING OTHER ADEQUATE IDENTIFICATION IN JOINT MILITARY SPECIFICATIONS. NOMENCLATURE ASSIGNMENTS REMAIN UNCHANGED REGARDLESS OF LATER CHANGES IN INSTALLATION AND/OR APPLICATION.

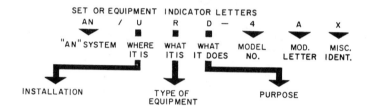

A -- AIRBORNE (INSTALLED AND OPERATED IN AIRCRAFT).
B -- UNDERWATER MOBILE, SUBMARINE.
C -- AIR TRANSPORTABLE (INACTIVATED, DO NOT USE).
D -- PILOTLESS CARRIER.
F -- FIXED.
G -- GROUND, GENERAL GROUND USE (INCLUDES TWO OR MORE GROUND-TYPE INSTALLATIONS).
K -- AMPHIBIOUS.
M -- GROUND, MOBILE (INSTALLED AS OPERATING UNIT IN A VEHICLE WHICH HAS NO FUNCTION OTHER THAN TRANSPORTING THE EQUIPMENT).
P -- PACK OR PORTABLE (ANIMAL OR MAN).
S -- WATER SURFACE CRAFT.
T -- GROUND, TRANSPORTABLE.
U -- GENERAL UTILITY (INCLUDES TWO OR MORE GENERAL INSTALLATION CLASSES, AIRBORNE, SHIPBOARD, AND GROUND).
V -- GROUND, VEHICULAR (INSTALLED IN VEHICLE DESIGNED FOR FUNCTIONS OTHER THAN CARRYING ELECTRONIC EQUIPMENT, ETC., SUCH AS TANKS).
W -- WATER SURFACE AND UNDERWATER.

A -- INVISIBLE LIGHT, HEAT RADIATION.
B -- PIGEON.
C -- CARRIER.
D -- RADIAC.
E -- NUPAC.
F -- PHOTOGRAPHIC.[1]
G -- TELEGRAPH OR TELETYPE.
I -- INTERPHONE AND PUBLIC ADDRESS.
J -- ELECTROMECHANICAL OR INERTIAL WIRE COVERED.
K -- TELEMETERING.
L -- COUNTERMEASURES.
M -- METEOROLOGICAL.
N -- SOUND IN AIR.
P -- RADAR.
Q -- SONAR AND UNDERWATER SOUND.
R -- RADIO
S -- SPECIAL TYPES, MAGNETIC, ETC., OR COMBINATIONS OF TYPES.
T -- TELEPHONE (WIRE).
V -- VISUAL AND VISIBLE LIGHT.
W -- ARMAMENT (PECULIAR TO ARMAMENT, NOT OTHERWISE COVERED).
X -- FACSIMILE OR TELEVISION.
Y -- DATA PROCESSING.

A -- AUXILIARY ASSEMBLIES (NOT COMPLETE OPERATING SETS USED WITH OR PART OF TWO OR MORE SETS OR SETS SERIES).
B -- BOMBING.
C -- COMMUNICATIONS (RECEIVING AND TRANSMITTING).
D -- DIRECTION FINDER, RECONNAISSANCE, AND/OR SURVEILLANCE.
E -- EJECTION AND/OR RELEASE.
G -- FIRE-CONTROL OR SEARCHLIGHT DIRECTING.
H -- RECORDING AND/OR REPRODUCING (GRAPHIC METEOROLOGICAL AND SOUND).
K -- COMPUTING.
L -- SEARCHLIGHT CONTROL (INACTIVATED, USE G).
M -- MAINTENANCE AND TEST ASSEMBLIES (INCLUDING TOOLS).
N -- NAVIGATIONAL AIDS (INCLUDING ALTIMETERS, BEACONS, COMPASSES, RACONS, DEPTH SOUNDING, APPROACH, AND LANDING).

P -- REPRODUCING (INACTIVATED, DO NOT USE).
Q -- SPECIAL, OR COMBINATION OF PURPOSES.
R -- RECEIVING, PASSIVE DETECTING.
S -- DETECTING AND/OR RANGE AND BEARING, SEARCH.
T -- TRANSMITTING.
W -- AUTOMATIC FLIGHT OR REMOTE CONTROL.
X -- IDENTIFICATION AND RECOGNITION.

[1] NOT FOR US USE EXCEPT FOR ASSIGNING SUFFIX LETTERS TO PREVIOUSLY NOMENCLATURED ITEMS.

20.484
Figure A5-1. — AN system.

APPENDIX VI

ELECTRONICS COLOR CODING

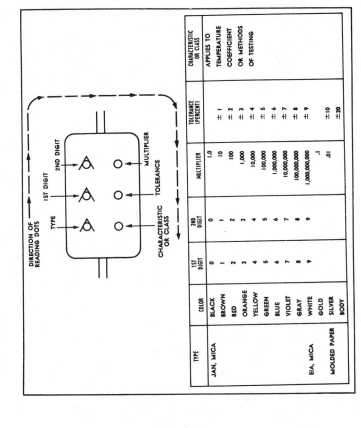

Figure A6-2.—6-Dot color code for mica and molded paper capacitors.

20.376

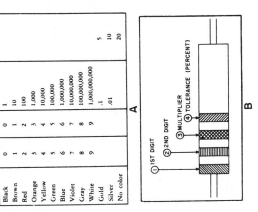

Figure A6-1.—Resistor color code.

20.373:.374

COLOR	1ST DIGIT	2ND DIGIT	MULTIPLIER	TOLERANCE (percent)
Black	0	0	1	
Brown	1	1	10	
Red	2	2	100	
Orange	3	3	1,000	
Yellow	4	4	10,000	
Green	5	5	100,000	
Blue	6	6	1,000,000	
Violet	7	7	10,000,000	
Gray	8	8	100,000,000	
White	9	9	1,000,000,000	
Gold			.1	5
Silver			.01	10
No color				20

TYPE	COLOR	1ST DIGIT	2ND DIGIT	MULTIPLIER	TOLERANCE (PERCENT)	CHARACTERISTIC OR CLASS
JAN, MICA	BLACK	0	0	1.0		APPLIES TO
	BROWN	1	1	10	± 1	TEMPERATURE
	RED	2	2	100	± 2	COEFFICIENT
	ORANGE	3	3	1,000	± 3	OR METHODS
	YELLOW	4	4	10,000	± 4	OF TESTING
	GREEN	5	5	100,000	± 5	
	BLUE	6	6	1,000,000	± 6	
	VIOLET	7	7	10,000,000	± 7	
	GRAY	8	8	100,000,000	± 8	
	WHITE	9	9	1,000,000,000	± 9	
EIA, MICA	GOLD			.1	±10	
	SILVER			.01	±20	
MOLDED PAPER	BODY					

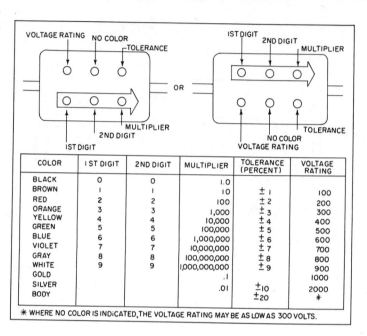

COLOR	I ST DIGIT	2ND DIGIT	MULTIPLIER	TOLERANCE (PERCENT)	VOLTAGE RATING
BLACK	0	0	1.0		
BROWN	1	1	10	± 1	100
RED	2	2	100	± 2	200
ORANGE	3	3	1,000	± 3	300
YELLOW	4	4	10,000	± 4	400
GREEN	5	5	100,000	± 5	500
BLUE	6	6	1,000,000	± 6	600
VIOLET	7	7	10,000,000	± 7	700
GRAY	8	8	100,000,000	± 8	800
WHITE	9	9	1,000,000,000	± 9	900
GOLD			.1		1000
SILVER			.01	± 10	2000
BODY				± 20	*

* WHERE NO COLOR IS INDiCATED,THE VOLTAGE RATING MAY BE AS LOW AS 300 VOLTS.

Figure A6-3. — 5-Dot color code for capacitors (dielectric not specified). 20.486

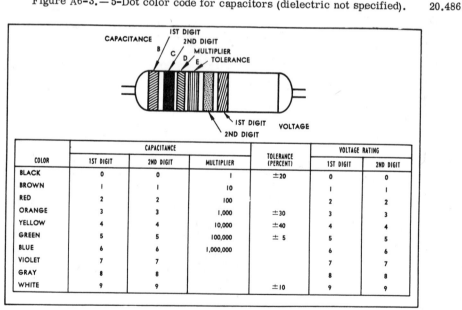

COLOR	CAPACITANCE			TOLERANCE (PERCENT)	VOLTAGE RATING	
	1ST DIGIT	2ND DIGIT	MULTIPLIER		1ST DIGIT	2ND DIGIT
BLACK	0	0	1	±20	0	0
BROWN	1	1	10		1	1
RED	2	2	100		2	2
ORANGE	3	3	1,000	±30	3	3
YELLOW	4	4	10,000	±40	4	4
GREEN	5	5	100,000	± 5	5	5
BLUE	6	6	1,000,000		6	6
VIOLET	7	7			7	7
GRAY	8	8			8	8
WHITE	9	9		±10	9	9

Figure A6-4. — 6-Band color code for tubular paper dielectric capacitors. 20.487

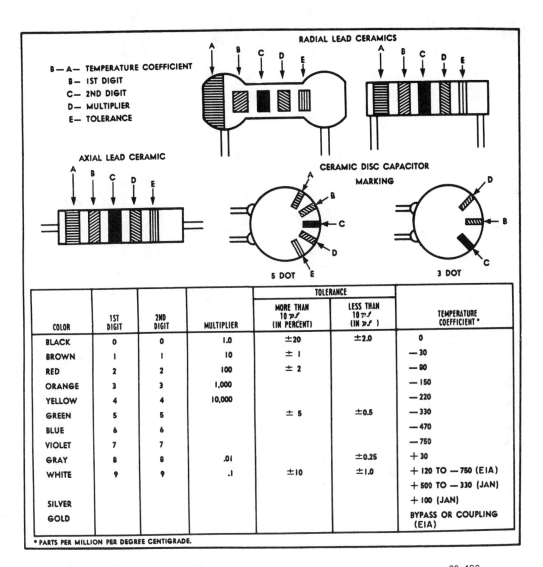

RADIAL LEAD CERAMICS

B — A — TEMPERATURE COEFFICIENT
B — 1ST DIGIT
C — 2ND DIGIT
D — MULTIPLIER
E — TOLERANCE

AXIAL LEAD CERAMIC

CERAMIC DISC CAPACITOR MARKING

5 DOT

3 DOT

COLOR	1ST DIGIT	2ND DIGIT	MULTIPLIER	TOLERANCE		TEMPERATURE COEFFICIENT*
				MORE THAN 10 $p f$ (IN PERCENT)	LESS THAN 10 $p f$ (IN $p f$)	
BLACK	0	0	1.0	±20	±2.0	0
BROWN	1	1	10	± 1		— 30
RED	2	2	100	± 2		— 80
ORANGE	3	3	1,000			— 150
YELLOW	4	4	10,000			— 220
GREEN	5	5		± 5	±0.5	— 330
BLUE	6	6				— 470
VIOLET	7	7				— 750
GRAY	8	8	.01		±0.25	+ 30
WHITE	9	9	.1	±10	±1.0	+ 120 TO — 750 (EIA)
						+ 500 TO — 330 (JAN)
SILVER						+ 100 (JAN)
GOLD						BYPASS OR COUPLING (EIA)

* PARTS PER MILLION PER DEGREE CENTIGRADE.

20.488

Figure A6-5. — Color code for ceramic capacitors having different configurations.

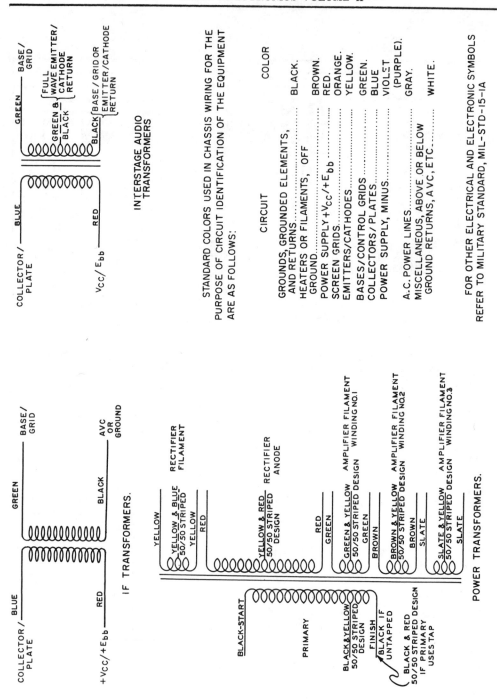

STANDARD COLORS USED IN CHASSIS WIRING FOR THE PURPOSE OF CIRCUIT IDENTIFICATION OF THE EQUIPMENT ARE AS FOLLOWS:

CIRCUIT	COLOR
GROUNDS, GROUNDED ELEMENTS, AND RETURNS	BLACK.
HEATERS OR FILAMENTS, OFF GROUND	BROWN.
POWER SUPPLY $+V_{cc}/+E_{bb}$	RED.
SCREEN GRIDS	ORANGE.
EMITTERS/CATHODES	YELLOW.
BASES/CONTROL GRIDS	GREEN.
COLLECTORS/PLATES	BLUE
POWER SUPPLY, MINUS	VIOLET (PURPLE).
A.C. POWER LINES	GRAY.
MISCELLANEOUS, ABOVE OR BELOW GROUND RETURNS, AVC, ETC	WHITE.

FOR OTHER ELECTRICAL AND ELECTRONIC SYMBOLS REFER TO MILITARY STANDARD, MIL–STD–15–1A

20.378:.380

Figure A6-6.—Color code for transformers.

APPENDIX VII

ELECTRONICS SYMBOLS

APPENDIX VII
ELECTRONICS SYMBOLS

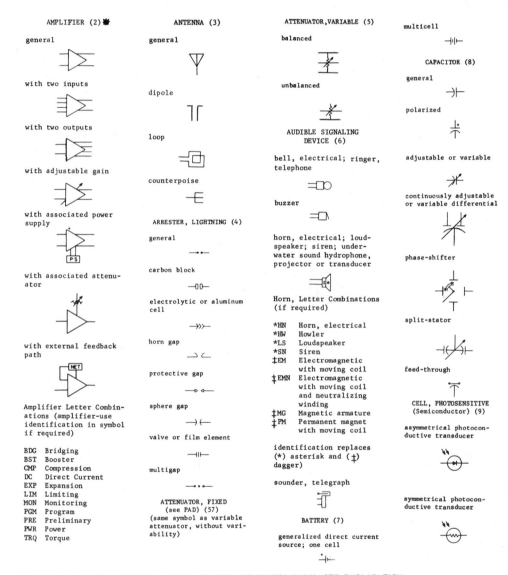

AMPLIFIER (2)✹

general

with two inputs

with two outputs

with adjustable gain

with associated power supply

with associated attenuator

with external feedback path

Amplifier Letter Combinations (amplifier-use identification in symbol if required)

BDG   Bridging
BST   Booster
CMP   Compression
DC   Direct Current
EXP   Expansion
LIM   Limiting
MON   Monitoring
PGM   Program
PRE   Preliminary
PWR   Power
TRQ   Torque

ANTENNA (3)

general

dipole

loop

counterpoise

ARRESTER, LIGHTNING (4)

general

carbon block

electrolytic or aluminum cell

horn gap

protective gap

sphere gap

valve or film element

multigap

ATTENUATOR, FIXED (see PAD) (57)
(same symbol as variable attenuator, without variability)

ATTENUATOR, VARIABLE (5)

balanced

unbalanced

AUDIBLE SIGNALING DEVICE (6)

bell, electrical; ringer, telephone

buzzer

horn, electrical; loudspeaker; siren; underwater sound hydrophone, projector or transducer

Horn, Letter Combinations (if required)

*HN   Horn, electrical
*HW   Howler
*LS   Loudspeaker
*SN   Siren
‡EM   Electromagnetic with moving coil
‡EMN   Electromagnetic with moving coil and neutralizing winding
‡MG   Magnetic armature
‡PM   Permanent magnet with moving coil

identification replaces (*) asterisk and (‡) dagger)

sounder, telegraph

BATTERY (7)

generalized direct current source; one cell

multicell

CAPACITOR (8)

general

polarized

adjustable or variable

continuously adjustable or variable differential

phase-shifter

split-stator

feed-through

CELL, PHOTOSENSITIVE (Semiconductor) (9)

asymmetrical photoconductive transducer

symmetrical photoconductive transducer

✹ NUMBER IN PARENTHESES INDICATES LOCATION OF SYMBOL IN MIL-STD PUBLICATION

Figure A7-1.— Electronics symbols.   13.5(179)A

photovoltaic transducer; solar cell

CIRCUIT BREAKER (11)

general

with magnetic overload

drawout type

CIRCUIT ELEMENT (12)

general

Circuit Element Letter Combinations (replaces (*) asterisk)

EG Equalizer
FAX Facsimile set
FL Filter
FL-BE Filter, band elimination
FL-BP Filter, band pass
FL-HP Filter, high pass
FL-LP Filter, low pass
PS Power supply
RG Recording unit
RU Reproducing unit
DIAL Telephone dial
TEL Telephone station
TPR Teleprinter
TTY Teletypewriter

Additional Letter Combinations (symbols preferred)

AR Amplifier
AT Attenuator
C Capacitor
CB Circuit breaker
HS Handset
I Indicating or switch board lamp
L Inductor
J Jack
LS Loudspeaker
MIC Microphone
OSC Oscillator
PAD Pad
P Plug
HT Receiver, headset
K Relay
R Resistor
S Switch or key switch
T Transformer
WR Wall receptacle

CLUTCH; BRAKE (14)

disengaged when operating means is de-energized

engaged when operating means is de-energized

COIL, RELAY and OPERATING (16)

semicircular dot indicates inner end of wiring

CONNECTOR (18)

assembly, movable or stationary portion; jack, plug, or receptacle

jack or receptacle

plug

separable connectors

two-conductor switchboard jack

two-conductor switchboard plug

jacks normalled through one way

jacks normalled through both ways

2-conductor nonpolarized, female contacts

2-conductor polarized, male contacts

waveguide flange

plain, rectangular

choke, rectangular

engaged 4-conductor; the plug has 1 male and 3 female contacts, individual contact designations shown

coaxial, outside conductor shown carried through

coaxial, center conductor shown carried through; outside conductor not carried through

mated choke flanges in rectangular waveguide

COUNTER, ELECTROMAGNETIC; MESSAGE REGISTER (26)

general

with a make contact

COUPLER, DIRECTIONAL (27) (common coaxial/waveguide usage)

(common coaxial/waveguide usage)

E-plane aperture-coupling, 30-decibel transmission loss

COUPLING (28)

by loop from coaxial to circular waveguide, direct-current grounds connected

CRYSTAL, PIEZO-ELECTRIC (62)

DELAY LINE (31)

general

tapped delay

bifilar slow-wave structure (commonly used in traveling-wave tubes)

(length of delay indication replaces (*) asterisk)

DETECTOR, PRIMARY; MEASURING TRANSDUCER (30) (see HALL GENERATOR and THERMAL CONVERTER)

DISCONTINUITY (33) (common coaxial/waveguide usage)

equivalent series element, general

capacitive reactance

inductive reactance

inductance-capacitance circuit, infinite reactance at resonance

Figure A7-1.— Electronics symbols—continued. 13.5(179)B

313

inductance-capacitance
circuit, zero reactance
at resonance

resistance

equivalent shunt element,
general

capacitive susceptance

inductive susceptance

inductance-capacitance
circuit, infinite sus-
ceptance at resonance

inductance-capacitance
circuit, zero suscept-
ance at resonance

ELECTRON TUBE (34)

triode

pentode, envelope connec-
ted to base terminal

twin triode, equipoten-
tial cathode

typical wiring figure to
show tube symbols placed
in any convenient position

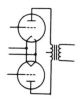

**rectifier; voltage regu-
lator
(see LAMP, GLOW)**

phototube, single and
multiplier

cathode-ray tube, electrostatic and magnetic de-
flection

mercury-pool tube,
ignitor and control
grid (see RECTIFIER)

resonant magnetron, co-
axial output and perma-
nent magnet

**reflex klystron, integral
cavity, aperture coupled**

**transmit-receive (TR)
tube gas filled, tunable
integral cavity, aperture
coupled, with starter**

**traveling-wave tube
(typical)**

**forward-wave traveling-
wave-tube amplifier shown
with four grids, having
slow-wave structure with
attenuation, magnetic
focusing by external
permanent magnet, rf in-
put and rf output cou-
pling each E-plane aper-
ture to external rect-
angular waveguide**

FERRITE DEVICES (100)

field polarization
rotator

field polarization
amplitude modulator

FUSE (36)

high-voltage primary
cutout, dry

**high-voltage primary cut-
out, oil**

**GOVERNOR (Contact-
making) (37)**

contacts shown here as
closed

HALL GENERATOR (39)

HANDSET (40)

general

operator's set with push-
to talk switch

HYBRID (41)

general

junction
(common coaxial/wave-
guide usage)

circular

(E, H or HE transverse
field indicators re-
place (*) asterisk)

rectangular waveguide and
coaxial coupling

INDUCTOR (42)

general

Figure A7-1.— Electronics symbols—continued.

13.5(179)C

314

Appendix VII — ELECTRONICS SYMBOLS

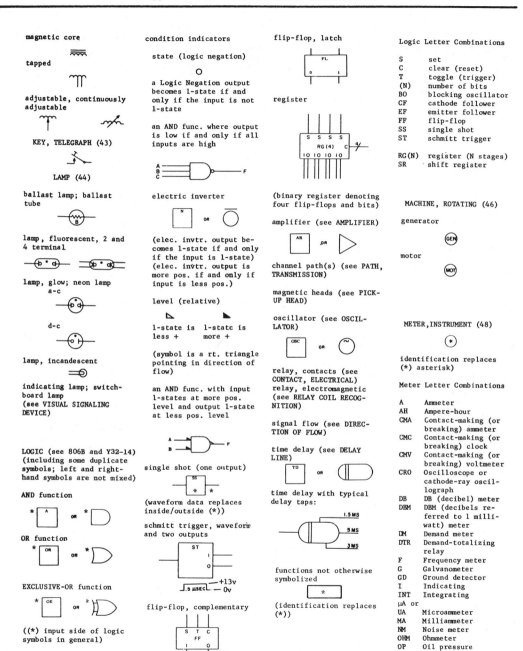

Figure A7-1.— Electronics symbols—continued.

13.5(179)D

MODE TRANSDUCER (53)

(common coaxial/waveguide usage)

transducer from rectangular waveguide to coaxial with mode suppression, direct-current grounds connected

MOTION, MECHANICAL (54)

rotation applied to a resistor

(identification replaces (*) asterisk)

NUCLEAR-RADIATION DE-TECTOR, gas filled; IONIZATION CHAMBER; PROPORTIONAL COUNTER TUBE; GEIGER-MULLER COUNTER TUBE (50) (see RADIATION-SENSITIVITY INDICATOR)

PATH, TRANSMISSION (58)

cable; 2-conductor, shield grounded and 5-conductor shielded

PICKUP HEAD (61)

general

writing; recording

reading; playback

erasing

writing, reading, and erasing

stereo

RECTIFIER (65)

semiconductor diode; metallic rectifier; electrolytic rectifier; asymmetrical varistor

mercury-pool tube power rectifier

fullwave bridge-type

RESISTOR (68)

general

tapped

heating

symmetrical varistor resistor, voltage sensitive (silicon carbide, etc.)

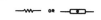

(identification marks replace (*) asterisk)

with adjustable contact

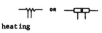

adjustable or continuously adjustable (variable)

(identification replaces (*) asterisk)

RESONATOR, TUNED CAVITY (71)

(common coaxial/waveguide usage)

resonator with mode suppression coupled by an E-plane aperture to a guided transmission path and by a loop to a coaxial path

tunable resonator with direct-current ground connected to an electron device and adjustably coupled by an E-plane aperture to a rectangular waveguide

ROTARY JOINT, RF (COU-PLER) (72)

general; with rectangular waveguide

(transmission path recognition symbol replaces (*) asterisk)

coaxial type in rectangular waveguide

circular waveguide type in rectangular waveguide

SEMICONDUCTOR DEVICE (73) (Two Terminal, diode)

semiconductor diode; rectifier

capacitive diode (also Varicap, Varactor, reactance diode, parametric diode)

breakdown diode, unidirectional (also backward diode, avalanche diode, voltage regulator diode, Zener diode, voltage reference diode)

breakdown diode, bidirectional and backward diode (also bipolar voltage limiter)

tunnel diode (also Esaki diode)

temperature-dependent diode

photodiode (also solar cell)

semiconductor diode, PNPN switch (also Shockley diode, four-layer diode and SCR).

(Multi-Terminal, transistor, etc.)

PNP transistor

NPN transistor

unijunction transistor, N-type base

Figure A7-1.— Electronics symbols—continued.   13.5(179)E

unijunction transistor,
P-type base

field-effect transistor,
N-type base

 OR

field-effect transistor,
P-type base

 OR

semiconductor triode,
PNPN-type switch

semiconductor triode,
NPNP-type switch

NPN transistor, trans-
verse-biased base

 OR

PNIP transistor, ohmic
connection to the intrin-
sic region

NPIN transistor, ohmic
connection to the in-
trinsic region

PNIN transistor, ohmic
connection to the in-
trinsic region

NPIP transistor, ohmic
connection to the intrin-
sic region

SQUIB (75)

explosive

igniter

sensing link; fusible
link operated

SWITCH (76)

push button, circuit closing (make)

push button, circuit op-
ening (break)

nonlocking; momentary
circuit closing (make)

OR

nonlocking; momentary
circuit opening (break)

OR

transfer

OR

locking, circuit closing
(make)

OR

locking, circuit opening
(break)

OR

transfer, 3-position

OFF

wafer

(example shown: 3-pole
3-circuit with 2 non-
shorting and 1 shorting
moving contacts)

safety interlock, circuit
opening and closing

2-pole field-discharge
knife, with terminals and
discharge resistor

(identification replaces
(*) asterisk)

SYNCHRO (78)

Synchro Letter Combina-
tions
CDX Control-differential
     transmitter
CT   Control transformer
CX   Control transmitter
TDR Torque-differential
     receiver
TDX Torque-differential
     transmitter
TR   Torque receiver
TX   Torque transmitter
RS   Resolver
B   Outer winding rotat-
     able in bearings

THERMAL ELEMENT (83)

actuating device

____ OR ____

thermal cutout; flasher

____ OR ____

thermal relay

OR

OR

OR

thermostat (operates on
rising temperature), con-
tact)

OR

thermostat, make contact

OR

thermostat, integral
heater and transfer
contacts

 OR

THERMISTOR; THERMAL
RESISTOR (84)

with integral heater

THERMOCOUPLE (85)

temperature-measuring

current-measuring,inte-
gral heater connected

current-measuring,inte-
gral heater insulated

temperature-measuring,
semiconductor

current-measuring, semi-
conductor

TRANSFORMER (86)

general

 OR

magnetic-core

one winding with adjust-
able inductance

separately adjustable
inductance

adjustable mutual induc-
tor, constant-current

Figure A7-1. — Electronics symbols — continued.   13.5(179)F

317

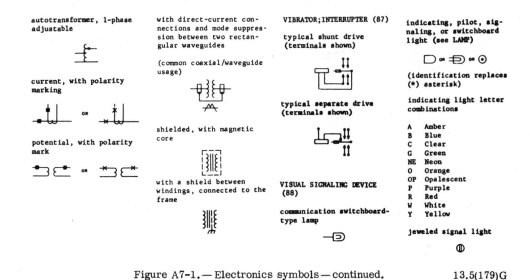

autotransformer, 1-phase adjustable

current, with polarity marking

potential, with polarity mark

with direct-current connections and mode suppression between two rectangular waveguides

(common coaxial/waveguide usage)

shielded, with magnetic core

with a shield between windings, connected to the frame

VIBRATOR;INTERRUPTER (87)

typical shunt drive (terminals shown)

typical separate drive (terminals shown)

VISUAL SIGNALING DEVICE (88)

communication switchboard-type lamp

indicating, pilot, signaling, or switchboard light (see LAMP)

(identification replaces (*) asterisk)

indicating light letter combinations

A   Amber
B   Blue
C   Clear
G   Green
NE   Neon
O   Orange
OP   Opalescent
P   Purple
R   Red
W   White
Y   Yellow

jeweled signal light

Figure A7-1.— Electronics symbols—continued.

13.5(179)G

INDEX

A CATALOGUE OF SELECTED DOVER BOOKS
IN ALL FIELDS OF INTEREST

A CATALOGUE OF SELECTED DOVER BOOKS
IN ALL FIELDS OF INTEREST

AMERICA'S OLD MASTERS, James T. Flexner. Four men emerged unexpectedly from provincial 18th century America to leadership in European art: Benjamin West, J. S. Copley, C. R. Peale, Gilbert Stuart. Brilliant coverage of lives and contributions. Revised, 1967 edition. 69 plates. 365pp. of text.

21806-6 Paperbound $3.00

FIRST FLOWERS OF OUR WILDERNESS: AMERICAN PAINTING, THE COLONIAL PERIOD, James T. Flexner. Painters, and regional painting traditions from earliest Colonial times up to the emergence of Copley, West and Peale Sr., Foster, Gustavus Hesselius, Feke, John Smibert and many anonymous painters in the primitive manner. Engaging presentation, with 162 illustrations. xxii + 368pp.

22180-6 Paperbound $3.50

THE LIGHT OF DISTANT SKIES: AMERICAN PAINTING, 1760-1835, James T. Flexner. The great generation of early American painters goes to Europe to learn and to teach: West, Copley, Gilbert Stuart and others. Allston, Trumbull, Morse; also contemporary American painters—primitives, derivatives, academics—who remained in America. 102 illustrations. xiii + 306pp. 22179-2 Paperbound $3.00

A HISTORY OF THE RISE AND PROGRESS OF THE ARTS OF DESIGN IN THE UNITED STATES, William Dunlap. Much the richest mine of information on early American painters, sculptors, architects, engravers, miniaturists, etc. The only source of information for scores of artists, the major primary source for many others. Unabridged reprint of rare original 1834 edition, with new introduction by James T. Flexner, and 394 new illustrations. Edited by Rita Weiss. 6⅝ x 9⅝.

21695-0, 21696-9, 21697-7 Three volumes, Paperbound $13.50

EPOCHS OF CHINESE AND JAPANESE ART, Ernest F. Fenollosa. From primitive Chinese art to the 20th century, thorough history, explanation of every important art period and form, including Japanese woodcuts; main stress on China and Japan, but Tibet, Korea also included. Still unexcelled for its detailed, rich coverage of cultural background, aesthetic elements, diffusion studies, particularly of the historical period. 2nd, 1913 edition. 242 illustrations. lii + 439pp. of text.

20364-6, 20365-4 Two volumes, Paperbound $6.00

THE GENTLE ART OF MAKING ENEMIES, James A. M. Whistler. Greatest wit of his day deflates Oscar Wilde, Ruskin, Swinburne; strikes back at inane critics, exhibitions, art journalism; aesthetics of impressionist revolution in most striking form. Highly readable classic by great painter. Reproduction of edition designed by Whistler. Introduction by Alfred Werner. xxxvi + 334pp.

21875-9 Paperbound $2.50

THE ARCHITECTURE OF COUNTRY HOUSES, Andrew J. Downing. Together with Vaux's *Villas and Cottages* this is the basic book for Hudson River Gothic architecture of the middle Victorian period. Full, sound discussions of general aspects of housing, architecture, style, decoration, furnishing, together with scores of detailed house plans, illustrations of specific buildings, accompanied by full text. Perhaps the most influential single American architectural book. 1850 edition. Introduction by J. Stewart Johnson. 321 figures, 34 architectural designs. xvi + 560pp.
22003-6 Paperbound $4.00

LOST EXAMPLES OF COLONIAL ARCHITECTURE, John Mead Howells. Full-page photographs of buildings that have disappeared or been so alteied as to be denatured, including many designed by major early American architects. 245 plates. xvii + 248pp. 7⅞ x 10¾. 21143-6 Paperbound $3.00

DOMESTIC ARCHITECTURE OF THE AMERICAN COLONIES AND OF THE EARLY REPUBLIC, Fiske Kimball. Foremost architect and restorer of Williamsburg and Monticello covers nearly 200 homes between 1620-1825. Architectural details, construction, style features, special fixtures, floor plans, etc. Generally considered finest work in its area. 219 illustrations of houses, doorways, windows, capital mantels. xx + 314pp. 7⅞ x 10¾. 21743-4 Paperbound $3.50

EARLY AMERICAN ROOMS: 1650-1858, edited by Russell Hawes Kettell. Tour of 12 rooms, each representative of a different era in American history and each furnished, decorated, designed and occupied in the style of the era. 72 plans and elevations, 8-page color section, etc., show fabrics, wall papers, arrangements, etc. Full descriptive text. xvii + 200pp. of text. 8⅜ x 11¼.
21633-0 Paperbound $5.00

THE FITZWILLIAM VIRGINAL BOOK, edited by J. Fuller Maitland and W. B. Squire. Full modern printing of famous early 17th-century ms. volume of 300 works by Morley, Byrd, Bull, Gibbons, etc. For piano or other modern keyboard instrument; easy to read format. xxxvi + 938pp. 8⅜ x 11.
21068-5, 21069-3 Two volumes, Paperbound $8.00

HARPSICHORD MUSIC, Johann Sebastian Bach. Bach Gesellschaft edition. A rich selection of Bach's masterpieces for the harpsichord: the six English Suites, six French Suites, the six Partitas (Clavierübung part I), the Goldberg Variations (Clavierübung part IV), the fifteen Two-Part Inventions and the fifteen Three-Part Sinfonias. Clearly reproduced on large sheets with ample margins; eminently playable. vi + 312pp. 8⅛ x 11. 22360-4 Paperbound $5.00

THE MUSIC OF BACH: AN INTRODUCTION, Charles Sanford Terry. A fine, nontechnical introduction to Bach's music, both instrumental and vocal. Covers organ music, chamber music, passion music, other types. Analyzes themes, developments, innovations. x + 114pp. 21075-8 Paperbound $1.25

BEETHOVEN AND HIS NINE SYMPHONIES, Sir George Grove. Noted British musicologist provides best history, analysis, commentary on symphonies. Very thorough, rigorously accurate; necessary to both advanced student and amateur music lover. 436 musical passages. vii + 407 pp. 20334-4 Paperbound $2.25

DESIGN BY ACCIDENT; A BOOK OF "ACCIDENTAL EFFECTS" FOR ARTISTS AND DESIGNERS, James F. O'Brien. Create your own unique, striking, imaginative effects by "controlled accident" interaction of materials: paints and lacquers, oil and water based paints, splatter, crackling materials, shatter, similar items. Everything you do will be different; first book on this limitless art, so useful to both fine artist and commercial artist. Full instructions. 192 plates showing "accidents," 8 in color. viii + 215pp. 8⅜ x 11¼.   21942-9 Paperbound $3.50

THE BOOK OF SIGNS, Rudolf Koch. Famed German type designer draws 493 beautiful symbols: religious, mystical, alchemical, imperial, property marks, runes, etc. Remarkable fusion of traditional and modern. Good for suggestions of timelessness, smartness, modernity. Text. vi + 104pp. 6⅛ x 9¼.
20162-7 Paperbound $1.25

HISTORY OF INDIAN AND INDONESIAN ART, Ananda K. Coomaraswamy. An unabridged republication of one of the finest books by a great scholar in Eastern art. Rich in descriptive material, history, social backgrounds; Sunga reliefs, Rajput paintings, Gupta temples, Burmese frescoes, textiles, jewelry, sculpture, etc. 400 photos. viii + 423pp. 6⅜ x 9¾.   21436-2 Paperbound $4.00

PRIMITIVE ART, Franz Boas. America's foremost anthropologist surveys textiles, ceramics, woodcarving, basketry, metalwork, etc.; patterns, technology, creation of symbols, style origins. All areas of world, but very full on Northwest Coast Indians. More than 350 illustrations of baskets, boxes, totem poles, weapons, etc. 378 pp.
20025-6 Paperbound $3.00

THE GENTLEMAN AND CABINET MAKER'S DIRECTOR, Thomas Chippendale. Full reprint (third edition, 1762) of most influential furniture book of all time, by master cabinetmaker. 200 plates, illustrating chairs, sofas, mirrors, tables, cabinets, plus 24 photographs of surviving pieces. Biographical introduction by N. Bienenstock. vi + 249pp. 9⅞ x 12¾.   21601-2 Paperbound $4.00

AMERICAN ANTIQUE FURNITURE, Edgar G. Miller, Jr. The basic coverage of all American furniture before 1840. Individual chapters cover type of furniture— clocks, tables, sideboards, etc.—chronologically, with inexhaustible wealth of data. More than 2100 photographs, all identified, commented on. Essential to all early American collectors. Introduction by H. E. Keyes. vi + 1106pp. 7⅞ x 10¾.
21599-7, 21600-4 Two volumes, Paperbound $11.00

PENNSYLVANIA DUTCH AMERICAN FOLK ART, Henry J. Kauffman. 279 photos, 28 drawings of tulipware, Fraktur script, painted tinware, toys, flowered furniture, quilts, samplers, hex signs, house interiors, etc. Full descriptive text. Excellent for tourist, rewarding for designer, collector. Map. 146pp. 7⅞ x 10¾.
21205-X Paperbound $2.50

EARLY NEW ENGLAND GRAVESTONE RUBBINGS, Edmund V. Gillon, Jr. 43 photographs, 226 carefully reproduced rubbings show heavily symbolic, sometimes macabre early gravestones, up to early 19th century. Remarkable early American primitive art, occasionally strikingly beautiful; always powerful. Text. xxvi + 207pp. 8⅜ x 11¼.   21380-3 Paperbound $3.50

VISUAL ILLUSIONS: THEIR CAUSES, CHARACTERISTICS, AND APPLICATIONS, Matthew Luckiesh. Thorough description and discussion of optical illusion, geometric and perspective, particularly; size and shape distortions, illusions of color, of motion; natural illusions; use of illusion in art and magic, industry, etc. Most useful today with op art, also for classical art. Scores of effects illustrated. Introduction by William H. Ittleson. 100 illustrations. xxi + 252pp.
21530-X Paperbound $2.00

A HANDBOOK OF ANATOMY FOR ART STUDENTS, Arthur Thomson. Thorough, virtually exhaustive coverage of skeletal structure, musculature, etc. Full text, supplemented by anatomical diagrams and drawings and by photographs of undraped figures. Unique in its comparison of male and female forms, pointing out differences of contour, texture, form. 211 figures, 40 drawings, 86 photographs. xx + 459pp. 5⅜ x 8⅜.
21163-0 Paperbound $3.50

150 MASTERPIECES OF DRAWING, Selected by Anthony Toney. Full page reproductions of drawings from the early 16th to the end of the 18th century, all beautifully reproduced: Rembrandt, Michelangelo, Dürer, Fragonard, Urs, Graf, Wouwerman, many others. First-rate browsing book, model book for artists. xviii + 150pp. 8⅜ x 11¼.
21032-4 Paperbound $2.50

THE LATER WORK OF AUBREY BEARDSLEY, Aubrey Beardsley. Exotic, erotic, ironic masterpieces in full maturity: Comedy Ballet, Venus and Tannhauser, Pierrot, Lysistrata, Rape of the Lock, Savoy material, Ali Baba, Volpone, etc. This material revolutionized the art world, and is still powerful, fresh, brilliant. With *The Early Work,* all Beardsley's finest work. 174 plates, 2 in color. xiv + 176pp. 8⅛ x 11.
21817-1 Paperbound $3.00

DRAWINGS OF REMBRANDT, Rembrandt van Rijn. Complete reproduction of fabulously rare edition by Lippmann and Hofstede de Groot, completely reedited, updated, improved by Prof. Seymour Slive, Fogg Museum. Portraits, Biblical sketches, landscapes, Oriental types, nudes, episodes from classical mythology—All Rembrandt's fertile genius. Also selection of drawings by his pupils and followers. "Stunning volumes," *Saturday Review.* 550 illustrations. lxxviii + 552pp. 9⅛ x 12¼.
21485-0, 21486-9 Two volumes, Paperbound $7.00

THE DISASTERS OF WAR, Francisco Goya. One of the masterpieces of Western civilization—83 etchings that record Goya's shattering, bitter reaction to the Napoleonic war that swept through Spain after the insurrection of 1808 and to war in general. Reprint of the first edition, with three additional plates from Boston's Museum of Fine Arts. All plates facsimile size. Introduction by Philip Hofer, Fogg Museum. v + 97pp. 9⅜ x 8¼.
21872-4 Paperbound $2.00

GRAPHIC WORKS OF ODILON REDON. Largest collection of Redon's graphic works ever assembled: 172 lithographs, 28 etchings and engravings, 9 drawings. These include some of his most famous works. All the plates from *Odilon Redon: oeuvre graphique complet,* plus additional plates. New introduction and caption translations by Alfred Werner. 209 illustrations. xxvii + 209pp. 9⅛ x 12¼.
21966-8 Paperbound $4.00

ALPHABETS AND ORNAMENTS, Ernst Lehner. Well-known pictorial source for decorative alphabets, script examples, cartouches, frames, decorative title pages, calligraphic initials, borders, similar material. 14th to 19th century, mostly European. Useful in almost any graphic arts designing, varied styles. 750 illustrations. 256pp. 7 x 10. 21905-4 Paperbound $4.00

PAINTING: A CREATIVE APPROACH, Norman Colquhoun. For the beginner simple guide provides an instructive approach to painting: major stumbling blocks for beginner; overcoming them, technical points; paints and pigments; oil painting; watercolor and other media and color. New section on "plastic" paints. Glossary. Formerly *Paint Your Own Pictures.* 221pp. 22000-1 Paperbound $1.75

THE ENJOYMENT AND USE OF COLOR, Walter Sargent. Explanation of the relations between colors themselves and between colors in nature and art, including hundreds of little-known facts about color values, intensities, effects of high and low illumination, complementary colors. Many practical hints for painters, references to great masters. 7 color plates, 29 illustrations. x + 274pp. 20944-X Paperbound $2.50

THE NOTEBOOKS OF LEONARDO DA VINCI, compiled and edited by Jean Paul Richter. 1566 extracts from original manuscripts reveal the full range of Leonardo's versatile genius: all his writings on painting, sculpture, architecture, anatomy, astronomy, geography, topography, physiology, mining, music, etc., in both Italian and English, with 186 plates of manuscript pages and more than 500 additional drawings. Includes studies for the Last Supper, the lost Sforza monument, and other works. Total of xlvii + 866pp. 7⅞ x 10¾. 22572-0, 22573-9 Two volumes, Paperbound $10.00

MONTGOMERY WARD CATALOGUE OF 1895. Tea gowns, yards of flannel and pillow-case lace, stereoscopes, books of gospel hymns, the New Improved Singer Sewing Machine, side saddles, milk skimmers, straight-edged razors, high-button shoes, spittoons, and on and on . . . listing some 25,000 items, practically all illustrated. Essential to the shoppers of the 1890's, it is our truest record of the spirit of the period. Unaltered reprint of Issue No. 57, Spring and Summer 1895. Introduction by Boris Emmet. Innumerable illustrations. xiii + 624pp. 8½ x 11⅝. 22377-9 Paperbound $6.95

THE CRYSTAL PALACE EXHIBITION ILLUSTRATED CATALOGUE (LONDON, 1851). One of the wonders of the modern world—the Crystal Palace Exhibition in which all the nations of the civilized world exhibited their achievements in the arts and sciences—presented in an equally important illustrated catalogue. More than 1700 items pictured with accompanying text—ceramics, textiles, cast-iron work, carpets, pianos, sleds, razors, wall-papers, billiard tables, beehives, silverware and hundreds of other artifacts—represent the focal point of Victorian culture in the Western World. Probably the largest collection of Victorian decorative art ever assembled—indispensable for antiquarians and designers. Unabridged republication of the Art-Journal Catalogue of the Great Exhibition of 1851, with all terminal essays. New introduction by John Gloag, F.S.A. xxxiv + 426pp. 9 x 12. 22503-8 Paperbound $4.50

PLANETS, STARS AND GALAXIES: DESCRIPTIVE ASTRONOMY FOR BEGINNERS, A. E. Fanning. Comprehensive introductory survey of astronomy: the sun, solar system, stars, galaxies, universe, cosmology; up-to-date, including quasars, radio stars, etc. Preface by Prof. Donald Menzel. 24pp. of photographs. 189pp. 5¼ x 8¼.
21680-2 Paperbound $1.50

TEACH YOURSELF CALCULUS, P. Abbott. With a good background in algebra and trig, you can teach yourself calculus with this book. Simple, straightforward introduction to functions of all kinds, integration, differentiation, series, etc. "Students who are beginning to study calculus method will derive great help from this book." Faraday House Journal. 308pp.
20683-1 Clothbound $2.00

TEACH YOURSELF TRIGONOMETRY, P. Abbott. Geometrical foundations, indices and logarithms, ratios, angles, circular measure, etc. are presented in this sound, easy-to-use text. Excellent for the beginner or as a brush up, this text carries the student through the solution of triangles. 204pp.
20682-3 Clothbound $2.00

TEACH YOURSELF ANATOMY, David LeVay. Accurate, inclusive, profusely illustrated account of structure, skeleton, abdomen, muscles, nervous system, glands, brain, reproductive organs, evolution. "Quite the best and most readable account,' *Medical Officer.* 12 color plates. 164 figures. 311pp. 4¾ x 7.
21651-9 Clothbound $2.50

TEACH YOURSELF PHYSIOLOGY, David LeVay. Anatomical, biochemical bases; digestive, nervous, endocrine systems; metabolism; respiration; muscle; excretion; temperature control; reproduction. "Good elementary exposition," *The Lancet.* 6 color plates. 44 illustrations. 208pp. 4¼ x 7.
21658-6 Clothbound $2.50

THE FRIENDLY STARS, Martha Evans Martin. Classic has taught naked-eye observation of stars, planets to hundreds of thousands, still not surpassed for charm, lucidity, adequacy. Completely updated by Professor Donald H. Menzel, Harvard Observatory. 25 illustrations. 16 x 30 chart. x + 147pp.
21099-5 Paperbound $1.25

MUSIC OF THE SPHERES: THE MATERIAL UNIVERSE FROM ATOM TO QUASAR, SIMPLY EXPLAINED, Guy Murchie. Extremely broad, brilliantly written popular account begins with the solar system and reaches to dividing line between matter and nonmatter; latest understandings presented with exceptional clarity. Volume One: Planets, stars, galaxies, cosmology, geology, celestial mechanics, latest astronomical discoveries; Volume Two: Matter, atoms, waves, radiation, relativity, chemical action, heat, nuclear energy, quantum theory, music, light, color, probability, antimatter, antigravity, and similar topics. 319 figures. 1967 (second) edition. Total of xx + 644pp.
21809-0, 21810-4 Two volumes, Paperbound $5.00

OLD-TIME SCHOOLS AND SCHOOL BOOKS, Clifton Johnson. Illustrations and rhymes from early primers, abundant quotations from early textbooks, many anecdotes of school life enliven this study of elementary schools from Puritans to middle 19th century. Introduction by Carl Withers. 234 illustrations. xxxiii + 381pp.
21031-6 Paperbound $2.50

CATALOGUE OF DOVER BOOKS

TWO LITTLE SAVAGES; BEING THE ADVENTURES OF TWO BOYS WHO LIVED AS INDIANS AND WHAT THEY LEARNED, Ernest Thompson Seton. Great classic of nature and boyhood provides a vast range of woodlore in most palatable form, a genuinely entertaining story. Two farm boys build a teepee in woods and live in it for a month, working out Indian solutions to living problems, star lore, birds and animals, plants, etc. 293 illustrations. vii + 286pp.

20985-7 Paperbound $2.50

PETER PIPER'S PRACTICAL PRINCIPLES OF PLAIN & PERFECT PRONUNCIATION. Alliterative jingles and tongue-twisters of surprising charm, that made their first appearance in America about 1830. Republished in full with the spirited woodcut illustrations from this earliest American edition. 32pp. 4½ x 6⅜.

22560-7 Paperbound $1.00

SCIENCE EXPERIMENTS AND AMUSEMENTS FOR CHILDREN, Charles Vivian. 73 easy experiments, requiring only materials found at home or easily available, such as candles, coins, steel wool, etc.; illustrate basic phenomena like vacuum, simple chemical reaction, etc. All safe. Modern, well-planned. Formerly *Science Games for Children*. 102 photos, numerous drawings. 96pp. 6⅛ x 9¼.

21856-2 Paperbound $1.25

AN INTRODUCTION TO CHESS MOVES AND TACTICS SIMPLY EXPLAINED, Leonard Barden. Informal intermediate introduction, quite strong in explaining reasons for moves. Covers basic material, tactics, important openings, traps, positional play in middle game, end game. Attempts to isolate patterns and recurrent configurations. Formerly *Chess*. 58 figures. 102pp.   (USO) 21210-6 Paperbound $1.25

LASKER'S MANUAL OF CHESS, Dr. Emanuel Lasker. Lasker was not only one of the five great World Champions, he was also one of the ablest expositors, theorists, and analysts. In many ways, his Manual, permeated with his philosophy of battle, filled with keen insights, is one of the greatest works ever written on chess. Filled with analyzed games by the great players. A single-volume library that will profit almost any chess player, beginner or master. 308 diagrams. xli x 349pp.

20640-8 Paperbound $2.75

THE MASTER BOOK OF MATHEMATICAL RECREATIONS, Fred Schuh. In opinion of many the finest work ever prepared on mathematical puzzles, stunts, recreations; exhaustively thorough explanations of mathematics involved, analysis of effects, citation of puzzles and games. Mathematics involved is elementary. Translated by F. Göbel. 194 figures. xxiv + 430pp.   22134-2 Paperbound $3.00

MATHEMATICS, MAGIC AND MYSTERY, Martin Gardner. Puzzle editor for Scientific American explains mathematics behind various mystifying tricks: card tricks, stage "mind reading," coin and match tricks, counting out games, geometric dissections, etc. Probability sets, theory of numbers clearly explained. Also provides more than 400 tricks, guaranteed to work, that you can do. 135 illustrations. xii + 176pp.

20338-2 Paperbound $1.50

AMERICAN FOOD AND GAME FISHES, David S. Jordan and Barton W. Evermann. Definitive source of information, detailed and accurate enough to enable the sportsman and nature lover to identify conclusively some 1,000 species and sub-species of North American fish, sought for food or sport. Coverage of range, physiology, habits, life history, food value. Best methods of capture, interest to the angler, advice on bait, fly-fishing, etc. 338 drawings and photographs. 1 + 574pp. 6⅝ x 9⅜.
22383-1 Paperbound $4.50

THE FROG BOOK, Mary C. Dickerson. Complete with extensive finding keys, over 300 photographs, and an introduction to the general biology of frogs and toads, this is the classic non-technical study of Northeastern and Central species. 58 species; 290 photographs and 16 color plates. xvii + 253pp.
21973-9 Paperbound $4.00

THE MOTH BOOK: A GUIDE TO THE MOTHS OF NORTH AMERICA, William J. Holland. Classical study, eagerly sought after and used for the past 60 years. Clear identification manual to more than 2,000 different moths, largest manual in existence. General information about moths, capturing, mounting, classifying, etc., followed by species by species descriptions. 263 illustrations plus 48 color plates show almost every species, full size. 1968 edition, preface, nomenclature changes by A. E. Brower. xxiv + 479pp. of text. 6½ x 9¼.
21948-8 Paperbound $5.00

THE SEA-BEACH AT EBB-TIDE, Augusta Foote Arnold. Interested amateur can identify hundreds of marine plants and animals on coasts of North America; marine algae; seaweeds; squids; hermit crabs; horse shoe crabs; shrimps; corals; sea anemones; etc. Species descriptions cover: structure; food; reproductive cycle; size; shape; color; habitat; etc. Over 600 drawings. 85 plates. xii + 490pp.
21949-6 Paperbound $3.50

COMMON BIRD SONGS, Donald J. Borror. 33⅓ 12-inch record presents songs of 60 important birds of the eastern United States. A thorough, serious record which provides several examples for each bird, showing different types of song, individual variations, etc. Inestimable identification aid for birdwatcher. 32-page booklet gives text about birds and songs, with illustration for each bird.
21829-5 Record, book, album. Monaural. $2.75

FADS AND FALLACIES IN THE NAME OF SCIENCE, Martin Gardner. Fair, witty appraisal of cranks and quacks of science: Atlantis, Lemuria, hollow earth, flat earth, Velikovsky, orgone energy, Dianetics, flying saucers, Bridey Murphy, food fads, medical fads, perpetual motion, etc. Formerly "In the Name of Science." x + 363pp.
20394-8 Paperbound $2.00

HOAXES, Curtis D. MacDougall. Exhaustive, unbelievably rich account of great hoaxes: Locke's moon hoax, Shakespearean forgeries, sea serpents, Loch Ness monster, Cardiff giant, John Wilkes Booth's mummy, Disumbrationist school of art, dozens more; also journalism, psychology of hoaxing. 54 illustrations. xi + 338pp.
20465-0 Paperbound $2.75

JIM WHITEWOLF: THE LIFE OF A KIOWA APACHE INDIAN, Charles S. Brant, editor. Spans transition between native life and acculturation period, 1880 on. Kiowa culture, personal life pattern, religion and the supernatural, the Ghost Dance, breakdown in the White Man's world, similar material. 1 map. xii + 144pp.
22015-X Paperbound $1.75

THE NATIVE TRIBES OF CENTRAL AUSTRALIA, Baldwin Spencer and F. J. Gillen. Basic book in anthropology, devoted to full coverage of the Arunta and Warramunga tribes; the source for knowledge about kinship systems, material and social culture, religion, etc. Still unsurpassed. 121 photographs, 89 drawings. xviii + 669pp.
21775-2 Paperbound $5.00

MALAY MAGIC, Walter W. Skeat. Classic (1900); still the definitive work on the folklore and popular religion of the Malay peninsula. Describes marriage rites, birth spirits and ceremonies, medicine, dances, games, war and weapons, etc. Extensive quotes from original sources, many magic charms translated into English. 35 illustrations. Preface by Charles Otto Blagden. xxiv + 685pp.
21760-4 Paperbound $4.00

HEAVENS ON EARTH: UTOPIAN COMMUNITIES IN AMERICA, 1680-1880, Mark Holloway. The finest nontechnical account of American utopias, from the early Woman in the Wilderness, Ephrata, Rappites to the enormous mid 19th-century efflorescence; Shakers, New Harmony, Equity Stores, Fourier's Phalanxes, Oneida, Amana, Fruitlands, etc. "Entertaining and very instructive." *Times Literary Supplement*. 15 illustrations. 246pp.
21593-8 Paperbound $2.00

LONDON LABOUR AND THE LONDON POOR, Henry Mayhew. Earliest (c. 1850) sociological study in English, describing myriad subcultures of London poor. Particularly remarkable for the thousands of pages of direct testimony taken from the lips of London prostitutes, thieves, beggars, street sellers, chimney-sweepers, street-musicians, "mudlarks," "pure-finders," rag-gatherers, "running-patterers," dock laborers, cab-men, and hundreds of others, quoted directly in this massive work. An extraordinarily vital picture of London emerges. 110 illustrations. Total of lxxvi + 1951pp. 6⅝ x 10.
21934-8, 21935-6, 21936-4, 21937-2 Four volumes, Paperbound $14.00

HISTORY OF THE LATER ROMAN EMPIRE, J. B. Bury. Eloquent, detailed reconstruction of Western and Byzantine Roman Empire by a major historian, from the death of Theodosius I (395 A.D.) to the death of Justinian (565). Extensive quotations from contemporary sources; full coverage of important Roman and foreign figures of the time. xxxiv + 965pp. 21829-5 Record, book, album. Monaural. $3.50

AN INTELLECTUAL AND CULTURAL HISTORY OF THE WESTERN WORLD, Harry Elmer Barnes. Monumental study, tracing the development of the accomplishments that make up human culture. Every aspect of man's achievement surveyed from its origins in the Paleolithic to the present day (1964); social structures, ideas, economic systems, art, literature, technology, mathematics, the sciences, medicine, religion, jurisprudence, etc. Evaluations of the contributions of scores of great men. 1964 edition, revised and edited by scholars in the many fields represented. Total of xxix + 1381pp. 21275-0, 21276-9, 21277-7 Three volumes, Paperbound $7.75

INCIDENTS OF TRAVEL IN YUCATAN, John L. Stephens. Classic (1843) exploration of jungles of Yucatan, looking for evidences of Maya civilization. Stephens found many ruins; comments on travel adventures, Mexican and Indian culture. 127 striking illustrations by F. Catherwood. Total of 669 pp.

20926-1, 20927-X Two volumes, Paperbound $5.00

INCIDENTS OF TRAVEL IN CENTRAL AMERICA, CHIAPAS, AND YUCATAN, John L. Stephens. An exciting travel journal and an important classic of archeology. Narrative relates his almost single-handed discovery of the Mayan culture, and exploration of the ruined cities of Copan, Palenque, Utatlan and others; the monuments they dug from the earth, the temples buried in the jungle, the customs of poverty-stricken Indians living a stone's throw from the ruined palaces. 115 drawings by F. Catherwood. Portrait of Stephens. xii + 812pp.

22404-X, 22405-8 Two volumes, Paperbound $6.00

A NEW VOYAGE ROUND THE WORLD, William Dampier. Late 17-century naturalist joined the pirates of the Spanish Main to gather information; remarkably vivid account of buccaneers, pirates; detailed, accurate account of botany, zoology, ethnography of lands visited. Probably the most important early English voyage, enormous implications for British exploration, trade, colonial policy. Also most interesting reading. Argonaut edition, introduction by Sir Albert Gray. New introduction by Percy Adams. 6 plates, 7 illustrations. xlvii + 376pp. 6½ x 9¼.

21900-3 Paperbound $3.00

INTERNATIONAL AIRLINE PHRASE BOOK IN SIX LANGUAGES, Joseph W. Bátor. Important phrases and sentences in English paralleled with French, German, Portuguese, Italian, Spanish equivalents, covering all possible airport-travel situations; created for airline personnel as well as tourist by Language Chief, Pan American Airlines. xiv + 204pp.

22017-6 Paperbound $2.00

STAGE COACH AND TAVERN DAYS, Alice Morse Earle. Detailed, lively account of the early days of taverns; their uses and importance in the social, political and military life; furnishings and decorations; locations; food and drink; tavern signs, etc. Second half covers every aspect of early travel; the roads, coaches, drivers, etc. Nostalgic, charming, packed with fascinating material. 157 illustrations, mostly photographs. xiv + 449pp.

22518-6 Paperbound $4.00

NORSE DISCOVERIES AND EXPLORATIONS IN NORTH AMERICA, Hjalmar R. Holand. The perplexing Kensington Stone, found in Minnesota at the end of the 19th century. Is it a record of a Scandinavian expedition to North America in the 14th century? Or is it one of the most successful hoaxes in history. A scientific detective investigation. Formerly *Westward from Vinland*. 31 photographs, 17 figures. x + 354pp.

22014-1 Paperbound $2.75

A BOOK OF OLD MAPS, compiled and edited by Emerson D. Fite and Archibald Freeman. 74 old maps offer an unusual survey of the discovery, settlement and growth of America down to the close of the Revolutionary war: maps showing Norse settlements in Greenland, the explorations of Columbus, Verrazano, Cabot, Champlain, Joliet, Drake, Hudson, etc., campaigns of Revolutionary war battles, and much more. Each map is accompanied by a brief historical essay. xvi + 299pp. 11 x 13¾.

22084-2 Paperbound $6.00

ADVENTURES OF AN AFRICAN SLAVER, Theodore Canot. Edited by Brantz Mayer. A detailed portrayal of slavery and the slave trade, 1820-1840. Canot, an established trader along the African coast, describes the slave economy of the African kingdoms, the treatment of captured negroes, the extensive journeys in the interior to gather slaves, slave revolts and their suppression, harems, bribes, and much more. Full and unabridged republication of 1854 edition. Introduction by Malcom Cowley. 16 illustrations. xvii + 448pp. 22456-2 Paperbound $3.50

MY BONDAGE AND MY FREEDOM, Frederick Douglass. Born and brought up in slavery, Douglass witnessed its horrors and experienced its cruelties, but went on to become one of the most outspoken forces in the American anti-slavery movement. Considered the best of his autobiographies, this book graphically describes the inhuman treatment of slaves, its effects on slave owners and slave families, and how Douglass's determination led him to a new life. Unaltered reprint of 1st (1855) edition. xxxii + 464pp. 22457-0 Paperbound $2.50

THE INDIANS' BOOK, recorded and edited by Natalie Curtis. Lore, music, narratives, dozens of drawings by Indians themselves from an authoritative and important survey of native culture among Plains, Southwestern, Lake and Pueblo Indians. Standard work in popular ethnomusicology. 149 songs in full notation. 23 drawings, 23 photos. xxxi + 584pp. 6⅝ x 9⅜. 21939-9 Paperbound $4.50

DICTIONARY OF AMERICAN PORTRAITS, edited by Hayward and Blanche Cirker. 4024 portraits of 4000 most important Americans, colonial days to 1905 (with a few important categories, like Presidents, to present). Pioneers, explorers, colonial figures, U. S. officials, politicians, writers, military and naval men, scientists, inventors, manufacturers, jurists, actors, historians, educators, notorious figures, Indian chiefs, etc. All authentic contemporary likenesses. The only work of its kind in existence; supplements all biographical sources for libraries. Indispensable to anyone working with American history. 8,000-item classified index, finding lists, other aids. xiv + 756pp. 9¼ x 12¾. 21823-6 Clothbound $30.00

TRITTON'S GUIDE TO BETTER WINE AND BEER MAKING FOR BEGINNERS, S. M. Tritton. All you need to know to make family-sized quantities of over 100 types of grape, fruit, herb and vegetable wines; as well as beers, mead, cider, etc. Complete recipes, advice as to equipment, procedures such as fermenting, bottling, and storing wines. Recipes given in British, U. S., and metric measures. Accompanying booklet lists sources in U. S. A. where ingredients may be bought, and additional information. 11 illustrations. 157pp. 5⅝ x 8⅛. (USO) 22090-7 Clothbound $3.50

GARDENING WITH HERBS FOR FLAVOR AND FRAGRANCE, Helen M. Fox. How to grow herbs in your own garden, how to use them in your cooking (over 55 recipes included), legends and myths associated with each species, uses in medicine, perfumes, etc.—these are elements of one of the few books written especially for American herb fanciers. Guides you step-by-step from soil preparation to harvesting and storage for each type of herb. 12 drawings by Louise Mansfield. xiv + 334pp. 22540-2 Paperbound $2.50

THE PHILOSOPHY OF THE UPANISHADS, Paul Deussen. Clear, detailed statement of upanishadic system of thought, generally considered among best available. History of these works, full exposition of system emergent from them, parallel concepts in the West. Translated by A. S. Geden. xiv + 429pp.

21616-0 Paperbound $3.00

LANGUAGE, TRUTH AND LOGIC, Alfred J. Ayer. Famous, remarkably clear introduction to the Vienna and Cambridge schools of Logical Positivism; function of philosophy, elimination of metaphysical thought, nature of analysis, similar topics. "Wish I had written it myself," Bertrand Russell. 2nd, 1946 edition. 160pp.

20010-8 Paperbound $1.35

THE GUIDE FOR THE PERPLEXED, Moses Maimonides. Great classic of medieval Judaism, major attempt to reconcile revealed religion (Pentateuch, commentaries) and Aristotelian philosophy. Enormously important in all Western thought. Unabridged Friedländer translation. 50-page introduction. lix + 414pp.

(USO) 20351-4 Paperbound $2.50

OCCULT AND SUPERNATURAL PHENOMENA, D. H. Rawcliffe. Full, serious study of the most persistent delusions of mankind: crystal gazing, mediumistic trance, stigmata, lycanthropy, fire walking, dowsing, telepathy, ghosts, ESP, etc., and their relation to common forms of abnormal psychology. Formerly *Illusions and Delusions of the Supernatural and the Occult.* iii + 551pp.   20503-7 Paperbound $3.50

THE EGYPTIAN BOOK OF THE DEAD: THE PAPYRUS OF ANI, E. A. Wallis Budge. Full hieroglyphic text, interlinear transliteration of sounds, word for word translation, then smooth, connected translation; Theban recension. Basic work in Ancient Egyptian civilization; now even more significant than ever for historical importance, dilation of consciousness, etc. clvi + 377pp. 6½ x 9¼.

21866-X Paperbound $3.95

PSYCHOLOGY OF MUSIC, Carl E. Seashore. Basic, thorough survey of everything known about psychology of music up to 1940's; essential reading for psychologists, musicologists. Physical acoustics; auditory apparatus; relationship of physical sound to perceived sound; role of the mind in sorting, altering, suppressing, creating sound sensations; musical learning, testing for ability, absolute pitch, other topics. Records of Caruso, Menuhin analyzed. 88 figures. xix + 408pp.

21851-1 Paperbound $2.75

THE I CHING (THE BOOK OF CHANGES), translated by James Legge. Complete translated text plus appendices by Confucius, of perhaps the most penetrating divination book ever compiled. Indispensable to all study of early Oriental civilizations. 3 plates. xxiii + 448pp.   21062-6 Paperbound $3.00

THE UPANISHADS, translated by Max Müller. Twelve classical upanishads: Chandogya, Kena, Aitareya, Kaushitaki, Isa, Katha, Mundaka, Taittiriyaka, Brhadaranyaka, Svetasvatara, Prasna, Maitriyana. 160-page introduction, analysis by Prof. Müller. Total of 826pp.   20398-0, 20399-9 Two volumes, Paperbound $5.00

POEMS OF ANNE BRADSTREET, edited with an introduction by Robert Hutchinson. A new selection of poems by America's first poet and perhaps the first significant woman poet in the English language. 48 poems display her development in works of considerable variety—love poems, domestic poems, religious meditations, formal elegies, "quaternions," etc. Notes, bibliography. viii + 222pp.

22160-1 Paperbound $2.00

THREE GOTHIC NOVELS: THE CASTLE OF OTRANTO BY HORACE WALPOLE; VATHEK BY WILLIAM BECKFORD; THE VAMPYRE BY JOHN POLIDORI, WITH FRAGMENT OF A NOVEL BY LORD BYRON, edited by E. F. Bleiler. The first Gothic novel, by Walpole; the finest Oriental tale in English, by Beckford; powerful Romantic supernatural story in versions by Polidori and Byron. All extremely important in history of literature; all still exciting, packed with supernatural thrills, ghosts, haunted castles, magic, etc. xl + 291pp.

21232-7 Paperbound $2.00

THE BEST TALES OF HOFFMANN, E. T. A. Hoffmann. 10 of Hoffmann's most important stories, in modern re-editings of standard translations: Nutcracker and the King of Mice, Signor Formica, Automata, The Sandman, Rath Krespel, The Golden Flowerpot, Master Martin the Cooper, The Mines of Falun, The King's Betrothed, A New Year's Eve Adventure. 7 illustrations by Hoffmann. Edited by E. F. Bleiler. xxxix + 419pp.

21793-0 Paperbound $2.50

GHOST AND HORROR STORIES OF AMBROSE BIERCE, Ambrose Bierce. 23 strikingly modern stories of the horrors latent in the human mind: The Eyes of the Panther, The Damned Thing, An Occurrence at Owl Creek Bridge, An Inhabitant of Carcosa, etc., plus the dream-essay, Visions of the Night. Edited by E. F. Bleiler. xxii + 199pp.

20767-6 Paperbound $1.50

BEST GHOST STORIES OF J. S. LEFANU, J. Sheridan LeFanu. Finest stories by Victorian master often considered greatest supernatural writer of all. Carmilla, Green Tea, The Haunted Baronet, The Familiar, and 12 others. Most never before available in the U. S. A. Edited by E. F. Bleiler. 8 illustrations from Victorian publications. xvii + 467pp.

20415-4 Paperbound $2.50

THE TIME STREAM, THE GREATEST ADVENTURE, AND THE PURPLE SAPPHIRE— THREE SCIENCE FICTION NOVELS, John Taine (Eric Temple Bell). Great American mathematician was also foremost science fiction novelist of the 1920's. *The Time Stream,* one of all-time classics, uses concepts of circular time; *The Greatest Adventure,* incredibly ancient biological experiments from Antarctica threaten to escape; The *Purple Sapphire,* superscience, lost races in Central Tibet, survivors of the Great Race. 4 illustrations by Frank R. Paul. v + 532pp.

21180-0 Paperbound $3.00

SEVEN SCIENCE FICTION NOVELS, H. G. Wells. The standard collection of the great novels. Complete, unabridged. *First Men in the Moon, Island of Dr. Moreau, War of the Worlds, Food of the Gods, Invisible Man, Time Machine, In the Days of the Comet.* Not only science fiction fans, but every educated person owes it to himself to read these novels. 1015pp.

20264-X Clothbound $5.00

JOHANN SEBASTIAN BACH, Philipp Spitta. One of the great classics of musicology, this definitive analysis of Bach's music (and life) has never been surpassed. Lucid, nontechnical analyses of hundreds of pieces (30 pages devoted to St. Matthew Passion, 26 to B Minor Mass). Also includes major analysis of 18th-century music. 450 musical examples. 40-page musical supplement. Total of xx + 1799pp.
(EUK) 22278-0, 22279-9 Two volumes, Clothbound $15.00

MOZART AND HIS PIANO CONCERTOS, Cuthbert Girdlestone. The only full-length study of an important area of Mozart's creativity. Provides detailed analyses of all 23 concertos, traces inspirational sources. 417 musical examples. Second edition. 509pp. (USO) 21271-8 Paperbound $3.50

THE PERFECT WAGNERITE: A COMMENTARY ON THE NIBLUNG'S RING, George Bernard Shaw. Brilliant and still relevant criticism in remarkable essays on Wagner's Ring cycle, Shaw's ideas on political and social ideology behind the plots, role of Leitmotifs, vocal requisites, etc. Prefaces. xxi + 136pp.
21707-8 Paperbound $1.50

DON GIOVANNI, W. A. Mozart. Complete libretto, modern English translation; biographies of composer and librettist; accounts of early performances and critical reaction. Lavishly illustrated. All the material you need to understand and appreciate this great work. Dover Opera Guide and Libretto Series; translated and introduced by Ellen Bleiler. 92 illustrations. 209pp.
21134-7 Paperbound $1.50

HIGH FIDELITY SYSTEMS: A LAYMAN'S GUIDE, Roy F. Allison. All the basic information you need for setting up your own audio system: high fidelity and stereo record players, tape records, F.M. Connections, adjusting tone arm, cartridge, checking needle alignment, positioning speakers, phasing speakers, adjusting hums, trouble-shooting, maintenance, and similar topics. Enlarged 1965 edition. More than 50 charts, diagrams, photos. iv + 91pp. 21514-8 Paperbound $1.25

REPRODUCTION OF SOUND, Edgar Villchur. Thorough coverage for laymen of high fidelity systems, reproducing systems in general, needles, amplifiers, preamps, loudspeakers, feedback, explaining physical background. "A rare talent for making technicalities vividly comprehensible," R. Darrell, *High Fidelity.* 69 figures. iv + 92pp. 21515-6 Paperbound $1.00

HEAR ME TALKIN' TO YA: THE STORY OF JAZZ AS TOLD BY THE MEN WHO MADE IT, Nat Shapiro and Nat Hentoff. Louis Armstrong, Fats Waller, Jo Jones, Clarence Williams, Billy Holiday, Duke Ellington, Jelly Roll Morton and dozens of other jazz greats tell how it was in Chicago's South Side, New Orleans, depression Harlem and the modern West Coast as jazz was born and grew. xvi + 429pp.
21726-4 Paperbound $2.50

FABLES OF AESOP, translated by Sir Roger L'Estrange. A reproduction of the very rare 1931 Paris edition; a selection of the most interesting fables, together with 50 imaginative drawings by Alexander Calder. v + 128pp. 6½x9¼.
21780-9 Paperbound $1.25

CATALOGUE OF DOVER BOOKS

THE RED FAIRY BOOK, Andrew Lang. Lang's color fairy books have long been children's favorites. This volume includes Rapunzel, Jack and the Bean-stalk and 35 other stories, familiar and unfamiliar. 4 plates, 93 illustrations x + 367pp.
21673-X Paperbound $2.50

THE BLUE FAIRY BOOK, Andrew Lang. Lang's tales come from all countries and all times. Here are 37 tales from Grimm, the Arabian Nights, Greek Mythology, and other fascinating sources. 8 plates, 130 illustrations. xi + 390pp.
21437-0 Paperbound $2.50

HOUSEHOLD STORIES BY THE BROTHERS GRIMM. Classic English-language edition of the well-known tales — Rumpelstiltskin, Snow White, Hansel and Gretel, The Twelve Brothers, Faithful John, Rapunzel, Tom Thumb (52 stories in all). Translated into simple, straightforward English by Lucy Crane. Ornamented with head-pieces, vignettes, elaborate decorative initials and a dozen full-page illustrations by Walter Crane. x + 269pp.
21080-4 Paperbound $2.50

THE MERRY ADVENTURES OF ROBIN HOOD, Howard Pyle. The finest modern versions of the traditional ballads and tales about the great English outlaw. Howard Pyle's complete prose version, with every word, every illustration of the first edition. Do not confuse this facsimile of the original (1883) with modern editions that change text or illustrations. 23 plates plus many page decorations. xxii + 296pp.
22043-5 Paperbound $2.50

THE STORY OF KING ARTHUR AND HIS KNIGHTS, Howard Pyle. The finest children's version of the life of King Arthur; brilliantly retold by Pyle, with 48 of his most imaginative illustrations. xviii + 313pp. 6⅛ x 9¼.
21445-1 Paperbound $2.50

THE WONDERFUL WIZARD OF OZ, L. Frank Baum. America's finest children's book in facsimile of first edition with all Denslow illustrations in full color. The edition a child should have. Introduction by Martin Gardner. 23 color plates, scores of drawings. iv + 267pp.
20691-2 Paperbound $2.25

THE MARVELOUS LAND OF OZ, L. Frank Baum. The second Oz book, every bit as imaginative as the Wizard. The hero is a boy named Tip, but the Scarecrow and the Tin Woodman are back, as is the Oz magic. 16 color plates, 120 drawings by John R. Neill. 287pp.
20692-0 Paperbound $2.50

THE MAGICAL MONARCH OF MO, L. Frank Baum. Remarkable adventures in a land even stranger than Oz. The best of Baum's books not in the Oz series. 15 color plates and dozens of drawings by Frank Verbeck. xviii + 237pp.
21892-9 Paperbound $2.00

THE BAD CHILD'S BOOK OF BEASTS, MORE BEASTS FOR WORSE CHILDREN, A MORAL ALPHABET, Hilaire Belloc. Three complete humor classics in one volume. Be kind to the frog, and do not call him names . . . and 28 other whimsical animals. Familiar favorites and some not so well known. Illustrated by Basil Blackwell. 156pp.
(USO) 20749-8 Paperbound $1.25

MATHEMATICAL PUZZLES FOR BEGINNERS AND ENTHUSIASTS, Geoffrey Mott-Smith. 189 puzzles from easy to difficult—involving arithmetic, logic, algebra, properties of digits, probability, etc.—for enjoyment and mental stimulus. Explanation of mathematical principles behind the puzzles. 135 illustrations. viii + 248pp.
20198-8 Paperbound $1.25

PAPER FOLDING FOR BEGINNERS, William D. Murray and Francis J. Rigney. Easiest book on the market, clearest instructions on making interesting, beautiful origami. Sail boats, cups, roosters, frogs that move legs, bonbon boxes, standing birds, etc. 40 projects; more than 275 diagrams and photographs. 94pp.
20713-7 Paperbound $1.00

TRICKS AND GAMES ON THE POOL TABLE, Fred Herrmann. 79 tricks and games— some solitaires, some for two or more players, some competitive games—to entertain you between formal games. Mystifying shots and throws, unusual caroms, tricks involving such props as cork, coins, a hat, etc. Formerly *Fun on the Pool Table*. 77 figures. 95pp.
21814-7 Paperbound $1.00

HAND SHADOWS TO BE THROWN UPON THE WALL: A SERIES OF NOVEL AND AMUSING FIGURES FORMED BY THE HAND, Henry Bursill. Delightful picturebook from great-grandfather's day shows how to make 18 different hand shadows: a bird that flies, duck that quacks, dog that wags his tail, camel, goose, deer, boy, turtle, etc. Only book of its sort. vi + 33pp. 6½ x 9¼. 21779-5 Paperbound $1.00

WHITTLING AND WOODCARVING, E. J. Tangerman. 18th printing of best book on market. "If you can cut a potato you can carve" toys and puzzles, chains, chessmen, caricatures, masks, frames, woodcut blocks, surface patterns, much more. Information on tools, woods, techniques. Also goes into serious wood sculpture from Middle Ages to present, East and West. 464 photos, figures. x + 293pp.
20965-2 Paperbound $2.00

HISTORY OF PHILOSOPHY, Julián Marias. Possibly the clearest, most easily followed, best planned, most useful one-volume history of philosophy on the market; neither skimpy nor overfull. Full details on system of every major philosopher and dozens of less important thinkers from pre-Socratics up to Existentialism and later. Strong on many European figures usually omitted. Has gone through dozens of editions in Europe. 1966 edition, translated by Stanley Appelbaum and Clarence Strowbridge. xviii + 505pp.
21739-6 Paperbound $3.00

YOGA: A SCIENTIFIC EVALUATION, Kovoor T. Behanan. Scientific but non-technical study of physiological results of yoga exercises; done under auspices of Yale U. Relations to Indian thought, to psychoanalysis, etc. 16 photos. xxiii + 270pp.
20505-3 Paperbound $2.50

Prices subject to change without notice.
Available at your book dealer or write for free catalogue to Dept. GI, Dover Publications, Inc., 180 Varick St., N. Y., N. Y. 10014. Dover publishes more than 150 books each year on science, elementary and advanced mathematics, biology, music, art, literary history, social sciences and other areas.